Oregano

Oregano

The genera *Origanum* and *Lippia*

Edited by
Spiridon E. Kintzios

Agricultural University of Athens, Athens, Greece

CRC PRESS

Boca Raton London New York Washington, D.C.

FIRST INDIAN REPRINT, 2012

This book contains information obtained from authentic and highly regarded sources. Reprinted material is quoted with permission, and sources are indicated. A wide variety of references are listed. Reasonable efforts have been made to publish reliable data and information, but the author and the publisher cannot assume responsibility for the validity of all materials or for the consequences of their use.

Neither this book nor any part may be reproduced or transmitted in any form or by any means, electronic or mechanical, including photocopying, microfilming, and recording, or by any information storage or retrieval system, without prior permission in writing from the publisher.

Direct all inquiries to CRC Press LLC, 2000 N.W. Corporate Blvd., Boca Raton, Florida 33431.

© 2002 Taylor & Francis Group, LLC CRC Press is an imprint of Taylor & Francis Group

Trademark Notice: Product or corporate names may be trademarks or registered trademarks, and are used only for identification and explanation, without intent to infringe.

Visit the CRC Press Web site at www.crcpress.com

Printed and bound in India by
Replika Press Pvt. Ltd.

ISBN 10 : 0-415-36943-6
ISBN 13 : 978-0-415-36943-5

FOR SALE IN SOUTH ASIA ONLY.

Dedicated to my mother, Stavroula

Contents

Preface to the series — ix
Acknowledgements — xi
List of contributors — xii

PART 1
Introduction — 1

1 Profile of the multifaceted prince of the herbs — 3
SPIRIDON E. KINTZIOS

PART 2
Botany — 9

2 Structural features of *Origanum* sp. — 11
ARTEMIOS M. BOSABALIDIS

PART 3
Taxonomy and chemistry — 65

3 The taxonomy and chemistry of *Origanum* — 67
MELPOMENI SKOULA AND JEFFREY B. HARBORNE

4 The Turkish *Origanum* species — 109
K. HÜSNÜ CAN BASER

5 The chemistry of the genus *Lippia* (Verbenaceae) — 127
CESAR A.N. CATALAN AND MARINA E.P. DE LAMPASONA

PART 4
Cultivation and breeding — 151

6 Cultivation of Oregano — 153
OLGA MAKRI

7 Breeding of Oregano — 163
CHLODWIG FRANZ AND JOHANNES NOVAK

PART 5
Pharmacology 175

8 The biological/pharmacological activity of the *Origanum* Genus 177
   DEA BARIČEVIČ AND TOMAŽ BARTOL

PART 6
Uses of Oregano in the food industry 215

9 Processing, effects and use of Oregano and marjoram
   in foodstuffs and in food preparation 217
   SEIJA MARJATTA MÄKINEN AND KIRSTI KAARINA PÄÄKKÖNEN

PART 7
Biotechnology 235

10 The biotechnology of Oregano (*Origanum* sp. and *Lippia* sp.) 237
   SPIRIDON E. KINTZIOS

PART 8
Miscellaneous 243

11 Bibliometric analysis of agricultural and biomedical
   bibliographic databases with regard to medicinal plants
   genera *Origanum* and *Lippia* in the period 1981–1998 245
   TOMAŽ BARTOL AND DEA BARIČEVIČ

   Index 269

Preface to the series

There is increasing interest in industry, academia and the health sciences in medicinal and aromatic plants. In passing from plant production to the eventual product used by the public, many sciences are involved. This series brings together information which is currently scattered through an ever increasing number of journals. Each volume gives an in-depth look at one plant genus, about which an area specialist has assembled information ranging from the production of the plant to market trends and quality control.

Many industries are involved such as forestry, agriculture, chemical, food, flavour, beverage, pharmaceutical, cosmetic and fragrance. The plant raw materials are roots, rhizomes, bulbs, leaves, stems, barks, wood, flowers, fruits and seeds. These yield gums, resins, essential (volatile) oils, fixed oils, waxes, juices, extracts and spices for medicinal and aromatic purposes. All these commodities are traded worldwide. A dealer's market report for an item may say 'Drought in the country of origin has forced up prices'.

Natural products do not mean safe products and account of this has to be taken by the above industries, which are subject to regulation. For example, a number of plants which are approved for use in medicine must not be used in cosmetic products.

The assessment of safe to use starts with the harvested plant material which has to comply with an official monograph. This may require absence of, or prescribed limits of, radioactive material, heavy metals, aflatoxin, pesticide residue, as well as the required level of active principle. This analytical control is costly and tends to exclude small batches of plant material. Large scale contracted mechanized cultivation with designated seed or plantlets is now preferable.

Today, plant selection is not only for the yield of active principle, but for the plant's ability to overcome disease, climatic stress and the hazards caused by mankind. Such methods as *in vitro* fertilization, meristem cultures and somatic embryogenesis are used. The transfer of sections of DNA is giving rise to controversy in the case of some end-uses of the plant material.

Some suppliers of plant raw material are now able to certify that they are supplying organically-farmed medicinal plants, herbs and spices. The Economic Union directive (CVO/EU No 2092/91) details the specifications for the *obligatory* quality controls to be carried out at all stages of production and processing of organic products.

Fascinating plant folklore and ethnopharmacology leads to medicinal potential. Examples are the muscle relaxants based on the arrow poison, curare, from species of *Chondrodendron*, and the anti-malarials derived from species of *Cinchona* and *Artemisia*. The methods of detection of pharmacological activity have become increasingly reliable and specific, frequently involving enzymes in bioassays and avoiding the use of laboratory animals. By using bioassay linked fractionation of crude plant juices or extracts, compounds can be specifically targeted which, for example, inhibit blood platelet aggregation, or have antitumour, or antiviral, or any other required activity. With the assistance of robotic devices, all the members of a genus may be readily screened. However, the plant material must be *fully* authenticated by a specialist.

The medicinal traditions of ancient civilisations such as those of China and India have a large armamentaria of plants in their pharmacopoeias which are used throughout south-east Asia. A similar situation exists in Africa and South America. Thus, a very high percentage of the World's population relies on medicinal and aromatic plants for their medicine. Western medicine is also responding. Already in Germany all medical practitioners have to pass an examination in phytotherapy before being allowed to practise. It is noticeable that throughout Europe and the USA, medical, pharmacy and health related schools are increasingly offering training in phytotherapy.

Multinational pharmaceutical companies have become less enamoured of the single compound magic bullet cure. The high costs of such ventures and the endless competition from me too compounds from rival companies often discourage the attempt. Independent phytomedicine companies have been very strong in Germany. However, by the end of 1995, eleven (almost all) had been acquired by the multinational pharmaceutical firms, acknowledging the lay public's growing demand for phytomedicines in the Western World.

The business of dietary supplements in the Western World has expanded from the Health Store to the pharmacy. Alternative medicine includes plant-based products. Appropriate measures to ensure the quality, safety and efficacy of these either already exist or are being answered by greater legislative control by such bodies as the Food and Drug Administration of the USA and the recently created European Agency for the Evaluation of Medicinal Products, based in London.

In the USA, the Dietary Supplement and Health Education Act of 1994 recognized the class of phytotherapeutic agents derived from medicinal and aromatic plants. Furthermore, under public pressure, the US Congress set up an Office of Alternative Medicine and this office in 1994 assisted the filing of several Investigational New Drug (IND) applications, required for clinical trials of some Chinese herbal preparations. The significance of these applications was that each Chinese preparation involved several plants and yet was handled as a *single* IND. A demonstration of the contribution to efficacy, of *each* ingredient of *each* plant, was not required. This was a major step forward towards more sensible regulations in regard to phytomedicines.

My thanks are due to the staffs of Harwood Academic Publishers and Taylor & Francis who have made this series possible and especially to the volume editors and their chapter contributors for the authoritative information.

Roland Hardman

Acknowledgements

This book would never have been completed in time without the active support of the following persons, who I warmly acknowledge:

Dr Roland Hardman for his guidance, his suggestions and the frequent provision of literature references and valuable data that kept me updated on several aspects of the genera *Origanum* and *Lippia* during the compilation of this volume.

Mrs Claire Redhead and Mrs Rowena Millan for their technical assistance on editorial matters and their incredible patience on dealing with my strangest queries.

My associate Mrs Olga Makri, and a contributor in this volume for undertaking the burden of compiling reference data and checking each contribution for format and spelling errors.

Mr John Konstas for assisting me in the vast literature survey that was essential for the realization of the present effort.

Finally, I would like to thank all the contributors to this volume, whom it was an honour and a pleasure to collaborate with.

It is possible that in the course of editing this volume, some reference material may have been used without knowledge, on my behalf, of a copyright ownership of it. In such a case, I apologize to any copyright holder whose rights may have been unwittingly infringed.

Contributors

Dea Baričevič
University of Ljubljana
Biotechnical Faculty
Agronomy Department
Jamnikarjeva 101
61000 Ljubljana
Slovenia

Tomaž Bartol
Slovenian National AGRIS Centre
Biotechnical Faculty
University of Ljubljana
Jamnikarjeva 101
1111 Ljubljana
Slovenia

K. Hüsnü Can Baser
Anadolu University
Medicinal and Aromatic Plant and
   Drug Research Centre (TBAM)
Yunus Emre Kampusu
26 470 Eskisehir
Turkey

Artemios M. Bosabalidis
Department of Botany
School of Biology, Faculty of Sciences
Aristotle University of Thessaloniki
Thessaloniki 54006, Greece

Cesar A.N. Catalan
Instituto de Quimica Organica. Facultad
de Bioquimica, Quimica y Farmacia
Universidad Nacional de Tucuman
Ayacucho 471. S.M. de Tucuman 4000
Argentina

Chlodwig Franz
Institute for Applied Botany
University of Veterinary Medicine
Veterinärplatz 1
A-1210 Wien, Austria

Jeffrey B. Harborne
Department of Botany
University of Reading
Whiteknights
P.O. Box 221
Reading RG6 6AS
United Kingdom

Spiridon E. Kintzios
Laboratory of Plant Physiology
Faculty of Agricultural Biotechnology
Agricultural University of Athens
Iera Odos 75
11141 Athens
Greece

Marina E.P. de Lampasona
Instituto de Quimica Organica. Facultad
de Bioquimica, Quimica y Farmacia
Universidad Nacional de Tucuman
Ayacucho 471. S.M. de Tucuman 4000
Argentina

Seija Marjatta Mäkinen
Department of Applied Chemistry
   and Microbiology, Division
   of Nutrition
P.O. Box 27, FIN-00014
University of Helsinki
Finland

Olga Makri
Laboratory of Plant Physiology
Faculty of Agricultural Biotechnology
Agricultural University of Athens
Iera Odos 75
11141 Athens
Greece

Johannes Novak
Institute for Applied Botany
University of Veterinary Medicine
Veterinärplatz 1
A-1210 Wien
Austria

Kirsti Kaarina Pääkkönen
Department of Food Technology,
P.O. Box 27, FIN-00014
University of Helsinki
Finland

Melpomeni Skoula
Department of Natural
 Products
Mediterranean Agronomical
 Institute of Chania (MAICh)
P.O. Box 85
73 100 Chania-Crete
Greece

Part 1

Introduction

1 Profile of the multifaceted prince of the herbs

Spiridon E. Kintzios

The scope of this volume is to offer an updated and analytic review on the currently available technical knowledge and market information of the world's commercially most valued spice – oregano. In addition, the book treats in detail various aspects of practical significance for the crop's industrialization – such as optimizing germplasm selection and utilization, novel cultivation methods and product processing, blending and uses in different countries, along with other market-related issues never included in previous reviews.

Oregano is the common name for a general aroma and flavor primarily derived from more than 60 plant species used all over the world as a spice. The majority of them belong to the Lamiaceae and Verbenaceae families, while a large distinction is made between the European (*Origanum* sp.) and Mexican (*Lippia* sp.) oregano.

European oregano is used as a flavoring in meat and sausage products, salads, stews, sauces, and soups. Prior to the introduction of hops, oregano was used to flavor ale and beer. The essential oil and oleoresin, used extensively in place of the plant material, are found in food products, cosmetics, and alcoholic liqueurs. Oregano is also a good salt replacement in tomato-containing recipes. Mexican oregano is used predominantly in flavoring Mexican foods, pizza, and barbecue sauces. Mexican oregano has a somewhat sharper and more pungent flavor than European oregano. The reader will find detailed information on the dietary properties of the spice and the various ways of adding it to foodstuff and beverage preparations. Issues of chemical stability and compatibility to diversified demand specifications will also be examined.

Most widely used is the genus *Origanum* (family Lamiaceae) (from the Greek words *oros* – mountain and hill and *ganos* – ornament). The taxonomy of the genus is rather complicated and a current issue of debate: indeed, *Origanum* sp. is characterized by a large (and still little investigated) morphological and chemical diversity resulting in the distinction of 49 taxa and 42 species. Respecting Ietswaart taxonomic revision there exist ten sections (*Amaracus* Bentham, *Anatolicon* Bentham, *Brevifilamentum* Ietswaart, *Longitubus* Ietswaart, *Chilocalyx* Ietswaart, *Majorana* Bentham, *Campanula ticalyx* Ietswaart, *Elongatispica* Ietswaart, *Origanum* Ietswaart, *Prolaticorolla* Ietswaart). Since Ietswaart's publication, five more species and one more hybrid have been recognized, raising the number of species to 43 and the number of hybrids to 18.

More than 300 scientific names have been given, during the last 150 years, to not more than 70 presently recognized *Origanum* species, subspecies, varieties and hybrids. This plethora of different names reflects the extent of morphological variation the genus exhibits in nature. The overwhelming majority of the taxa are locally distributed within the Mediterranean region, with nine species being located in Greece and 21 in

Turkey. Sixty per cent of all *Origanum* taxa are recorded to grow in Turkey, indicating this country as the gene center of *Origanum*. In addition, 17 hybrids between different species have been described, some of which are known only from artificial crosses. Very complex in their taxonomy, *Origanum* biotypes vary in respect of either the content of essential oil in aerial parts of the plant or essential oil composition. Essential oil 'rich' taxa with essential oil content of more than 2 per cent (most commercially known oregano plants), are mainly characterized either by the dominant occurrence of carvacrol and/or thymol (together with considerable amounts of γ-terpinene and p-cymene) or by linalool, terpinene-4-ol and sabinene hydrate as main components. The two most well commercially known 'oregano' species are *O. vulgare* subsp. *hirtum* (Greek oregano) (as well as winter sweet marjoram or pot marjoram, which is derived from *O. heracleoticum* and *O. onites* (Turkish oregano), each having an essential oil content of more than 2 per cent.

The genus *Lippia* (family Verbenaceae) is the most well known of several plants in Mexico that bear a resemblance to the Mediterranean oregano in terms of flavor and aroma, and the leaves of which have long been established by trade practice to be oregano (curly leaf oregano, Mexican sage, origan, oregamon, wild marjoram, Mexican marjoram, or Mexican wild sage). The genus *Lippia* Houst. consists of approximately 200 species of which 46 have been chemically examined.

The species mainly used are either *Lippia graveolens* or *L. berlandieri*. Because most of the species are aromatic, the studies on the chemistry of this genus are mostly related with the composition of the essential oils and only a very few ones devoted to the non-volatile constituents. An outstanding feature of *Lippia* is the difference observed in the essential oil composition reported for the same species from different geographic origins. The mono- and sesquiterpenoids found in the essential oils for all but two of the *Lippia* species investigated so far are quite common and widespread in the plant kingdom, the exceptions being *L. integrifolia* (Gris.) Hieron. which produces ketones based on the unique sesquiterpene skeletons named lippifoliane and integrifoliane and *L. dulcis* Trev. which contains (+)-hernandulcin, a sesquiterpenoid 1500 times sweeter than sucrose. Iridoids glucosides, phenylpropanoids, naphthoquinoids and flavonoids are the four types of significant non-volatile secondary metabolites reported in *Lippia*.

The *Origanum* species are perennial herbs native to the dry, rocky calcareous soils in the mountainous areas of southern Europe and south-west Asia, and the Mediterranean countries. The perennial, erect plants reach a height of 0.8 to 1 m and have pubescent stems, ovate, dark green leaves, and white or purple flowers. The root structure of oregano is such that it binds the soil and keeps it from washing away on steep slopes. European oregano is primarily produced in Greece, Italy, Spain, Turkey, and the United States. The *Lippia* species are small shrubs with larger leaves than the *Origanum* species and come primarily from Mexico.

The reported life zone for *Origanum vulgare* is 5–28 °C with an annual precipitation of 0.4–2.7 m and a soil pH of 4.5–8.7. Although much of the commercial material is collected from wild plants, fields can be seeded or established from transplants on light, dry, well-drained soils that are somewhat alkaline. Harvesting can take place two to six times per year. The *Lippia* species are predominantly collected as wild plants in Mexico.

Information is provided, in a very detailed manner, on structural dynamics of *Origanum* in order for the reader to get a picture about the construction and operation of the specific tissues composing the plant fundamental organs (leaf, stem, root).

Particular attention is given to glandular (peltate, capitate Type I and II) hairs and nonglandular hairs, their ontogenetic patterns from a protodermal initial cell and the topology of essential oil biosynthesis. Such data (e.g. knowing the structural components participating in a certain metabolic process as well as their temporal alterations and modifications) can be utilized for other fields of *Origanum* research.

There are a number of publications referring to the chemisty of *Origanum* which is known widely in the world of herbs and spices for its volatile oils. Oregano is the commercial name of those species that are rich in the phenolic monoterpenoids, mainly carvacrol, occasionally thymol, while marjoram is the commercial name of those that are rich in bicyclic monoterpenoids *cis-* and *trans*-sabinene hydrate. Besides the qualitative variation of the volatile compounds at the infragencric level, there is considerable quantitative variation at the infraspecific level. Remarkable chemical variations have not only been observed between but also within populations and accessions. The seasonal variation of essential oil yield and composition has received less attention. There is yet limited knowledge in the biosynthesis of the essential oil compounds and their inheritance, which would be useful for a more effective selection and establishing a targeted breeding program. For all these reasons, wild collection accompanied with quality and species maintaining assurance systems (sustainability, GHP) and/or field production of reliable genotypes are the future methods of choice for quality products. The enormous inter- and infraspecific chemical polymorphism of *Oreganum* sp. offers a wide range for selection towards the production of specific monoterpenes as fine chemicals, new odor and flavor profiles a.s.o. In the present book, a detailed chemotaxonomic account of *Origanum* taxa is presented by group and by section. *Origanum* species are rich in other compounds as well, such as various phenolics, lipids and fatty acids, flavonoids and anthocyanids. In a separate chapter, main breeding targets and quality criteria are set out and various breeding methods are discussed.

It is generally accepted that the Greek oregano has the best essential oil quality, the main constituents of which are carvacrol (the compound responsible for characterizing a plant as of the oregano type) and/or thymol, accompanied by *p*-cymene and γ-terpinene. Mexican oregano oil contains approximately equal amounts of carvacrol and thymol and smaller amounts of 1,8-cineole and other compounds. The basic composition of the oil varies with the plant source and geographical growth area.

Although oregano has been known and used for centuries, it gained only lately mass popularity, largely due to its relationship to marjoram (*O. marjorana*), the popular and botanical terms for both species having long been confused. And while sweet marjoram was one of the most popular herbs during the Middle Ages, oregano was scarcely cultivated, probably due to the plant's tendency to compete against other plants growing nearby. On the other hand, wild oregano has been traditionally collected in Mediterranean countries and in Mexico for use in many of the favorite dishes (e.g. for tomato-based sauces, lamb, seafood, chili peppers and almost any garlic flavored dish). The rest of the world discovered oregano after Second World War, with the expansion of pizza consumption (and in a lesser-degree, Mexican-style foods). Oregano consumption boomed from almost nil to a consumption volume of over 500 000 tons, demonstrating a per capita increase of importation into the United States of 3800 per cent from 1940 to 1985. Product prices depend heavily on quality. The overall market of oregano is expanding becoming thus by far the largest selling herb today. Latest estimates put worldwide production of oregano at about 10 000 tons. Turkey has a dominant position in the worldwide trade of oregano (over 2/3 of the total production), followed by Mexico, Greece and other Mediterranean countries. Greece has long been a leading source and its product has traditionally commanded the highest prices, nevertheless not always sufficiently meeting demand. Though Italy harvests large amounts of oregano, most of it is consumed domestically. The Mediterranean-type of product, as compared to the Mexican, is a smaller leaf of somewhat lighter green color and milder, sweeter flavor. As compared to sweet marjoram, however, it is much strongly flavored. The harvesting and processing of oregano is similar in Mediterranean and Mexican areas. In a separate chapter, a comparative demonstration is given of cultivation practices of oregano in different producing countries, such as Turkey, Hungary, Germany, Israel, Slovenia, the Federal Republic of Yugoslavia and Albania.

Cultivation of oregano can be a profitable business. Growers currently enjoy increased market prices due to the limited product availability, as a result of the exhaustion of wild oregano populations due to intensive collection. A recent survey in Greece (Papanagiotou *et al.*, 2001) indicated that for a given average yield of 1850 kg per hectare and an average product price of 4.1 € per kg, the net profit for the grower is 2500 € per hectare, a value considerably higher then for most crop and horticultural species. Labor (1260 man-hours/hectare) was estimated to reach 64 per cent of the total production cost.

The herb is often sold by mesh size, indicating average particle size. In United States oregano imports are roughly equal from both Mediterranean and Mexican species. The Mexican Oregano is a much stronger, more robustly, 'wild' flavored oregano. After cleaning, the leaves of Mediterranean oregano come into a size of 30 or 60 mesh, with larger leaf particles giving the choicest, more refined appearance. In Mexico, shippers often refer to their most refined product as 'Greek cut.' In the United States the herb is offered as ground or whole leaf oregano (although not always in the original whole form). Beyond that, various mesh sizes may also be available, each being the most appropriate choice for a particular use. Other important species collected and marketed as European oregano include *Thymus capitatus* (Spanish oregano), *Origanum syriacum* (*Origanum maru* – Syrian marjoram or zatar) and *Origanum virens*. Additional species used in Mexican oregano include *Lippia palmeri* and *Lippia origanoides*.

The original fresh material is the essential factor determining the quality of the dried herbs. Nevertheless, the drying method, type of packaging, and storage conditions also has clear effects on the microbiological quality of the herbs. An exemplary investigation is the Finnish Herb Study, presented in detail in this volume.

As a medicinal plant, European oregano has traditionally been used as a carminative, diaphoretic, expectorant, emmenagogue, stimulant, stomachic, and tonic. In addition, it has been used as a folk remedy against colic, coughs, headaches, nervousness, toothaches, and irregular menstrual cycles. Turkish villagers have traditionally used *kekik water*, the aromatic water obtained after removing essential oil from the distillate of oregano herbs, which has in recent years become a commercial commodity. *Origanum* oil is a powerful disinfectant, and carvacrol and thymol are considered to be anthelmintic and antifungal agents. The documented insecticidal and allelopathic properties of the species suggest its potential value as a biological control agent. Antibacterial/fungicidal properties are reviewed in detailed as is the use of the genus as a food ingredient, either for its antioxidant properties or its distinctive flavor. Especially, the monoterpenic phenol carvacrol, the main constituent of commercial oreganos should be given special emphasis since its low toxicity and surprisingly high and diverse biological activities render this simple molecule a promising lead for the development of novel medicines not only for humans but also for animals and plants. Carvacrol-rich oils are also used for their high potency especially as antibacterial and antifungal agents. The use of carvacrol-rich oils as ingredients in animal feed and for the preservation of food against bacterial or fungal spoilage and as antioxidants appears to increase. The *Lippia* species of oregano are generally recognized as safe for human consumption as natural flavorings/seasonings, and the *Origanum* species are generally recognized as safe as natural extracts/essential oils. However, oregano is one of the most common foodstuffs, which have the ability to cause aversions during pregnancy. In addition, and according to some reports, oregano intake can cause systemic allergic reactions. These items are particularly treated with.

Although the monograph documentation of *O. vulgare* was submitted to German Ministry of Health, the stuff responsible for phytotherapeutic medicinal domain – Commission E evaluated *Origani vulgaris herba* negatively (Banz. Nr. 122 from 6th July 1988), because of lack of scientific proofs for a number of indication areas (Blumenthal, 1998). Nevertheless, many of the studies confirmed benefit effects of oregano for human health. In this volume, extensive reference is provided on the bioactive/pharmacological properties of the *Origanum* genus, such as antioxidant and antimicrobial action and its use for the treatment of a vast list of ailments, such as respiratory tract disorders like cough or bronchial catarrh (as expectorant and spasmolitic agent), in gastrointestinal disorders (as choleretic, digestive, eupeptic and spasmolitic agent), as an oral antiseptic, in urinary tract disorders (as diuretic and antiseptic) and in dermatological affections (alleviation of itching, healing crusts, insect stings), viral infections and even cancer. Oregano, its essential oil or isolated compounds have been studied also from the point of possible applications in plant protection, in post-harvest crop/fruit protection or in apiculture, where species specific fungi endanger the production systems. Finally, *Origanum* taxa, especially those that are rich in essential oils, have been extensively studied for their insect-pollinating or nectar yielding effects.

In spite of its commercial importance *Origanum* sp. is still an underutilized plant species and very few commercial oregano cultivars exist. Even today, oregano production from cultivation is less than production from the wild, which results to an apparent danger of genetic erosion. This situation can probably change due to the increased knowledge on cultivation technology (such as fertilization, irrigation, pest management and hydroponics) and the exploitation of the natural variability, either towards the selection and introduction to cultivation of new biotypes or the creation of new cultivars through breeding. For this purpose, a deeper, updated knowledge of the plant's biosynthetic pathways, in relation to its ecophysiological behavior and biology is essential. Equally important is the construction of appropriate germplasm databases, an updated account of which is presented in this volume. Finally, the possible use of waste products from oregano processing, such as *cyclone powder* (Oregano lipids) in shampoos and other cosmetic preparations is evaluated.

Several *Origanum* species are cultivated as culinary herbs and as garden or medicinal plants. Hereby most of the genetic diversity of the genus is concentrated in collections of individual growers, which contain about 600 accessions. Different approaches have been undertaken for the conservation of oregano germplasm, and the activity of international seed- and germplasm banks and affiliated organizations (such as the Oregano Genetic Resources Network) is reviewed.

Both *Origanum* and *Lippia* species are yet a novel target for biotechnology, though some interesting progress has been achieved towards the micropropagation of certain species, such as *O. vulgare*, *O. bastetanum* and *L. junelliana*. The current focus of such applications is related to the establishment of clonal populations, which could offer a means to overcome the problems associated with germplasm variability. Further perspectives include the use of cell cultures for the scaled-up production of useful pharmaceuticals and the enhancement of breeding activities (e.g. *in vitro* selection for disease resistance, exploitation of somaclonal variation).

Monographs on specific life sciences related topics frequently forego certain aspects of methodology. When carrying out comprehensive Origanum- or Lippia-related searches, more than one database should be consulted. In Origanum some 54 per cent of all investigated references can retrieved by one (most inclusive) database what is not enough. By investigating several different bibliographic databases and references/records on genera *Lippia* and *Origanum*, one can identify a total of 1001 references published in 476 journals during 1981–1998. It is evident that single databases, or information sources, cannot serve as good examples to illustrate some principal directions in the field of medicinal plants. Prior to setting up a research and conducting database searches the end-users should gather comprehensive instructions on each particular information service and get familiar with its construction and maintenance characteristics, as extensively described in this book.

REFERENCE

Papanagiotou, E., Papanikolaou, K. and Zamanidis, S. (2001) The cultivation of aromatic and medicinal plants in Greece: 1. The economic dimension. *Agriculture* 1, 36–42 (in Greek).

Part 2
Botany

2 Structural features of *Origanum* sp.

Artemios M. Bosabalidis

INTRODUCTION

The genus *Origanum* (Lamiaceae) includes 39 species widely distributed in the Mediterranean region (Vokou *et al.*, 1993). The species *Origanum vulgare* predominates in occurrence, while *Origanum dictamnus* is endemic of the island of Crete in southern Greece. The plants are perennial herbs spontaneously growing in calcareous substrates. One of the striking morphological characteristics of the *Origanum* plants is the presence of glandular and nonglandular hairs covering the aerial organs. Both types of hairs originate from epidermal (protodermal) cells. Every epidermal cell, according to Netolitzky (1932), practically possesses the potentiality to develop into a hair. A number of external and internal factors, however, determine which ones of the typical epidermal cells will develop into hairs, and furthermore, which morphological pattern they will follow.

The glandular hairs are numerous on the vegetative organs (stems, leaves, bracts), while their density becomes reduced on the reproductive organs (calyces, corollas). On the stamens and the gynoecium of the flowers, no glandular hairs have been reported to occur (Werker *et al.*, 1985b). The glandular hairs secrete an essential oil with a characteristic odour, mainly due to the major components of the oil, the monoterpenes carvacrol and thymol. The essential oils of *Origanum* have been found to develop strong antioxidant, antimicrobial, insecticidal and genotoxic activities (Scheffer *et al.*, 1986; Sivropoulou *et al.*, 1996; Karpouhtsis *et al.*, 1998).

Though descriptions of the morphology of *Origanum* species are quite often given in the literature on plant taxonomy (Ietswaart, 1980; Kokkini *et al.*, 1991; Circella and D'Andrea, 1993), relevant reports on the anatomy and cytology of the various *Origanum* organs, are limited. In the present article, an attempt was made to partly bridge this gap by providing information on the structural features of *Origanum*.

STRUCTURE OF THE FUNDAMENTAL ORGANS

The leaf

Morphology of the leaf

The leaves of *Origanum* species may differ in size, shape and thickness, as well as in the density, type and size of the nonglandular and glandular hairs covering their surfaces. The leaf of *O. dictamnus* is nearly rounded with a mean side surface area of 358.75 mm^2 (Bosabalidis,

1990a). It is densely covered with branched nonglandular hairs which lend to it a velvety appearance (Figure 2.1). Hairs are more numerous on the lower leaf side, which, thus, becomes brighter. The distribution of the nonglandular hairs is uniform all over the leaf surface with no particular accumulation at a certain area. The leaf also bears glandular hairs which appear as small transparent droplets scattered throughout (Figure 2.1, arrows). Their population is remarkably lower than that of the nonglandular hairs.

The leaf of *Origanum* × *intercedens* (hybrid from *O. onites* and *O. vulgare* subsp. *hirtum*) is ovate in shape, petiolate, with a serrate margin. At full expansion, each leaf side was measured to display a mean surface area of 207.83 mm^2 and a thickness of 203.38 μm (Bosabalidis and Exarchou, 1995). The pubescence is composed of glandular and nonglandular hairs as well (Figure 2.2). The glandular hairs are uniformly distributed and their number and size are higher on the upper leaf side. The nonglandular hairs are multicellular uniseriate and their tips face the leaf apex. They are more numerous on the abaxial leaf side veins and their density and length increase toward the leaf base.

The leaf of *O. vulgare* subsp. *hirtum* does not significantly differ from that of *Origanum* × *intercedens*, as concerns the size, shape and thickness, as well as the density of the glandular and nonglandular hairs.

Anatomy of the leaf

Though a great number of publications on leaves of aromatic plants exists, most of them deal with the glandular hairs and their secretions, and only a few have proceeded with the study of leaf anatomy (Henderson *et al.*, 1970; Yamaura *et al.*, 1992; Werker *et al.*, 1993; Bosabalidis and Exarchou, 1995; Bosabalidis and Kokkini, 1997; Bosabalidis and Skoula, 1998b; Gavalas *et al.*, 1998).

In a cross section of a fully grown leaf of *O. vulgare* subsp. *hirtum*, the cells of the upper epidermis appear much more larger than those of the lower one and covered with a relatively thick cuticular layer (Figure 2.3). Stomata exist on both leaf epidermises, but their number is significantly higher on the lower epidermis (Table 2.5). The chlorenchyma cells of the palisade parenchyma are elongated and ordinarily arranged in one layer. The chloroplasts are numerous, usually located at the parietal cytoplasm. The cells of the spongy parenchyma mostly exhibit a round profile and they leave between them large intercellular spaces. At the positions on both leaf surfaces where peltate glandular hairs occur, the epidermis forms more or less deep depressions (not shown in Figure 2.3). Morphometric assessments conducted on leaf cross sections of *O. vulgare* subsp. *hirtum* resulted in the estimation of the relative volumes of the partial histological components (Table 2.1).

Table 2.1 Relative volumes (±standard deviation) of the leaf histological components in *Origanum vulgare* subsp. *hirtum*

Component	n	RV (%)
Upper epidermis	24	9.3 ± 1.8
Palisade parenchyma cells	24	27.4 ± 12.6
Palisade parenchyma intercellular spaces	24	6.8 ± 1.9
Spongy parenchyma cells	24	28.8 ± 3.1
Spongy parenchyma intercellular cells	24	15.1 ± 2.8
Lower epidermis	24	7.7 ± 1.6

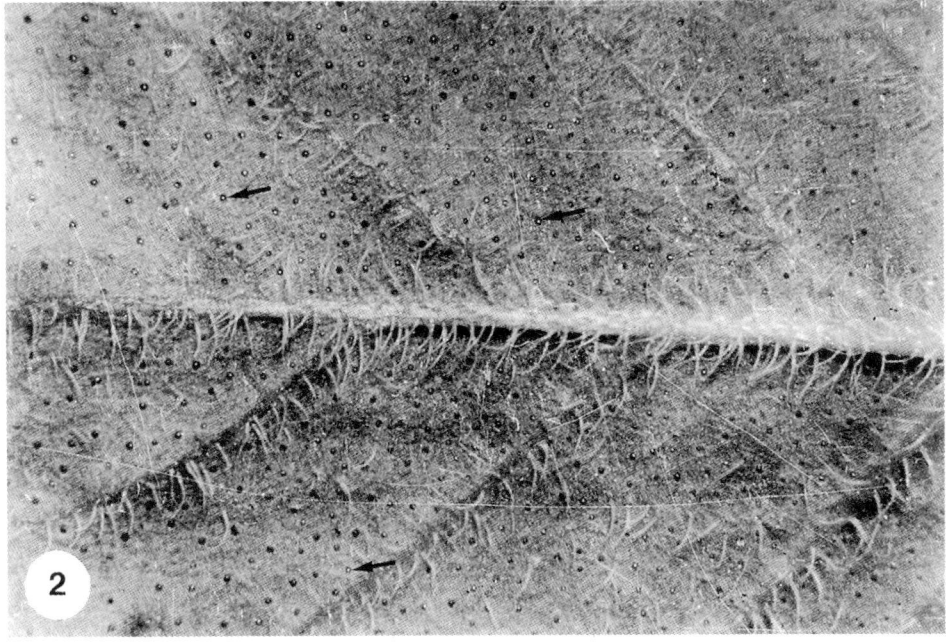

Figures 2.1 and 2.2 Partial view of the morphology of the leaf lower surface in *Origanum dictamnus* (Figure 2.1 [×19]) and *Origanum × intercedens* (Figure 2.2 [×11]). Arrows point to glandular hairs.

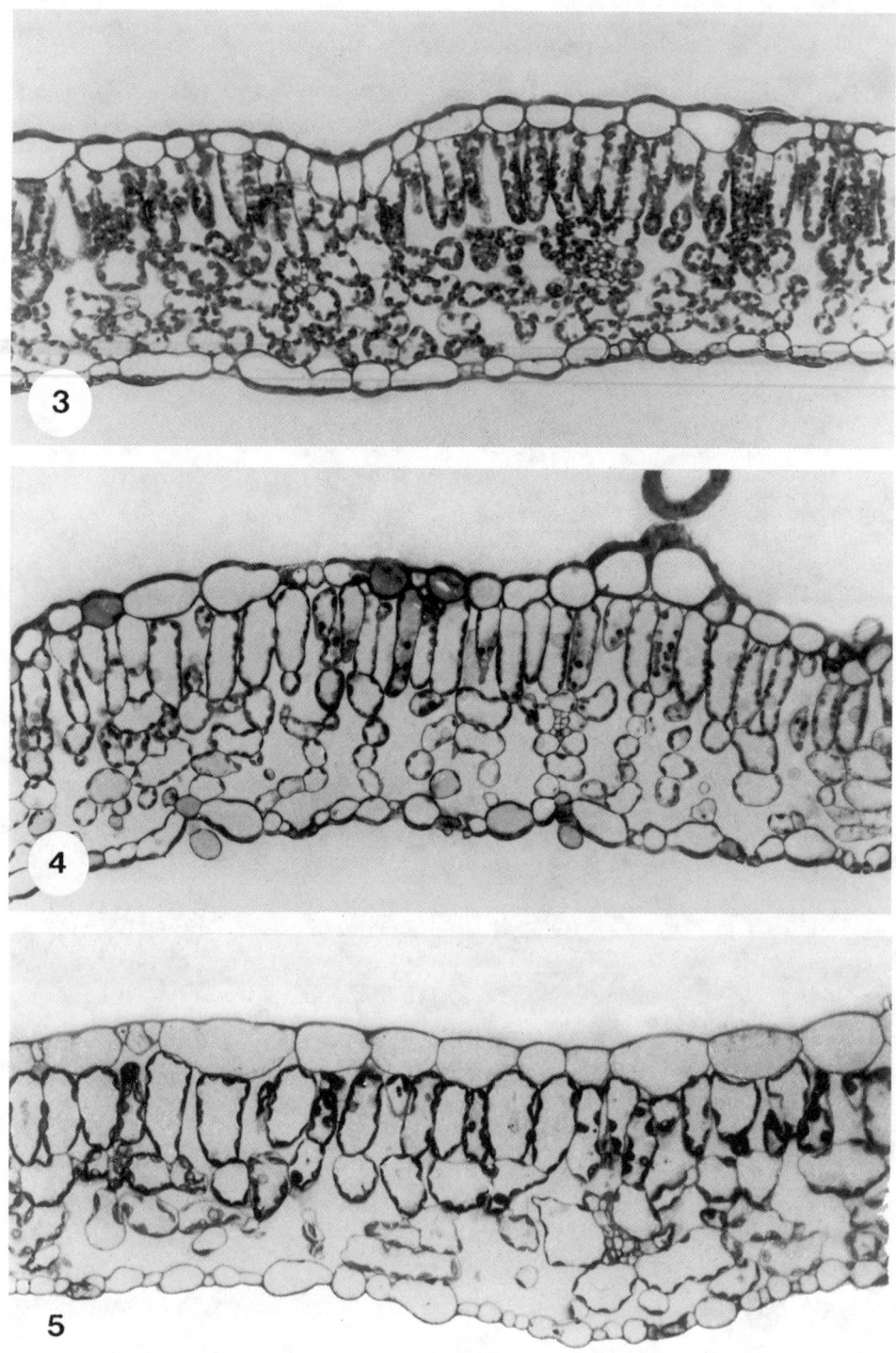

Figures 2.3–2.5 Cross sections of leaves from *Origanum vulgare* subsp. *hirtum* (Figure 2.3), *Origanum × intercedens* (Figure 2.4) and *Origanum dictamnus* (Figure 2.5) to show corresponding leaf anatomy (×200).

Cross sections of leaves of *Origanum* × *intercedens* (Figure 2.4) as well as those of *O. dictamnus* (Figure 2.5) have disclosed no significant anatomical differences, compared to *O. vulgare* subsp. *hirtum*. Worthy differences mainly comprise the length and width of the palisade parenchyma cells and the density of chloroplasts.

The epidermis

TYPICAL EPIDERMAL CELLS

In leaf cross sections of *Origanum* species, the typical epidermal cells appear to differ in size and number between the upper and lower leaf sides (Figures 2.3–2.5). Leaf paradermal sections reveal that epidermal cells are in close contact with each other leaving between no intercellular spaces (Figure 2.6). The cell outlines (anticlinal walls) are irregular in the form of successive foldings, which contribute to a better adhesion of the cells. This holds true for both the upper and lower leaf surfaces (Figures 2.6–2.7).

At early development, the epidermal cells appear in ultrathin sections rich in protoplasm with large nuclei, ordinarily located at the proximal cell region (Figure 2.8). Vacuoles are relatively small and they occupy the distal cell region. None of the typical cell organelles proliferates in volume or number. The leucoplasts have an irregular shape and they contain large globular inclusions which are amorphous and of high electron density (Figures 2.8 arrows, 10). Similar intraplastidal inclusions have been also encountered in the initial cells of the glandular hairs (peltate and capitate), as well as in the basal cells of the differentiated glandular hairs.

At advanced development, the epidermal cells become occupied at their greatest portion by a large vacuole with an electron translucent content (Figure 2.9) (Bosabalidis, 1987b). The cytoplasm is limited at the periphery of the cells forming a thin perietal layer in which plastids, small mitochondria, short elements of the rough endoplasmic reticulum and a few dictyosomes, are embedded. Quantitative estimations on the relative volumes of the cytoplasm, the central vacuole and the nucleus of dittany leaf epidermal cells, are presented in Table 2.2. The leucoplasts still contain large globular inclusions, which, however, this time have a fine granular appearance (Figures 2.9 arrow, 11). The inclusions are bordered by a single membrane and they occupy about 54 per cent of the plastidal volume (Table 2.3).

Morphometric assessments conducted on the epidermal cell plastids (and their inclusions) gave specific values associated with volume and surface area (Table 2.4) (Bosabalidis, 1987b). Leucoplast membrane-bound inclusions with a circular outline and an amorphous internal structure, have been also observed in the cambial cells of *Fraxinus americana* (Srivastava, 1966), the tissue culture of *Daucus carota* (Israel and Steward, 1967), the vascular parenchyma of *Callitriche stagnalis* (Favali and Patrignani, 1973), the mesophyll of *Phaseolus aureus* (Hurkman and Kennedy, 1976), the laticifers of *Papaver somniferum* (Nessler and Mahlberg, 1979), etc.

The chemical nature of the plastidal inclusions is proteinaceous, as cytochemical tests with pepsin incubation have shown (Bosabalidis, 1987a). Pepsin resulted in a progressive digestion of the inclusion bodies (Figures 2.12, 2.13). Similar results for plastidal membrane-limited inclusions have been also obtained in the chlorenchyma cells of palms (Gailhofer and Thaler, 1974) and of tobacco (Ames and Pivorun, 1974) by using the same proteolytic enzyme. Treatment of ultrathin sections with diaminobenzidine (DAB) further resulted in an intense staining of the inclusions (Figure 2.14). The great

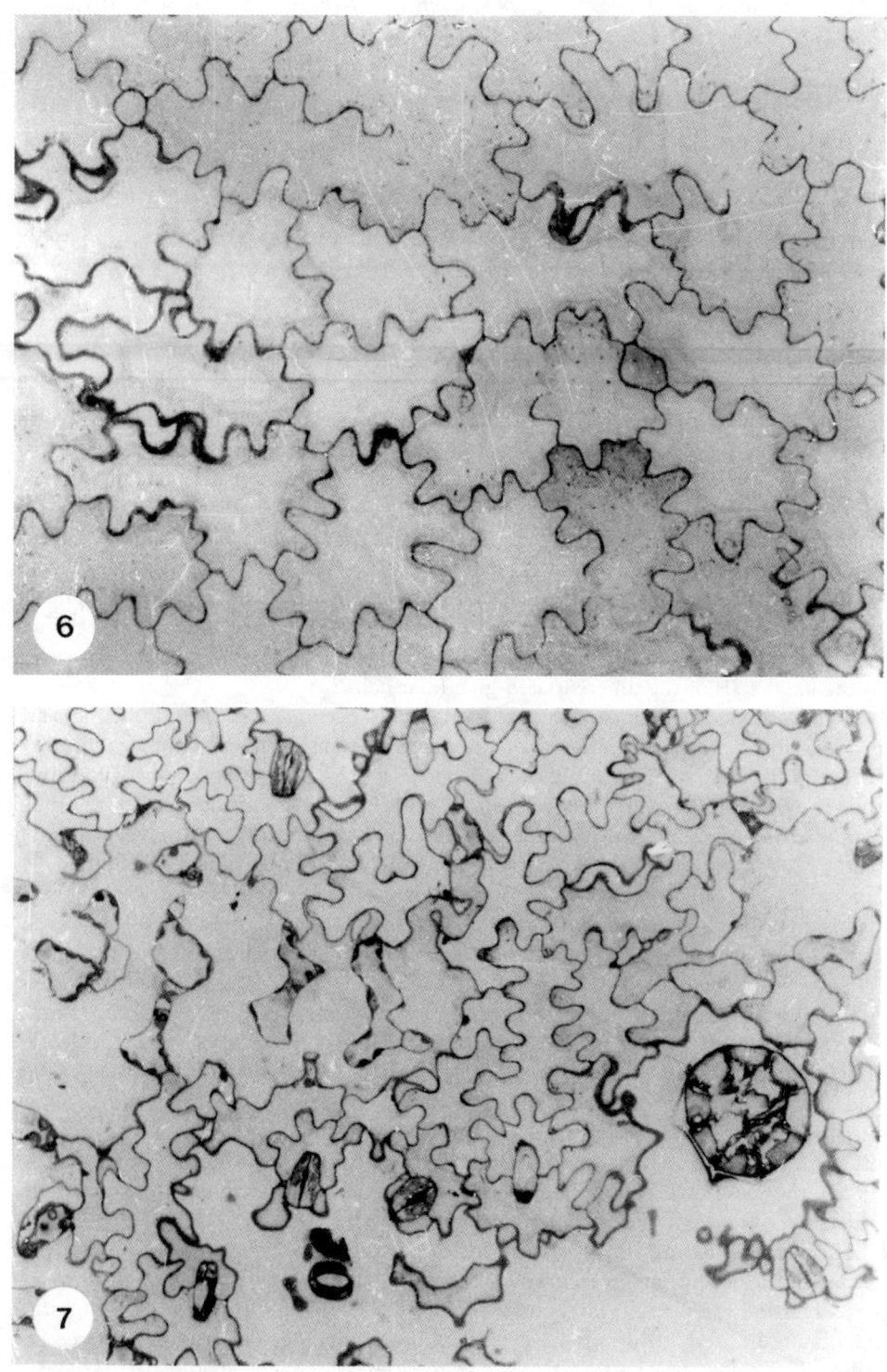

Figures 2.6–2.7 Surface views of the upper (Figure 2.6) and the lower (Figure 2.7) leaf epidermes of *Origanum*. The contours (anticlinal walls) of the epidermal cells are highly refolded (×320).

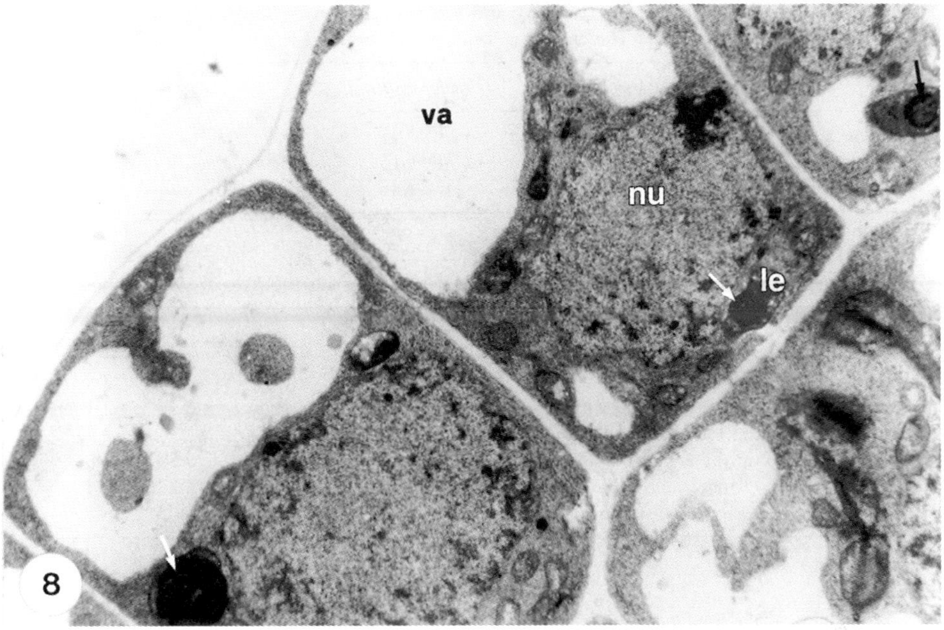

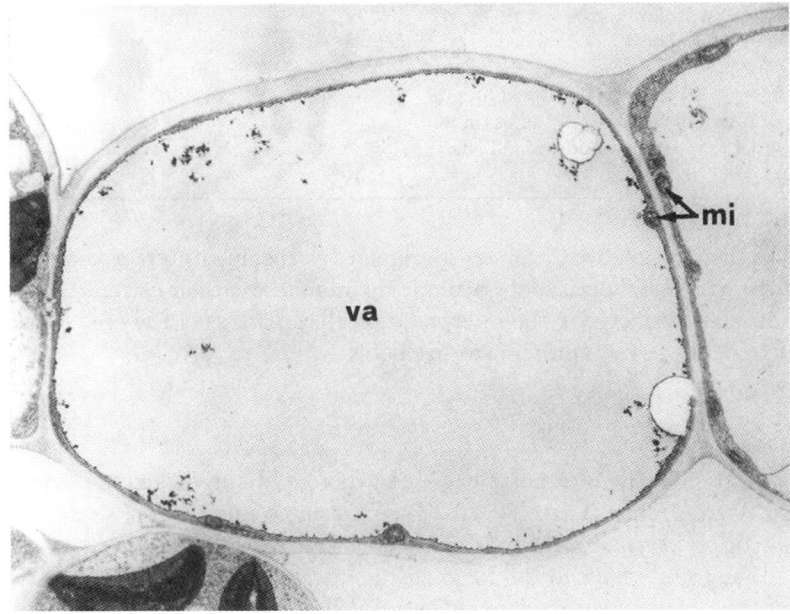

Figure 2.8 Epidermal cells of young leaves of *Origanum dictamnus*. Cells are rich in cytoplasm with large nuclei (nu). The leucoplasts (le) contain globular, electron dense inclusions (arrows). va, vacuole; mi, mitochondria (×8200).

Figure 2.9 Epidermal cells of developed leaves of *Origanum dictamnus*. Cells contain a voluminous central vacuole (va) limiting the cytoplasm to a thin parietal layer. Inclusions of plastids appear fine granular (arrow). mi, mitochondrion (×8000).

Table 2.2 Average actual volumes (in μm³) and percentage volume values of cell components in leaf epidermal cells of *Origanum dictamnus*

	Upper epidermis	Lower epidermis	%
Nucleus	3083.14	490.49	2.4
Central vacuole	118 187.14	18 801.15	92.0
Cytoplasm	7194.01	1144.48	5.6

Table 2.3 Morphometric characteristics of epidermal cell plastids in leaves of *Origanum dictamnus*

Parameters	Mean + SD
Volume fraction of plastids in epidermal cell cytoplasm	0.17 ± 0.03
Numerical density of plastids per μm³ of epidermal cell cytoplasm	0.16 ± 0.01
Volume fraction of inclusion bodies in plastids (estimated by point lattice overlay and phot. paper weighing)	0.54 ± 0.08
Surface density of inclusion bodies in plastids (μm²/μm³)	3.28 ± 0.75
Diameter of inclusion bodies, μm ($D = 4 d/\pi$)	1.05 ± 0.21

Table 2.4 Values of morphometric parameters concerning epidermal cell plastids in leaves of *Origanum dictamnus*

Parameters	Upper epidermis	Lower epidermis
Average total volume of plastids per epidermal cell (in μm³)	1222.98	194.56
Average number of plastids per epidermal cell	1151.04	183.12
Average volume of a plastid (in μm³)	1.06	1.06
Average total volume of plastid inclusion bodies per epidermal cell (in μm³)	656.09	104.38
Average volume of inclusion body per plastid (in μm³)	0.57	0.57
Average value of plastid inclusion body radius (in μm)	0.52	0.52
Average total surface area of plastid inclusion bodies per epidermal cell (in μm²)	4011.37	638.16

percentage of the intraplastidal volume occupied by the inclusions, their chemical nature, their compartmentalization by a single membrane and their existence throughout plastid lifetime, converge to the interpretation that plastids of leaf epidermal cells of *Origanum* probably serve as protein storing pools.

STOMATA

The structure and ontogeny of stomata have been thoroughly investigated in Lamiaceae (Inamdar and Bhatt, 1972; Cantino, 1990). The most often met types of stomatal complexes are the diacytic type (guard cells surrounded by two subsidiary cells with their common walls at right angles to the long axis of the stoma) and the diallelocytic one (guard cells surrounded by three or more C-shaped subsidiary cells of graded size, oriented perpendicularly to the guard cells). To a lesser extent, leaves of Lamiaceae bear the anisocytic type of stomatal complex (guard cells bordered by three subsidiary cells, one of

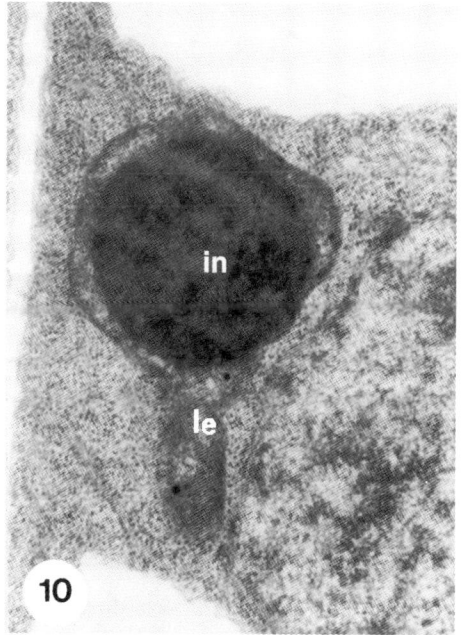

Figure 2.10 A large electron dense inclusion (in) in a leucoplast (le) of a young leaf epidermal cell (×23 500).

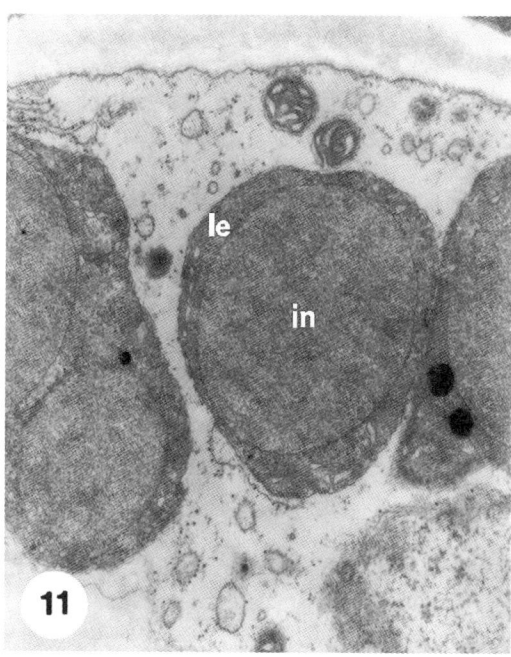

Figure 2.11 A membrane-bound, fine-granular inclusion (in) in a leucoplast (le) of a developed leaf epidermal cell (×24 000).

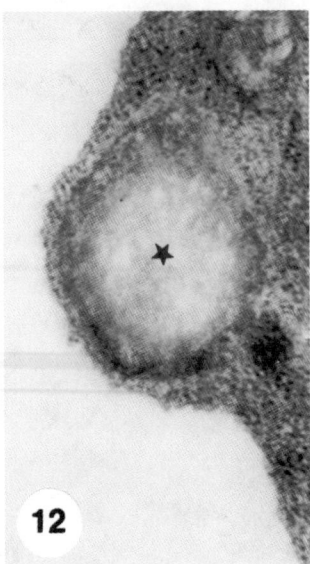

Figure 2.12 Short-time pepsin treatment of ultrathin sections. Plastid inclusion bodies appear partly digested (asterisk) (×22 500).

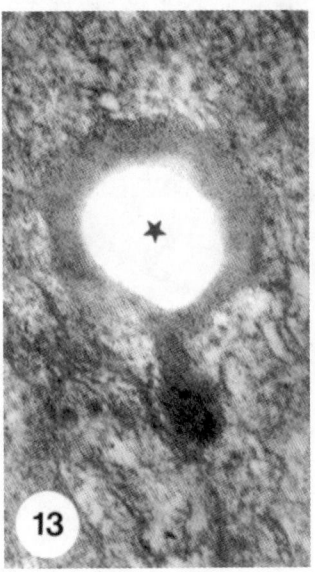

Figure 2.13 Long-time pepsin treatment of ultrathin sections. Plastid inclusion bodies have been thoroughly digested (asterisk) (×22 000).

which is markedly smaller than the other two) and the paracytic type (guard cells accompanied on either side by one or more subsidiary cells parallel to the long axis of the stoma).

Contrary to stomatal structure, studies on stomatal distribution and density in Lamiaceae are quite limited. Inamdar and Bhatt (1972) have reported in 33 species of

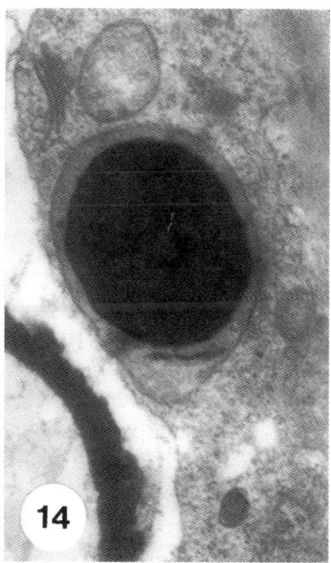

Figure 2.14 An electron opaque plastid inclusion strongly stained after diaminobenzidine (DAB) incubation (×22 000).

Table 2.5 Number of stomata and glandular hairs per mm² of leaf surface area in *Origanum* taxa (mean ± SD)

Taxon	Upper leaf surface		Lower leaf surface	
	Stomata	Gland. hairs	Stomata	Gland. hairs
Origanum vulgare subsp. *hirtum*[a]	97.6 ± 8.99	11.30 ± 0.92	206.32 ± 24.85	10.09 ± 1.77
Origanum vulgare subsp. *hirtum*[b]	103.08 ± 12.59	7.63 + 0.57	355.26 ± 52.51	5.54 ± 0.76
Origanum vulgare subsp. *vulgare*	43.31 ± 5.64	4.26 ± 0.29	283.08 ± 38.24	3.19 ± 0.50
Origanum vulgare subsp. *viridulum*	5.56 ± 0.63	1.48 ± 0.03	214.21 ± 33.07	1.16 ± 0.4
Origanum × *intercedens*	17.39 ± 4.57	6.42 ± 0.86	207.89 ± 16.08	5.15 ± 0.69
Origanum dictamnus		3.39 ± 0.75		4.11 ± 0.22

Notes
a from Crete.
b from Euboea.

the family (*Origanum* not included) that stomata are mostly localized on the lower leaf side only, with some exceptions where stomata occur on both leaf surfaces. Both authors have determined that the density of stomata ranges from 144 to 1200 per mm² of leaf surface area, depending upon the species. An analogous variability is also mentioned by Beerling and Kelly (1997) in several Lamiaceae plants. Bosabalidis and Kokkini (1997) and Bosabalidis and Skoula (1998b) have correlated the density (No/mm²) of stomata on both leaf sides of *Origanum* taxa with the corresponding density of glandular hairs (Table 2.5). Stomata were found to be significantly higher in number on the lower leaf side rather than on the upper one, in contrast to the glandular hairs which are more numerous on the upper leaf side. In Mentha longifolia, Mentha spicata and their hybrid Mentha × villoso-nervata, stomata as well as glandular hairs occur in higher numbers on the lower leaf side (Gavalas et al., 1998). Combined developmental studies on stomata and glandular hairs (both originating from leaf protodermal cells) could constitute a source of information for their simultaneous differentiation from cytologically similar initials into morphologically and functionally different structures. It would also be beneficial, to anatomically, physiologically and biochemically investigate the mechanisms mobilized by both types of epidermal structures, in order each of them to extend and mark its own territory area on the leaf surface. Whether in this allelopathic procedure, phytohormones or other primary and secondary substances play a decisive role, it is to be determined.

In O. vulgare subsp. hirtum, altitute and season do not appear to significantly affect the density of stomata, as concerns the lower leaf side (c. 430 stomata/mm^2) (Kofidis and Bosabalidis, unpublished). The relevant stomatal density for the upper leaf side also does not appear to alter with season (c. 100 stomata/mm^2), but it becomes higher with increasing altitute (c. from 20 to 200 stomata/mm^2).

TRICHOMES

The aerial parts of many plants are covered with trichomes, which may be glandular or nonglandular. The members of each category exhibit a broad spectrum of variability, as concerns their shape, size, anatomy, ultrastructure, function, origin, pattern of development, etc. More than a century ago, microscopists tried to investigate some of the above trichome characteristics (particularly the hair form) and used them as a criterion for plant classification (Weiss, 1867; de Bary, 1877). The importance of the various types of trichomes in systematics has been greatly acknowledged and many works on this subject have appeared since that time (Gupta and Bhambie, 1978; Metcalfe and Chalk, 1979; Cantino and Sanders, 1986; Karousou et al., 1992). Modern techniques, such as sophisticated microscopy (scanning and transmission electron microscopy), analytical chemistry, biochemistry, genetics, molecular biology, biotechnology, etc. have been use as fundamental means for the study and classification of plant trichomes (Wollenweber, 1984; Harborne et al., 1986; Walt van der and Demarne, 1988; Bini-Maleci and Servettaz, 1991; Berta et al., 1993). As to the functional and ecological roles of the plant pubescence, the nonglandular hairs were long time ago considered to have a significant contribution to the obstruction of the free movement of the water vapours from the stomata (transpiration), as well as to the reduction of leaf overheating (Gradmann, 1923; Staudermann, 1924). They have been also considered to create difficulties in the movement, feeding and oviposition of mites and aphids on the leaf surface (Walters et al., 1989; Goertzen and Small, 1993).

The glandular hairs seem to operate in different ways than the nonglandular ones. Their secretions (monoterpenes, sesquiterpenes, phenolics, sucrose esters, etc.) have been observed to have a repellent character on harmful insects (Levin, 1973; Werker, 1993), to be toxic when eaten (Klingauf et al., 1983; Bestmann et al., 1987), to inhibit egg hatching (Sharaby, 1988; Konstantopoulou et al., 1992) and to function as sticky traps (Kowalski et al., 1988). Diterpenes and triterpenes have been further found to be deterrent, toxic and severe skin irritants to herbivorous mammals (Rosenthal and Berenbaum, 1991). Of significance are, in addition, the antimicrobial and allelopathic properties of the terpenoid components of the essential oils secreted by the glandular hairs (Vokou and Margaris, 1986; Sivropoulou et al., 1996). Recently, an important role has been attributed to the glandular hairs and their secretions (phenolic compounds of the essential oils), as concerns their implication in the protection of the mesophyll cells from UV-B radiation (Fahn and Shimony, 1996; Bosabalidis and Skoula, 1998b; Manetas, 1999). Glandular hair products may not only have a repellent character for insects, but also an attractive one, as in the case of nectaries, where they play a decisive role in the process of pollination (Fahn, 1979; Sawidis et al., 1989).

Nonglandular hairs

In *Origanum* taxa, nonglandular hairs have been observed to occur on the vegetative and reproductive organs as well (Werker et al., 1985b). In the leaves, they are present on both surfaces, but their number is higher on the veins of the lower leaf surface (Figures 2.1, 2.2). Their density and length increase toward the base of the blade.

Structure

The nonglandular hairs of *Origanum* are usually falcate filiform, uniseriate, composed of 4–7 cells (branched configurations may also exist, as in the case of *O. dictamnus*). The most distal cell of each hair ordinarily points to the leaf tip (Werker et al., 1985b; Bosabalidis and Exarchou, 1995).

Apart from *Origanum*, this type of nonglandular hairs has been also observed in other genera of Lamiaceae, such as *Ocimum* (Werker et al., 1993), *Teucrium* (Bini-Maleci and Servettaz, 1991), *Nepeta* (Bourett et al., 1994), *Monarda* (Heinrich, 1973), *Thymus* (Economou-Amilli et al., 1982), *Satureja* (Bosabalidis, 1990b), *Salvia* (Serrato-Valenti et al., 1997), *Mentha* (Gavalas et al., 1998), *Plectranthus* (Ascensao et al., 1998), *Leonotis* (Ascensao et al., 1995), *Coridothymus*, *Majorana*, *Micromeria*, *Melissa* (Werker et al., 1985a), etc.

Though the falcate form of nonglandular hairs seems to be the dominant in Lamiaceae, other types, like flask-shaped in *Teucrium* (Bini-Maleci and Servettaz, 1991), conical (unicellular or bicellular) in *Salvia*, *Monarda* and *Thymus* (Heinrich, 1973; Economou-Amilli et al., 1982; Bisio et al., 1999) and capitate (uniseriate and multi-seriate) in *Ocimum* (Gupta and Bhambie, 1978) may also exist.

Ontogeny

The nonglandular hairs of *Origanum* initiate and develop very early before the mesophyll becomes differentiated into palisade and spongy parenchymas. They originate from a single protodermal cell which is much voluminous than its neighbouring cells (Figure 2.15). This initial cell contains a centrally located nucleus and it divides periclinally and asymmetrically to give two daughter cells i.e. a large vacuolated basal cell and a second dome-shaped cell seated above the former (Figure 2.16). The basal cell may further divide twice to produce four cells, or it may divide successively resulting in the formation of a pedestal. The apical daughter cell actually constitutes the mother cell of the hair and it undergoes a series of periclinal divisions to derive two (Figure 2.17), three (Figures 2.18–2.20), four, five (Figure 2.21), and so on, up to seven cells. The derivatives are initially short (Figures 2.17, 2.18), but they progressively increase in length (Figures 2.19, 2.20). The free end of the most distal cell of the hair becomes acute and in the progress of development it bends toward the leaf tip (Figure 2.20). Branching of nonglandular hairs (Figure 2.22) may be frequent (*O. dictamnus*) or rare (*O. vulgare*, *O. onites*, *Origanum × intercedens*, etc.). The cytological characteristics of the cells composing a nonglandular hair comprise a large central vacuole with no dark deposits, as well as a thin peripheral

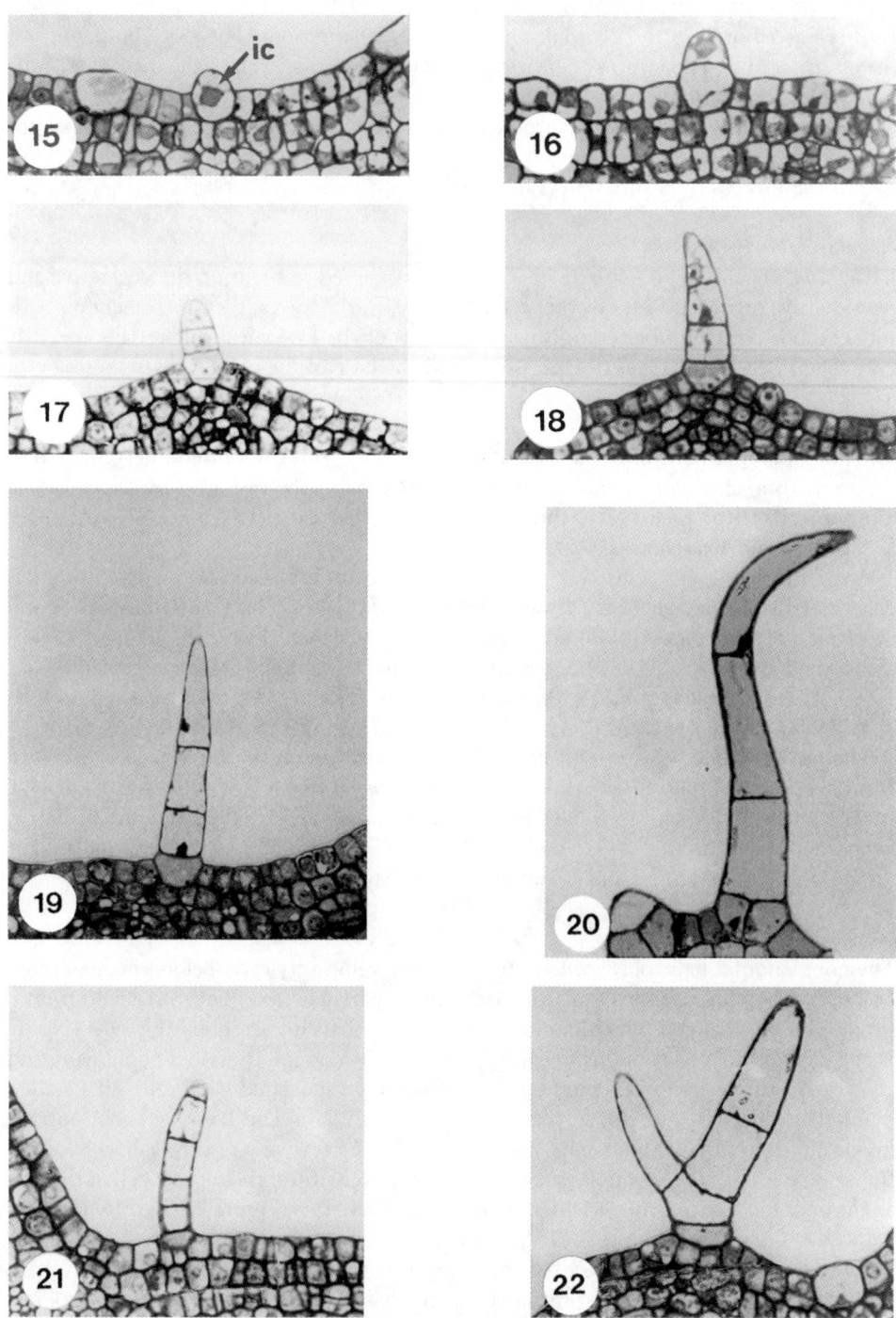

Figures 2.15–2.22 Ontogeny of nonglandular hairs of *Origanum vulgare* (× 450). *Figure 2.15* The initial cell of the hair (ic). *Figure 2.16* The 2-celled stage (one periclinal division). *Figure 2.17* The 3-celled stage (two periclinal divisions). *Figure 2.18* A young hair composed of one basal cell and three piled vacuolated cells above it. The most distal cell is pointed.

protoplasmic layer in which the nucleus and other organelles are embedded. The plastids are represented by leucoplasts. The cell walls (anticlinal and periclinal) are thin with primary texture and no deposition of cutin or suberin.

Glandular hairs
The glandular hairs are epidermal appendices, as the nonglandular hairs, but they are more complicated from the anatomical, ultrastructural and functional points of view. They biosynthesize and secrete a broad spectrum of substances, like essential oils, resins, gums, slimes, nectar, salty solutions, etc. In the family of Lamiaceae, glandular hairs principally produce essential oils and our report will be limited to only this sort of hairs.

Glandular hairs in members of Lamiaceae develop on all aerial plant organs. The majority of the relevant studies have been conducted on the leaf, but there is a remarkable number of observations made on the calyx, the corolla, the stamens and the carpels of the flower (Modenesi *et al.*, 1984; Werker *et al.*, 1985b; Servettaz *et al.*, 1994; Ascensao *et al.*, 1995; Corsi and Bottega, 1999). The glandular hairs initiate very early, when leaves are still at the primary stage. This situation seems to hold true not only for Lamiaceae (Bosabalidis and Tsekos, 1982a; Werker *et al.*, 1985b; Maffei *et al.*, 1986; Danilova and Kashina, 1989; Bourett *et al.*, 1994), but also for other families with trichomes secreting essential oils, such as the Compositae (Vermeer and Peterson, 1979a; Werker and Fahn, 1982; Spring and Bienert, 1987), Geraniaceae (Oosthuizen and Coetzee, 1984), Simarubaceae (Bory and Clair-Maczulajtys, 1981), Valerianaceae (Corsi and Pagni, 1990), Dioscoreaceae (Bruni *et al.*, 1987), etc. In the primary leaves, all stages of gland development can be found and consequently these leaves are the most appropriate for ontogenetic studies.

Microscopic investigations on aromatic plants have revealed that young leaves bear both developing and mature glandular hairs (Werker *et al.*, 1993; Duke and Paul, 1993; Ascensao *et al.*, 1995; Bosabalidis and Skoula, 1998b), whereas fully expanded leaves only mature glandular hairs (Amelunxen, 1965; Bory and Clair-Maczulajtys, 1981; Werker *et al.*, 1993; Ascensao *et al.*, 1999). Cases are reported, however, in which grown leaves possess both mature and young oil glands (Dell and McComb, 1975; Venkatachalam *et al.*, 1984; Oosthuizen and Coetzee, 1984; Bell and Curtis, 1985; Fahn and Shimony, 1998). The fact that in the same leaf (regardless of whether it is primary or grown) initiating, developing and senescing glandular hairs may simultaneously occur, reveals that the process of essential oil secretion does not proceed synchronously. The formation and pattern of development of the glandular hairs, as well as the biosynthesis of the terpenoids and the mechanism of their secretion seem to be governed by genes (Nielsen *et al.*, 1982; Rosenthal and Berenbaum, 1991). Environmental factors (light, water, nutrients, etc.) may though affect these parameters. An interesting point of consideration is the density of the glandular hairs on the leaf surface. In some aromatic plants [*Thymus* (Economou-Amilli *et al.*, 1982; Letchamo and Gosselin, 1996), *Rosmarinus* (Werker *et al.*, 1985a), *Mentha* (Maffei *et al.*, 1986; Gavalas

---

Figure 2.19 Advanced elongation of the hair cells. *Figure 2.20* A fully-developed hair. The basal region is composed of more than one cells, while the piled cells have undergone maximal elongation. The most distal cell is bent. *Figure 2.21* A nonglandular hair composed of the basal region and five piled cells. *Figure 2.22* A branched nonglandular hair.

et al., 1998), *Salvia* (Venkatachalam et al., 1984; Corsi and Bottega, 1999) *Plectranthus* (Ascensao et al., 1999), *Leonotis* (Ascensao et al., 1995), *Nepeta* (Bourett et al., 1994), *Rosmarinus* (Maffei et al., 1993), *Teucrium* (Bini-Maleci and Servettaz, 1991), etc.], glandular hairs are more numerous on the lower leaf side, whereas in some others [*Origanum* (Werker et al., 1985b; Bosabalidis and Kokkini, 1997), *Artemisia* (Corsi and Nencioni, 1995), *Tamus* (Bruni et al., 1987), *Ailanthus* (Bory and Clair-Maczulajtys, 1981), etc.] on the upper leaf side. The density of glandular hairs on the leaf surface is functionally associated with transpiration, leaf overheating, insect attack, UV-B radiation, etc.

During leaf development and expansion, the number of glandular hairs may remain more or less stable (Henderson et al., 1970; Werker et al., 1993; Ascensao et al., 1995), or it may change (Maffei et al., 1986). In several articles the view has been expressed that during leaf growth there is a continuous formation of new glandular hairs, so that old leaves are richer in glandular pubescence than young ones (Dell and McComb, 1975; Oothuizen and Coetzee, 1984; Bruni et al., 1987; Mahlberg et al., 1984). Goertzen and Small (1993) consider, however, that the density of glandular hairs decreases with leaf maturity, indicating that the young leaves have denser hair coverings. This could be an adaptive feature of the plant, wherein the young, most tender and appetizing to herbivores leaves, are given the highest level of protection (many secretions of glandular hairs were found to be deterrent or toxic to insects). A denser glandular hair covering in the young leaves has been also observed in *Inula* (Werker and Fahn, 1982), *Mentha* (Maffei et al., 1986), *Origanum* (Bosabalidis and Skoula, 1998b), *Pelargonium* (Walters et al., 1989), *Valeriana* (Corsi and Pagni, 1990), *Nicotiana* (Nielsen et al., 1991), etc.

The number of glandular hairs on the leaves of aromatic plants is linearly associated with the essential oil yield. Thus, the greater the number of glandular hairs on the leaves is, the higher the amount of essential oil derived from them by distillation (Bosabalidis and Kokkini, 1997; Gavalas et al., 1998). This is due to the fact that the glandular hairs are the exclusive leaf sites of essential oil biosynthesis, as biochemical studies with isolated glandular hairs have evidenced (Gershenzon et al., 1989; Mc Caskill and Croteau, 1995). No other leaf compartments, epidermal or mesophyllic, are able to metabolize terpenoids into essential oils. An indirect participation of the photosynthesizing leaf tissue in the provision of the secretory cells of the glandular trichomes with precursors is, however, acknowledged.

The glandular hairs on the upper and lower leaf surfaces may not only differ in number and density, but also in the qualitative and quantitative constitutions of the secretions they produce. Comparative studies of essential oil analyses separately for the hairs of the upper and the lower surfaces of mature leaves have shown that oil composition may either slightly fluctuate, as in the case of *Ocimum basilicum* (Werker et al., 1993) and *Origanum* × *intercedens* (Bosabalidis and Skoula, 1998b), or it may exhibit remarkable alterations, as in the case of *Mentha piperita* (Maffei et al., 1989). In the young leaves, the differences between the upper and lower leaf sides, as concerns the content and composition of the essential oil produced by the glandular hairs, are quite sharper (Werker et al., 1993).

The shape, density, size and position of the glandular hairs, as well as the oily property, the chemical constitution and the fragrant character of the essential oils, are parameters not accidentally created by nature on aromatic plants and much more investigation is needed to increase data on the structural, functional and ecological points of the glandular hairs and their secretions.

Types of glandular hairs

From the morphological studies conducted so far on members of Lamiaceae, it follows that on the aerial plant organs and particularly on the leaves, two types of glandular hairs mainly exist, the peltate and the capitate. These types co-occur to create a mixed population. An integrated glandular hair (peltate or capitate) has been described to be composed of a basal region (unicellular or multicellular), a stalk region (unicellular or multicellular) and a head region (unicellular or multicellular). In our opinion, however, a structurally and functionally integrated glandular hair, additionally to the above three regions, also contains the epidermal cells which radially surround the basal region. Such cells have been observed in the peltate hairs of *Origanum* (Bosabalidis and Tsekos, 1984; Bosabalidis and Exarchou, 1995; Bosabalidis and Kokkini, 1997), *Calamintha* (Hanlidou et al., 1991), *Ocimum* (Werker et al., 1993), *Satureja* (Bosabalidis, 1990b), *Mentha* (Gavalas et al., 1998), *Monarda* (Heinrich, 1973), etc. The beribasal cells do not operate like the typical epidermal cells, but they become induced by the glandular hair [presumably in a way analogous to that between the guard and subsidiary cells of stomatal complexes (Stebbins and Shah, 1960)] to constitute an accessory of the hair and serve the process of essential oil secretion. Thus, the large size of the beribasal cells, their shape, arrangement, vacuolation, density of plasmodesmata at the periclinal walls, etc, most probably contribute to the collection of the photosynthates from the chlorenchymatic mesophyll and to their further centripetal transport to the basal cell of the glandular hair. The latter cell in this way becomes a central pool in which the photosynthates are temporarily deposited, which is why it is very voluminous (significantly larger than the typical epidermal cells) and greatly vacuolated. The photosynthetic products become then moved to the stalk region of the glandular hair and finally to the head region, in which their elaboration into essential oil takes place through the enzymatic machinery exclusively possessed by the head secretory cells.

Capitate glandular hairs

Structure The capitate glandular hairs are much smaller than the peltate ones, occur in thicker populations and exhibit a greater morphological variability. Werker et al., (1985a) distinguish three main types of capitate hairs in Lamiaceae, i.e. Type I (short) with one basal cell, 1–2 stalk cells and 1–2 head cells (rounded, ovoid or pear-like), Type II (medium) with one basal cell, 1–2 stalk cells and one head cell (finger- or pestle-like) and Type III (long) with one basal cell, 2–5 stalk cells and one head cell (rounded). In Table 2.6, some morphological data from Lamiaceae members are summarized, as concerns the above-mentioned types of capitate glandular hairs occurring on the leaf surface. From these data it follows that the most frequent type of capitate hairs met in all species studied, is Type I. This type may be exclusive (*Satureja, Mentha, Thymus, Coridothymus, Rosmarinus, Nepeta*) or it may be accompanied either by Type II capitate trichomes (*Majorana, Micromeria, Melissa, Origanum, Ocimum, Calamintha*) or by Type III capitate trichomes (*Salvia, Plectranthus, Teucrium*).

In the leaves of *Sideritis syriaca* subsp. *syriaca*, another type of capitate glandular hair has been observed composed of four symmetric basal cells, three elongated stalk cells, a single plasma rich neck cell and four symmetric head cells (Karousou et al., 1992). Peculiar is also the anatomy of a sort of capitate glandular hair occurring on the leaves of *Leonotis leonurus* (Ascensao et al., 1995). This hair is constructed of an 8–32 celled pedestal base, a 2–3 celled stalk, an 1–2 celled neck and a four celled head. Both of

Table 2.6 Types of capitate glandular hairs in members of Lamiaceae

Taxon	Type I	Type II	Type III	References
Satureja thymbra	+			Werker et al., 1985a; Bosabalidis, 1990b
Mentha piperita	+			Maffei et al., 1989
Mentha viridis lavanduliodora	+			Maffei et al., 1986
Thymus vulgaris	+			Bruni and Modenesi, 1983
Coridothymus capitatus	+			Werker et al., 1985a
Rosmarinus officinalis	+			Werker et al., 1985a
Nepeta racemosa	+	+		Bourett et al., 1994
Majorana syriaca	+	+		Werker et al., 1985a
Micromeria fruticosa	+	+		Werker et al., 1985a
Melissa officinalis	+	+		Werker et al., 1985a
Origanum vulgare	+	+		Werker et al., 1985a
Ocimum basilicum	+	+		Werker et al., 1993
Calamintha menthifolia	+		+	Hanlidou et al., 1991
Salvia officinalis	+		+	Werker et al., 1985a; Bini-Maleci et al., 1983
Salvia aurea	+		+	Serato-Valenti et al., 1997
Ocimum basilicum	+		+	Gupta and Bhambie, 1978
Plectranthus madagascariensis	+		+	Ascensao et al., 1998
Plectanthrus ornatus	+		+	Ascensao et al., 1999
Teucrium sect. Chamaedrys	+		+	Bini-Maleci and Servettaz, 1991
Teucrium silicum	+			Servettaz et al., 1994

Notes

Type I (short) = 1 basal cell, 1–2 stalk cells, 1–2 head cells (rounder, ovoid or pear-like).
Type II (medium) = 1 basal cell, 1–2 stalk cells, 1 head cell (finger-like).
Type III (long) = 1 basal cell, 2–5 stalk cells, 1 head cell (rounded).

the above-mentioned types of capitate glandular hairs have not been earlier reported in Lamiaceae. In contrast to the peltate glandular hairs which are considered to exclusively be involved in essential oil production, the capitate hairs of Lamiaceae have been reported to secrete either essential oil (Amelunxen et al., 1969; Lovett and Speak, 1979; Bini-Maleci et al., 1983; Mc Caskill et al., 1992; Bisio et al., 1999) or mucilage (Modenesi et al., 1984; Danilova and Kashina, 1989; Bini-Maleci and Servettaz, 1991; Bourett et al., 1994; Ascensao and Pais, 1998). As to the question which one of both kinds of glandular hairs (peltate or capitate) develops earlier, Amelunxen (1964) and Brun et al., (1991) claim that the capitate hairs mature first followed by the peltate ones, while Ascensao et al. (1995) consider that capitate and peltate hairs mature more or less simultaneously.

In *O. vulgare*, two types of capitate glandular hairs have been recognized (Werker et al., 1985b). Hairs of Type I are short and consist of one basal cell, one stalk cell and one rounded or pear-like head cell. The stalk and head regions are not perpendicularly oriented to the epidermal surface, but they are ordinarily bent. Hairs of Type II are somehow longer and consist of one basal cell, one to three stalk cells and one finger-like or pestle-like head cell. These hairs have always a vertical orientation.

Ontogeny Studies on the ontogenetic pattern of the capitate glandular hairs in *O. vulgare* subsp. *hirtum* (Bosabalidis, unpublished) have shown, that hairs of Type I originate from a single protodermal cell when the leaf is still at the primary state (Figure 2.23). This initial cell progressively increases in volume protruding from the level of the epidermis. When it completes its growth, it undergoes an asymmetric periclinal division to give two daughter cells (Figure 2.24). The lower daughter cell does not divide any further and it constitutes the future basal cell of the capitate hair. The second daughter cell, which is seated above the basal cell, is ovoid, bent, and it redivides also periclinally and asymmetrically to give the stalk cell and the head cell (Figure 2.25).

The capitate hairs of Type II also originate from a single protodermal cell which is much larger than its neighbouring cells (Figure 2.26). This cell divides asymmetrically through a periclinal wall to give rise to two daughter cells differing in shape and protoplasmic content (Figure 2.27). The proximal cell is highly vacuolated (basal cell), whereas the distal one is plasma rich and dome-like. The latter undergoes a periclinal asymmetric division resulting in two cells, a flattened stalk cell and a pestle-like apical cell (Figure 2.28). The ontogenetic process becomes completed with an ultimate asymmetric periclinal division of the apical cell, from which a flattened neck cell and a pestle-like head cell are derived (Figure 2.29). Both cells contain a dense protoplast, contrary to the basal and stalk cells which have large vacuoles.

Similar to the above patterns of development in the capitate hairs (Types I and II) of *Origanum*, are those described for other members of Lamiaceae (Hanlidou *et al.*, 1991; Ascensao *et al.*, 1995), as well as for members of other families (Franceschi and Gianquinta, 1983; Oosthuizen and Coetzee, 1983). In the typical glandular hairs of

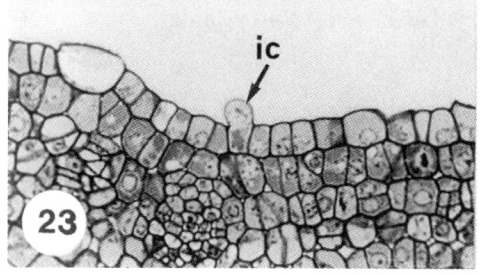
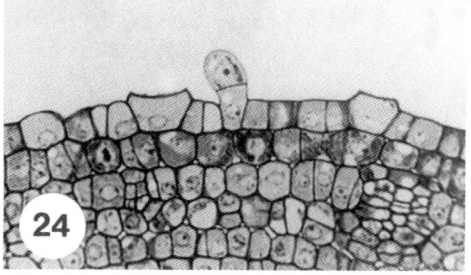
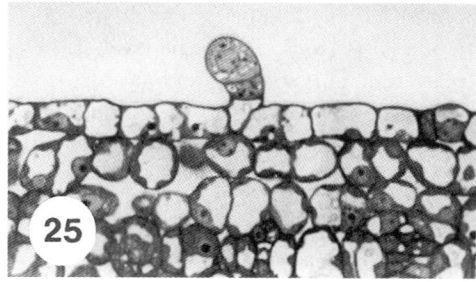

Figures 2.23–2.25 Ontogeny of Type I capitate glandular hairs in *Origanum vulgare* (×550). *Figure 2.23* The initial protodermal cell (ic). *Figure 2.24* The 2-celled stage after a periclinal division of the initial cell. *Figure 2.25* A fully-developed glandular hair composed of one basal cell, one stalk cell, and one plasma-rich head cell.

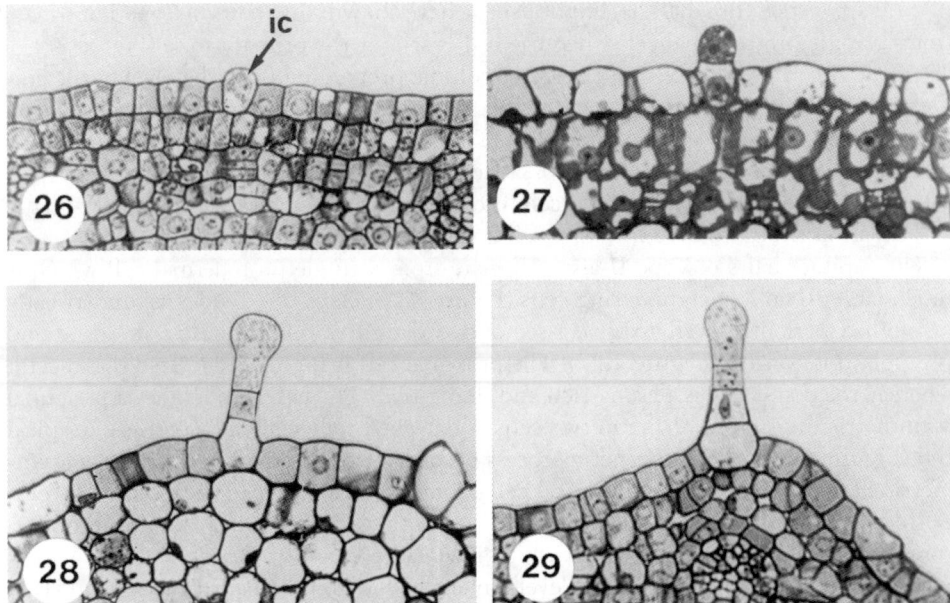

Figures 2.26–2.29 A series of light micrographs showing the course of development of Type II capitate glandular hairs in *Origanum vulgare* (×550). *Figure 2.26* The initial protodermal cell (ic). *Figure 2.27* A young glandular hair composed of one vacuolated basal cell and one apical plasma-rich cell. *Figure 2.28* A developing 3-celled glandular hair. The most distal cell is pestle-like. *Figure 2.29* A fully-developed glandular hair composed of four piled cells. The apical cell corresponds to the secretory head.

Asteraceae, the division of the initial cell is not an asymmetric periclinal one, but a symmetric anticlinal (Carlquist, 1958; Vermeer and Peterson, 1979a; Duke and Paul, 1993). Symmetric anticlinal is also the division of the initial cell in the capitate-stalked hairs of *Cannabis* (Mahlberg *et al.*, 1984). The following divisions run parallel to the leaf surface establishing first the head region and then the stalk and basal regions of the hairs (contrary to Lamiaceae in which the head region is the last to form).

Peltate (sessile) glandular hairs The peltate hairs are short, voluminous, and they are ordinarily composed of a large basal cell (at the level of the epidermis), a flattened stalk cell and a multicellular head. This sort of glandular hair is very frequent in the members of Lamiaceae. Depending upon the species, the number of the head cells may fluctuate, whereas the stalk and basal regions usually remain unicellular (Table 2.7). In a significant number of species, the number of the head cells ranges from 12 to 18. These cells are arranged in such a manner that four cells (small) occupy the centre of the head, while the remaining cells (large) are peripherally located. In some other species, the heads are composed of a reduced number of cells (4 or 8). The cells of the former category usually belong to the subfamily Nepetoideae and those of the latter to the Lamioideae. Since the head cells are the only cells of the peltate glandular hairs which are able to secrete the essential oil (Mc Caskill *et al.*, 1992; Bourett *et al.*, 1994), it easily follows that the species having peltate hairs with high numbers of head cells are producing higher amounts of essential oil compared to the species with a limited number of peltate hair

Table 2.7 Number of cells composing the regions of the head, the stalk, and the base of the peltate glandular hairs in members of Lamiaceae

Taxon	Head	Stalk	Base	References
Micromeria fruticosa	14–18	1	1	Werker et al., 1985a
Ocimum basilicum	16	2	2	Gupta and Bhambie, 1978
Coridothymus capitatus	14 (10 + 4)	1	1	Werker et al., 1985a
Thymus vulgaris	10–14	1	1	Bruni and Modenesi, 1983
Monarda fistulosa	10	1	1	Heinrich, 1973
Majorana syriaca	12 (8 + 4)	1	1	Werker et al., 1985a
Melissa officinalis	12 (8 + 4)	1	1	Werker et al., 1985a
Salvia officinalis	12 (8 + 4)	1	1	Werker et al., 1985a; Bini-Maleci et al., 1983; Cors and Bottega, 1999; Venkatachalam et al., 1984
Salvia fruticosa	12 (8 + 4)	1	1	Werker et al., 1985a
Satureja thymbra	12 (8 + 4)	1	1	Werker et al., 1985a; Bosabalidis, 1990b
Origanum dictamnus	12 (8 + 4)	1	1	Bosabalidis and Tsekos, 1982a
Origanum vulgare	12 (8 + 4)	1	1	Werker et al., 1985b
Origanum × intercedens	12 (8 + 4)	1	1	Bosabalidis and Exarchou, 1995
Calamintha menthifolia	12 (8 + 4)	1	1	Hanlidou et al., 1991
Mentha piperita	8	1	1	Amelunxen, 1965; Maffei et al., 1989
Salvia officinalis	8	1	1	Vezar-Petri and Then, 1975; Venkatachalam et al., 1984
Salvia aurea	8	1	1	Serrato-Valenti et al., 1997
Plectranthus madagascariensis	8	1	1	Ascensao et al., 1998
Plectranthus ornatus	8	1	1	Ascensao et al., 1999
Leonotis leonurus	8	1	1	Ascensao et al., 1995
Salvia blepharophylla	4	1	1	Bisio et al., 1999
Ocimum basilicum	4	1	1	Gupta and Bhambie, 1978; Werker et al., 1993
Teucrium sect. *Chamaedrys*	4	1	1	Bini-Maleci and Servetazz, 1991
Teucrium siculum	4	1	1	Servettaz et al., 1994
Sideritis syriaca	4	1	1	Karousou et al., 1992
Pogostemon cablin	4	1	1	Henderson et al., 1970
Nepeta racemosa	4	1	1	Bourett et al., 1994

head cells (Maffei et al., 1989; Karousou et al., 1992). Apparently, the density of the glandular hairs on the leaf surface should be considered, in the first place.

Peltate glandular hairs may differ not only in the number of the composing cells, but also in their size. Measurements made in some members of Lamiaceae showed that in leaf surface view, the diameter of the head along with the above subcuticular space may vary among the species and between the upper and lower leaf surfaces of the same species as well (Table 2.8).

The peltate hairs on the grown leaves of Lamiaceae are situated not on flat areas, but on local epidermal depressions. This observation is to some extent analogous to that of the *Citrus* fruit peel, in which the epidermis forms a local depression at the point where an oil gland exists in the exocarpic tissue. As to the expediency of the presence of such epidermal depressions, it might be that they probably serve to collect the essential oil, when the latter becomes released from the subcuticular space due to the action of various predators. In this way, thousands of small essential oil pools are formed on the organ

Table 2.8 Density and size (diameter in surface view) of the peltate glandular hairs in members of Lamiaceae

Taxon	Density (No/mm²)		Size (μm)		References
	Upper leaf side	Lower leaf side	Upper leaf side	Lower leaf side	
Origanum vulgare	<10				Werker et al., 1985b
Origanum vulgare subsp. vulgare	4	3	77	68	Bosabalidis and Kokkini, 1997
Origanum vulgare subsp. hirtum	8–11	6–10	80–103	71–93	Bosabalidis and Kokkini, 1997
Origanum vulgare subsp. viridulum	1.5	1.2	70	63	Bosabalidis and Kokkini, 1997
Origanum dictamnus	3.4	4.1		89.5	Bosabalidis, 1990a
Origanum × intercedens	6.4	5.2	86.9	84.3	Bosabalidis and Skoula, 1998
Mentha piperita	3.2	9.1	60		Maffei et al., 1989; Mc Caskill et al., 1992
Mentha longifolia	1.8	18.6			Gavalas et al., 1998
Mentha spicata	7.3	19.2			Gavalas et al., 1998
Mentha × villoso-nervata	4.3	11.3			Gavalas et al., 1998
Thymus vulgaris			47.3		Yamaura et al., 1992
Salvia reflexa			50		Lovett and Speak, 1979
Satureja thymbra	4.3		91.1		Bosabalidis, 1990b
Plectranthus ornattus			70		Ascensao et al., 1999
Leonotis leonurus			50		Ascensao et al., 1995
Monarda fistulosa	10.5		60		Heinrich, 1973
Pogostemon cablin	0.95				Henderson et al., 1970

surface, which in total create a shield protecting from microorganism attack and UV-B radiation.

Structure One of the most frequently occurring in Lamiaceae kind of peltate glandular hairs is that composed of a 12-celled head (Table 2.7). This kind of hair is represented in the species of *Origanum* (Bosabalidis and Tsekos, 1982a, 1984; Werker *et al.*, 1985b; Bosabalidis and Exarchou, 1995). At the developmental stage when cell divisions have been completed but secretion of the essential oil has not started yet (the oil-accumulating subcuticular space is not formed), the peltate glandular hairs appear in leaf cross section composed of a large vacuolated basal cell, a flattened stalk cell, and (most often) four plasma rich head cells arranged parallel to the epidermal surface (Figure 2.30). From the head cells, the two inner ones are smaller than the two lateral. In the fully-differentiated and secreting peltate hairs, the situation remains the same, as concerns the number of cells, and additionally a large subcuticular space filled with essential oil is formed (Figure 2.31). In leaf cross sections, thus, a mature peltate glandular hair usually appears composed of a four celled head, a unicellular stalk and also an unicellular base.

The anatomical picture of the *Origanum* peltate hair is quite different in leaf paradermal sections. In such sections, the head appears to consist of 12 cells, from which four are small and centrally located and eight are large and peripherally arranged (Figure 2.32). Serial paradermal sections cut toward the base of the peltate hair, disclose a progressive disappearance of the four central cells (the eight peripheral cells are still well distinguished), while simultaneously in the centre of the head the stalk region gradually emerges (Figures 2.33–2.35). This region becomes proved to be unicellular (Figure 2.36). At the most proximal level of section, the basal region appears also unicellular and radially surrounded by 12–18 elongated epidermal cells (Figure 2.37). Paradermal sections are thus necessary to elucidate the real number, as well as the manner of arrangement of the cells in all three hair regions. In conclusion, a typical peltate glandular hair in *Origanum* is composed of a 12-celled head, a unicellular stalk, a unicellular base and a peribasal crown of 12–18 radial epidermal cells.

Ontogeny The study of the ontogeny of the peltate glandular hairs is a key procedure elucidating obscure points related to the site of gland initiation, the sequence and manner of cell divisions, the final number and arrangement of the hair cells, the changes in the hair shape, the cytological alterations, the determination of the stage of secretion, the process of formation of the essential oil-accumulating chamber, the cell characteristics at the postsecretory and senescing phases, etc. Descriptions of peltate glandular hair ontogeny have been briefly or in detail reported for some Lamiaceae members, such as *Origanum* (Bosabalidis and Tsekos, 1982a, 1984; Bosabalidis and Exarchou, 1995), *Thymus* (Bruni and Modenesi, 1983), *Pogostemon* (Henderson *et al.*, 1970), *Salvia* (Venkatachalam *et al.*, 1984), *Sideritis* (Karousou *et al.*, 1992), *Leonotis* (Ascensao *et al.*, 1995), *Mentha* (Fischer and Hecht-Buchholz, 1985), *Calamintha* (Hanlidou *et al.*, 1991), *Monarda* (Heinrich, 1973), etc.

In *Origanum*, a peltate glandular hair originates from a single protodermal cell dominating in volume (Figure 2.38). In this cell, the centrally located nucleus progressively migrates toward the apical region, while the bottom region becomes occupied by vacuoles. The thus established morphological polarization results in an asymmetric periclinal division from which two unequally sized and shaped cells are derived (Figure 2.39). The proximal daughter cell is peg-like and it does not divide any further constituting the future basal cell of the peltate hair. The distal cell is dome-like, much larger and it

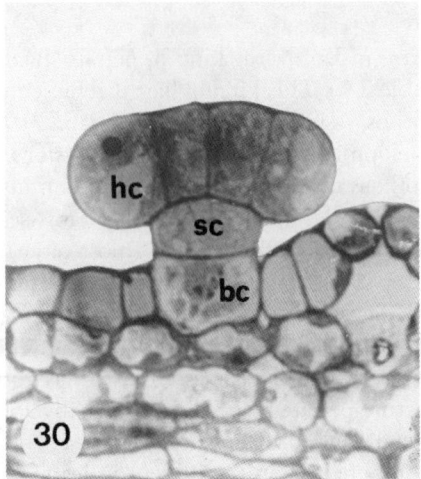

Figure 2.30 Longitudinal section of a peltate glandular hair after conclusion of divisions (subcuticular space not yet formed). The hair appears in the section composed of one basal cell (bc), one stalk cell (sc) and four head cells (hc) (×750).

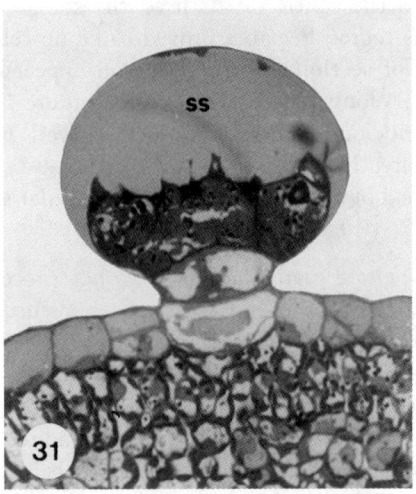

Figure 2.31 A fully-developed peltate glandular hair of *Origanum dictamnus*. The basal and stalk cells are highly vacuolated, whereas the head cells are plasma rich. The cuticle becomes greatly detached from the apical cell walls forming a large subcuticular space (ss) (×700).

redivides asymmetrically and periclinally to give rise to a lower flattened cell (future stalk cell of the hair) and an upper rounded cell (Figure 2.40). The latter corresponds to the mother cell of the head and it undergoes a series of anticlinal divisions to ultimately form the head of the peltate hair (Figures 2.41, 2.42).

Integration of cell divisions in the head of the peltate hair is followed by the stage of essential oil secretion. The secreted material becomes released into a space formed at the tip of the hair by detachment of the cuticle from the apical walls of the head cells. This

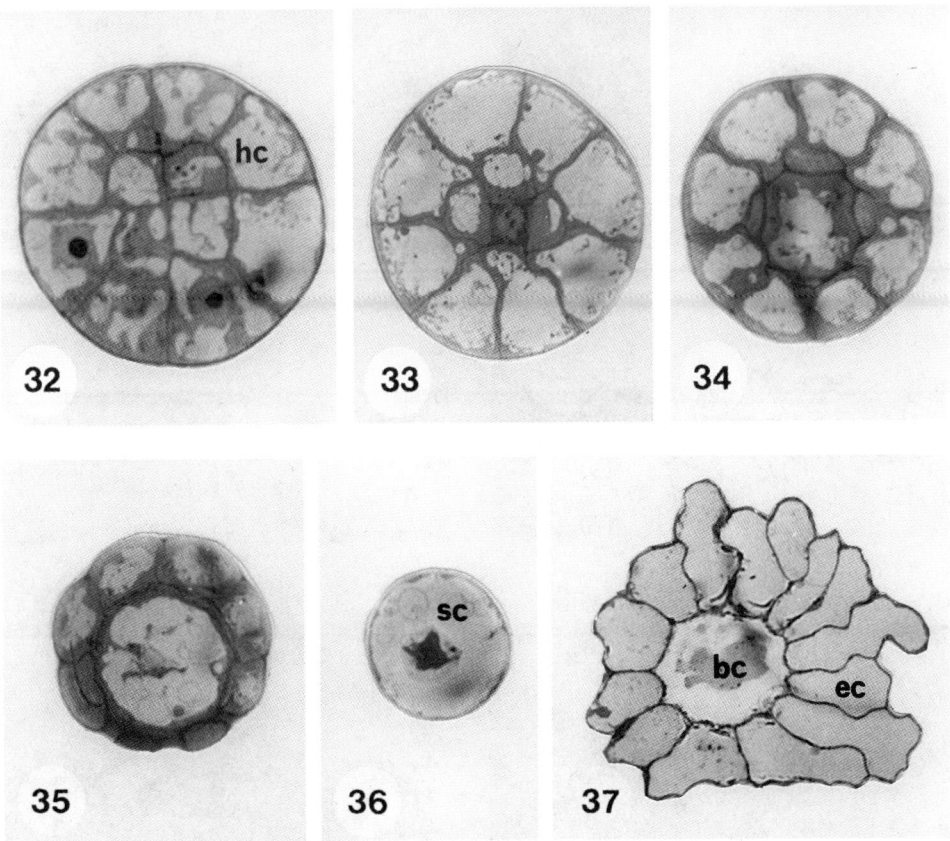

Figures 2.32–2.37 Leaf paradermal sections serially cut from the region of the head to the region of the base of a peltate glandular hair (×700). *Figure 2.32* A fully-developed head composed of four small central cells and eight large peripheral cells (hc). *Figure 2.33* A section at a lower level. In the centre of the head emerges the stalk region. *Figure 2.34* Even lower section. From the four central cells, only small portions can be discerned, whereas the eight peripheral cells are still intact. *Figure 2.35* The stalk region of the peltate hair surrounded by the eight peripheral cells in reduced volume (the four central cells become disappeared). *Figure 2.36* The stalk region composed of one cell (sc). The nucleus is centrally located. *Figure 2.37* The basal region, also composed of one cell (bc). This cell is radially surrounded by a number of epidermal cells (ec).

oil-containing subcuticular space initially is small (Figure 2.43) and later greatly increases in volume (Figure 2.44). After secretion is concluded, the peltate hair disintegrates (Figure 2.45). During this last ontogenetic stage, the cuticle does not remain any longer stretched and ballooned, but it sediments down to locally touch the head apical walls.

To follow the development of the head of the peltate glandular hair, leaf paradermal sections have been used (Bosabalidis and Tsekos, 1984). In such sections, the mother cell of the head (Figure 2.46) appears to divide twice symmetrically and anticlinally resulting in two (Figure 2.47) and then four (Figure 2.48) equally-sized daughter cells. Each of these cells redivides asymmetrically through a bent wall, to give a head with

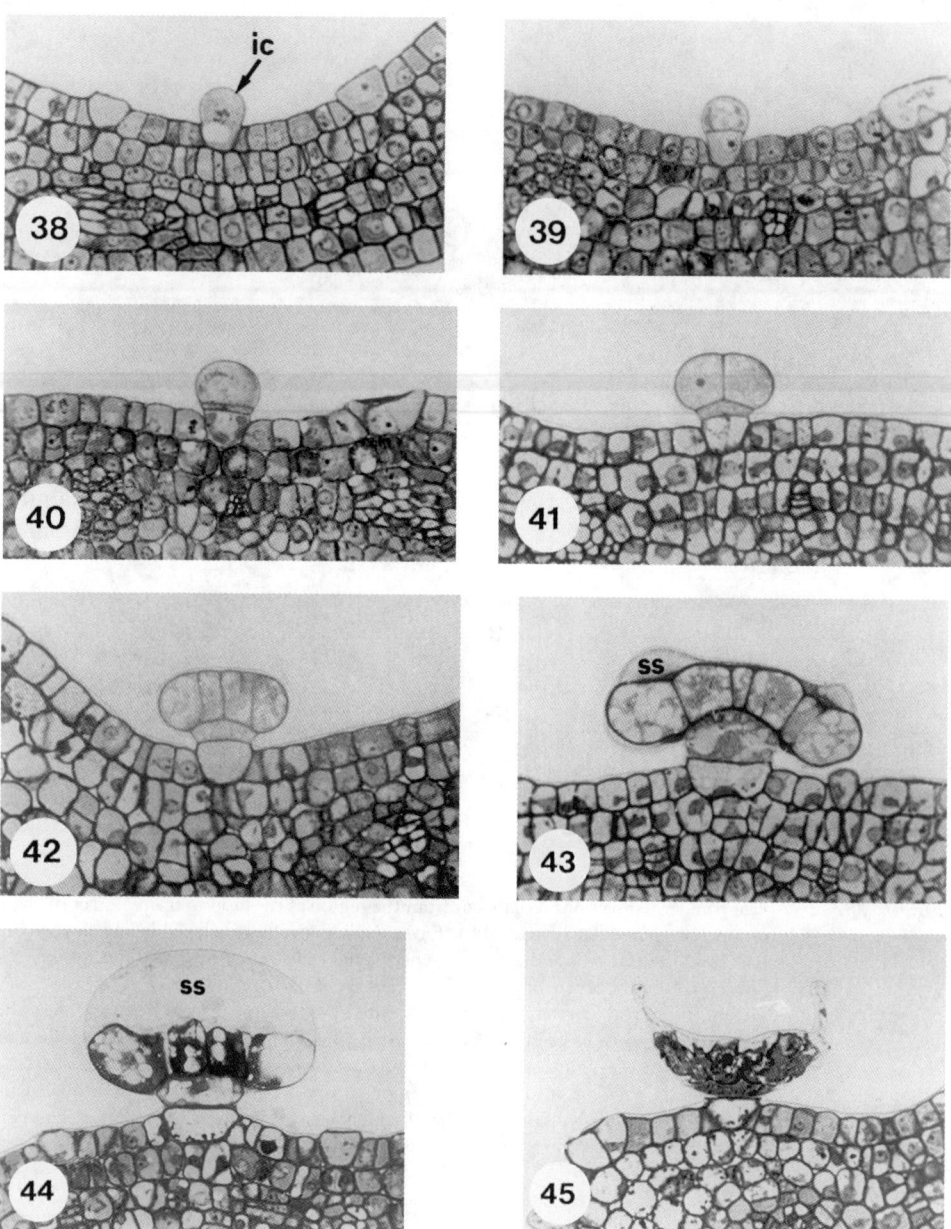

Figures 2.38–2.45 Ontogeny of peltate glandular hairs in *Origanum vulgare* (×500). *Figure 2.38* The initial cell of the peltate hair (ic). *Figure 2.39* First periclinal division of the initial cell to give the future basal cell and an apical dome-like cell. *Figure 2.40* Second periclinal division resulting in the basal cell, stalk cell and mother cell of the head. *Figure 2.41* First anticlinal divisions in the head region. *Figure 2.42* Integration of anticlinal divisions in the head region. *Figure 2.43* Slight detachment of the cuticle from the apical walls of the head cells to form a small subcuticular space (ss). *Figure 2.44* Advanced detachment of the cuticle creating a large subcuticular space (ss) in which the essential oil will be accumulated. *Figure 2.45* Disorganized head cells in a senescent peltate hair.

four small lateral cells and four larger inner cells (Figure 2.49). The last developmental stage involves a morphological polarization and subsequent asymmetric division of the inner cells, so that finally the head is composed of four small central cells surrounded by eight large peripheral ones (Figure 2.50). In consideration of the above manner of head cell origination, the speculation could be expressed that the four central cells of the head produce different essential oil constituents from the eight peripheral cells and that the final mixture takes place within the subcuticular chamber.

The ontogeny of the peltate glandular hairs of *Origanum* was further studied at the submicroscopic level, in order to draw detailed information about the intracellular alterations taking place throughout hair lifetime (Bosabalidis and Tsekos, 1982a,b; Bosabalidis and Kofidis, unpublished). The ontogenetic course can be roughly distinguished into three main phases, i.e. the presecretory phase, the phase of active secretion and the phase of senescence.

Presecretory phase The initial cell of a peltate hair appears ultrastructurally largely occupied by a central nucleus (Figure 2.51). The cytoplasmic organelles are quite limited in number (Bosabalidis, 1990a). They are principally represented by inclusion-containing leucoplasts, mitochondria, rough endoplasmic reticulum elements and dictyosomes. The ground plasm contains densely arranged ribosomes and small vacuoles with a

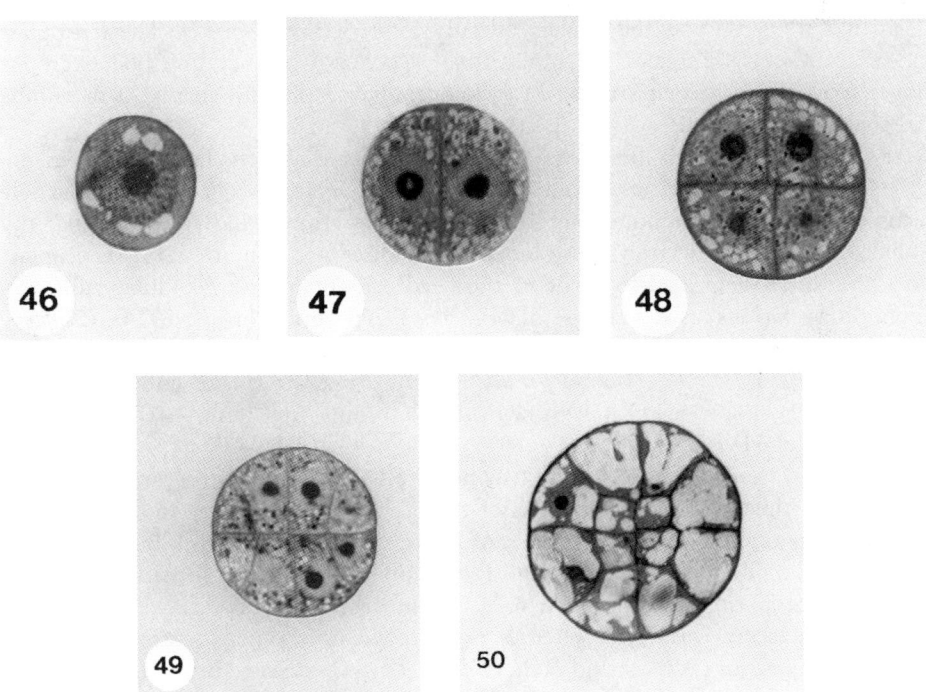

Figures 2.46–2.50 Successive stages of head development in the peltate glandular hairs (leaf surface sections) (×700). *Figure 2.46* The mother cell of the head. *Figure 2.47* First symmetric anticlinal division of the mother cell. *Figure 2.48* Second symmetric anticlinal division to give four equal cells. *Figure 2.49* An eight-celled head arisen after an asymmetric division of the four cells. *Figure 2.50* Final stage of head development. The head is composed of 12 cells, i.e. four small central cells and eight large peripheral cells.

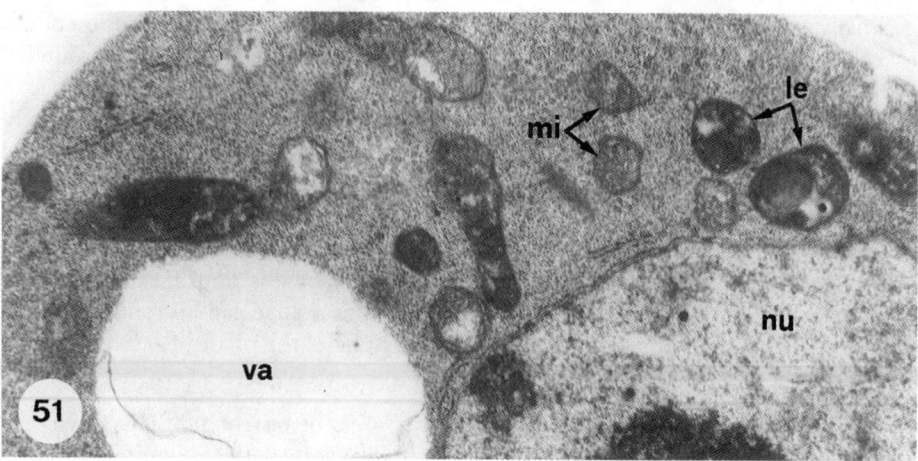

Figure 2.51 Ultrastructure of the initial cell of a peltate glandular hair. The nucleus (nu) is large and centrally located. Among organelles, leucoplasts (le) and mitochondria (mi) proliferate. Vacuoles (va) are small and limited in number (×15 000).

translucent content. When the head of the peltate hair becomes 8-celled, the ground plasm is still dense and the nuclei centrally located (Figure 2.52). The leucoplasts appear scattered throughout the intercellular space and no inclusion bodies are any longer recognized in their stroma. Mitochondria have somehow increased in number and contain many sacculi.

At the stage when cell divisions have been completed and the head is composed of 12 cells, the nucleus becomes significantly smaller, whereas concomitantly vacuoles become larger and more numerous. In the leucoplast stroma, individual annular thylakoids and stacked electron dense lamellae are discerned (Figure 2.53). Leucoplasts have been often described to occur in the head cells of peltate glandular hairs with lipophilic secretions, as in the case of *Viscaria* (Tsekos and Schnepf, 1974), *Comptonia* (Bell and Curtis, 1985), *Mentha* (Fischer and Hecht-Buchholz, 1985), *Monarda* (Heinrich, 1973), *Nepeta* (Bourett *et al.*, 1994), *Artemisia* (Duke and Paul, 1993) *Calceolaria* (Schnepf, 1969b), *Newcastelia* (Dell and McComb, 1975), *Cannabis* (Mahlberg *et al.*, 1984), etc.

In *Origanum*, leucoplasts of peltate hair head cells do not ultrastructurally appear to directly participate in the biosynthesis of the essential oil, contrary to other types of secretory structures, like the oil cavities of Rutaceae (Heinrich, 1969; Bosabalidis and Tsekos, 1982c; Gleizes *et al.*, 1983) and the oil ducts of Apiaceae (Bosabalidis, 1996), in which leucoplasts are considered to constitute the principal sites of essential oil biosynthesis (accumulations of secretory droplets within the plastid stroma). In some species [*Inula viscosa* (Werker and Fahn, 1982), *Thymus vulgaris* (Bruni and Modenesi, 1983), *Nicotiana tabacum* (Nielsen *et al.*, 1991), *Chrysanthemum morifolium* (Vermeer and Peterson, 1979b), etc.], the head cells of the peltate glandular hairs have been observed to contain instead of leucoplasts, chloroplasts, which, however, do not have a well-developed grana system.

The stalk cell of the peltate hair of *Origanum* exhibits at early development significant ultrastructural similarities with the head cells. It, thus, contains a centrally located large nucleus, a reduced number of cytoplasmic organelles and a dense ground plasm

Structural features of Origanum sp. 39

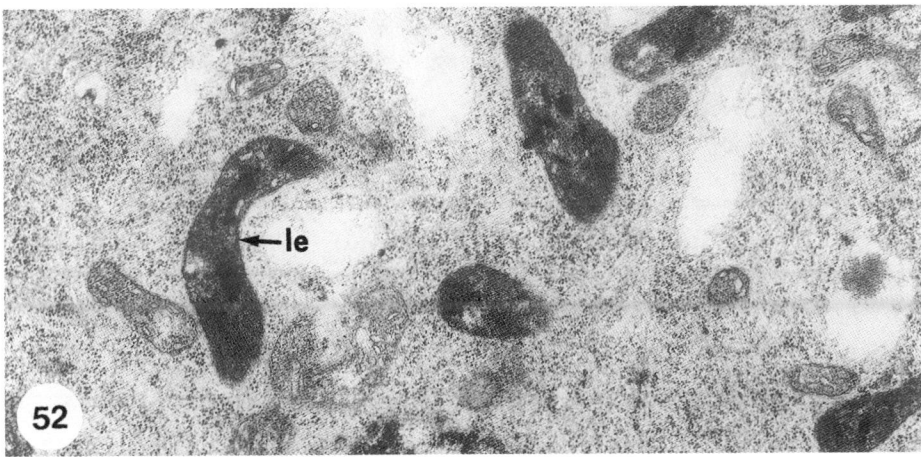

Figure 2.52 A cytoplasmic portion of a head cell at the 8-celled stage of head development. The leucoplasts (le) are scattered throughout the cytoplasm and they do not contain any longer globular inclusions (×23 000).

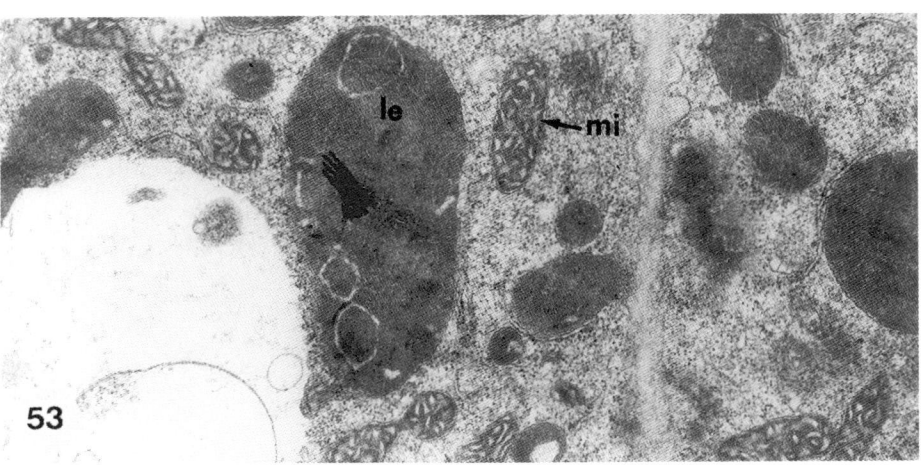

Figure 2.53 Head cells at the stage of integration of cell divisions. The leucoplasts (le) exhibit in their stroma annular thylakoids and stacked electron dense lamellae. The mitochondria (mi) have significantly increased in number (×23 000).

(Figure 2.54). The periclinal walls facing the head cells and the basal cell, are thin, not cutinized and they are transversed by numerous plasmodesmata. The lateral (anticlinal) walls of the stalk cell are thicker than the periclinal ones and they remain uncutinized until the head is four-celled. Afterwards, they gradually become impregnated with cutin up to the stage the head obtains its final size (Figure 2.55). The process of progressive cutinization becomes evident by a centripetal increase of the electron density of the wall matrix, which is presumably due to a polymerization of the cutin precursors by the air oxygen (Frey-Wyssling and Mühlethaler, 1959).

Cutinization of the lateral walls of the peltate hair stalk cell has been also reported to occur in other Lamiaceae members (Bruni and Modenesi, 1983; Mc Caskill *et al.*, 1992;

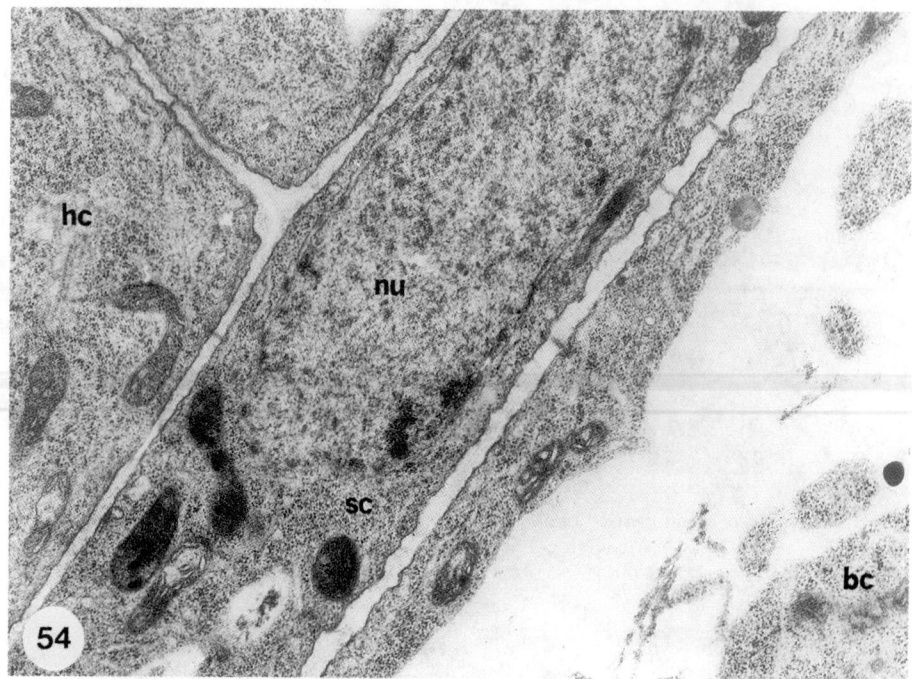

Figure 2.54 Early development of the peltate hair stalk cell (sc). The nucleus (nu) is flattened and centrally located. The cell walls between the stalk cell and head cells (hc), as well as between the stalk cell and the basal cell (bc) bear numerous plasmodesmata (×22 500).

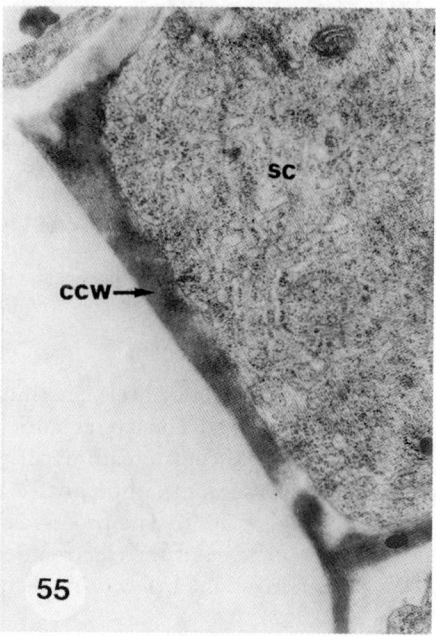

Figure 2.55 Advanced cutinization of the lateral wall of the stalk cell (sc). ccw, cutinized cell wall (×15 000).

Ascensao et al., 1995, 1999; Corsi and Bottega, 1999) and it has been interpreted to prevent an uncontrolled by the protoplasm apoplastic movement of substances from the mesophyll cells to the hair head cells (Dell and McComb, 1977). At the same time, a similar movement of the essential oil to the mesophyll cells, where it might have a deleterious effect, is avoided (Wollenweber and Schnepf, 1970). According to Dolzmann (1964), such a cutinization contributes to a better support of the voluminous head which becomes heavy during accumulation of the essential oil within the subcuticular space. On this subject, Amelunxen (1965) claims that the cutinized lateral walls of the stalk cell constitute a cutin source for the raising cuticle, so that the latter does not rupture. In the next steps of stalk cell differentiation, the vacuole remarkably increases and the cytoplasm along with the nucleus become restricted to the parietal cell zone.

The basal cell of the *Origanum* peltate hairs is very large and highly vacuolated, even from early development (Figure 2.56). It contains a limited number of organelles (particularly mitochondria) compared to a secretory head cell. The plastids are quite similar to those of the typical epidermal cells and each of them contains a voluminous membrane-bordered inclusion with a fine-granular appearance (Figure 2.56). The upper periclinal wall facing the stalk cell, as well as the lower one facing the mesophyll parenchyma, are transversed by numerous plasmodesmata (Figure 2.54). This event, along with the large size and the great vacuole of the basal cell favour the suggestion that this cell serves as a collector of the mesophyll photosynthates, which through the apoplast and the symplast are led to the stalk cell and finally to the head cells to constitute precursors for essential oil biosynthesis. It should be recalled that the secretory

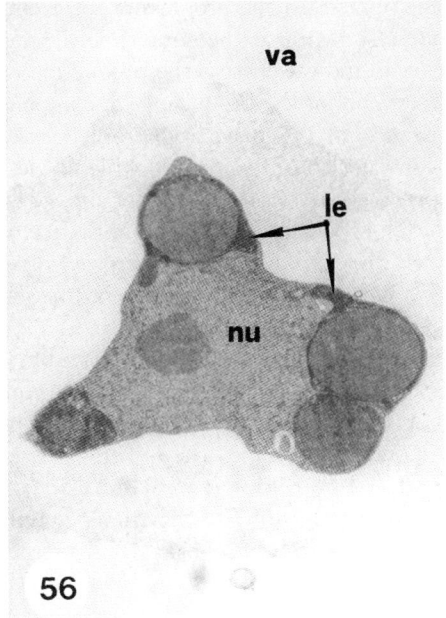

Figure 2.56 Section of a cytoplasmic strand transversing the large central vacuole (va) of the peltate hair basal cell. In the strand, the nucleus (nu) surrounded by several inclusion-containing leucoplasts (le), are embedded (×9500).

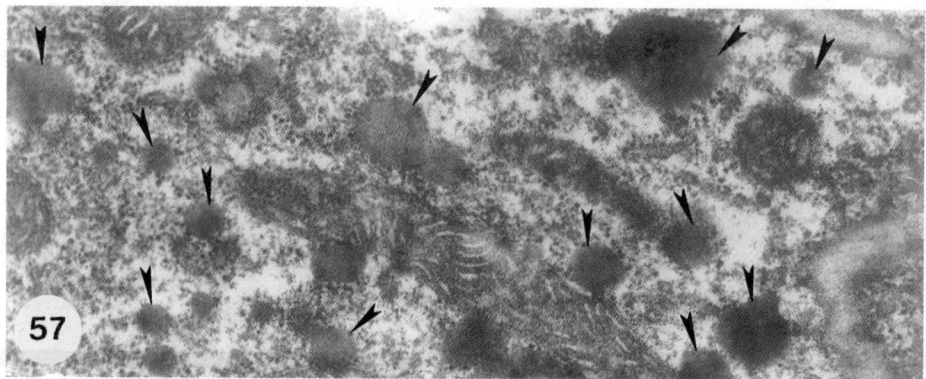

Figure 2.57 Variously-sized osmiophilic droplets of essential oil (arrowheads) in the ground plasm of a head cell during the phase of early secretion (×32 500).

head cells possess leucoplasts (not chloroplasts) and they, thus, are unable to produce the photosynthates they need.

Phase of active secretion   After cell divisions in the head region of the peltate glandular hairs of *Origanum* have been concluded and the cells have obtained their final size, secretion of the essential oil starts (Bosabalidis and Tsekos, 1982a,b). As previously mentioned, the only cells able to produce essential oil are those of the head of the hair. The first ultrastructural sign of secretion is the appearance of many small secretory droplets within the ground plasm of the head cells (Figure 2.57). These droplets are osmiophilic, round in shape and they do not possess a boundary membrane. In the progress of secretion, the size and number of the droplets gradually increase, while at the same time the density of the ground plasm becomes reduced. This is an indication that the production of the secretory material is likely performed at the expense of the ground plasm components. Presence of a substance identical in appearance to the plasmatic droplets was not observed within the plastids or the endoplasmic reticulum elements, which have been considered in the glandular hairs of other species to constitute the major sites of essential oil biosynthesis (Schnepf, 1969a; Tsekos, 1974; Bell and Curtis, 1985; Bourett *et al.*, 1994). Ultrastructural revelation that essential oil is produced in the ground plasm of the head cells in the form of droplets, further comes from observations on the peltate glandular hairs of *Monarda* (Heinrich, 1973), *Mentha* (Amelunxen, 1965), and several other Lamiaceae members (Werker *et al.*, 1985a). Mc Caskill *et al.*, (1992) have observed and specifically stained many droplets of essential oil within the ground plasm of the head cells of *Mentha* peltate hairs. These authors by incubating isolated heads with radio-labelled [^{14}C] sucrose have noticed an incorporation of it into the oil droplets, a fact suggesting that photosynthetic sucrose (introduced from the mesophyll cells) becomes glycolysed and converted to terpenoid precursors (pyruvate). The enzymatic equipment for this conversion is possessed by the head cells. A ground plasmic origin of the terpenoid precursor isopentenyl pyrophosphate (IPP) is further discussed by Gray (1987).

The view that ground plasm droplets in the head cells of *Origanum* peltate hairs most probably correspond to the secreted essential oil and not to cytoplasmic liposomes, is based on the facts that: (a) they appear only at the stage of secretion, (b) their appearance is combined with a reduction of the ground plasm density, (c) they progressively

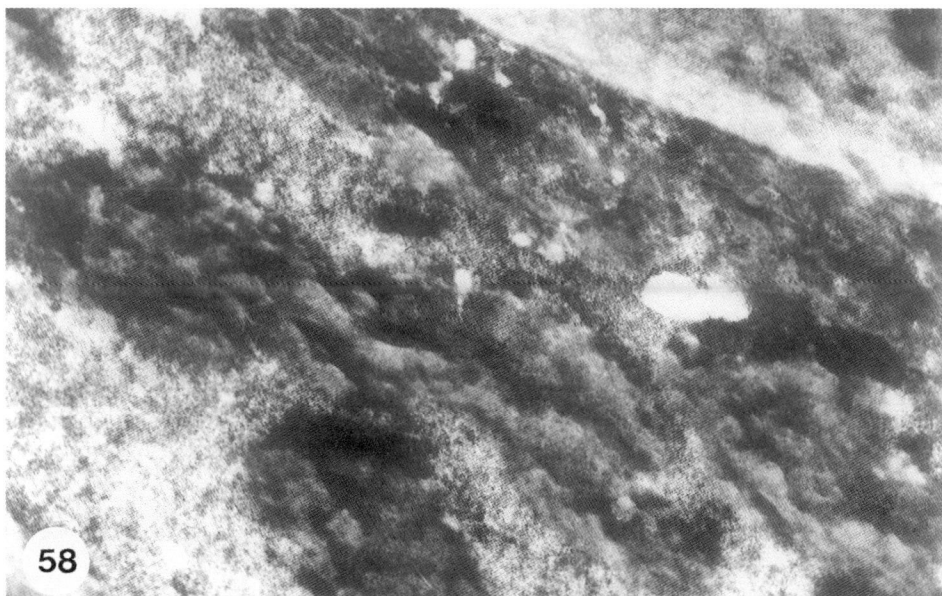

Figure 2.58 Irregular accumulations of essential oil in the ground plasm of a head cell during the phase of late secretion (×33 000).

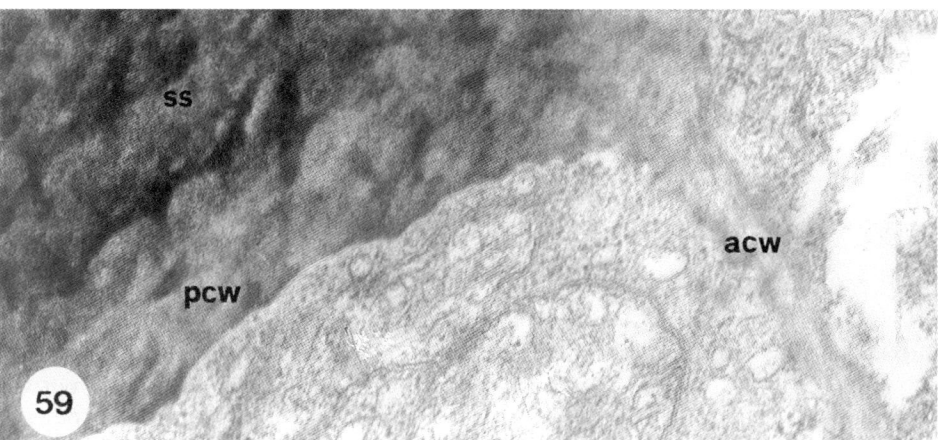

Figure 2.59 Osmiophilic essential oil secreted within the subcuticular space (ss) formed at the tip of the head. The apical periclinal walls of the head cells (pcw) are impregnated with the essential oil, whereas the anticlinal walls (acw) are free of oil. Note the apparent resemblance of the subcuticular space content to the irregular accumulations in Figure 2.58 (×36 500).

increase in number and size during secretion, and (d) they exhibit an intense osmiophilia. The latter observation is in accordance with relevant ones on glandular hairs of other species, in which the osmiophilic secretory product is also considered as essential oil (Amelunxen, 1965; Schnepf, 1969a; Heinrich, 1973; Bell and Curtis, 1985; Bosabalidis, 1990a).

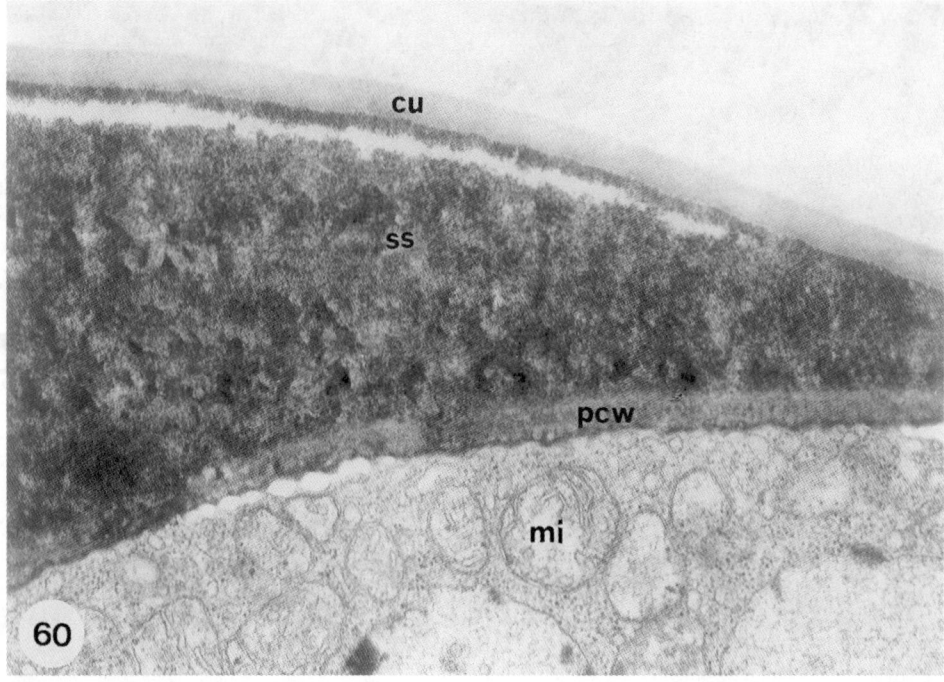

Figure 2.60 Osmiophilic essential oil accumulated in the subcuticular space (ss) formed by detachment of the cuticle (cu) from the apical walls of the head cells (pcw). Numerous mitochondria (mi) accumulate along these walls (×32 500).

In the progress of secretion, the essential oil droplets fuse with each other to create large intracellular accumulations (Figure 2.58). The secreted material does not remain stored in the head cells, but it migrates toward the tip of the head. There, it penetrates the apical walls (by impregnation of the microfibril capillaries) to become finally released into a space formed by separation of the cuticle from the walls (subcuticular space). The movement of the oil seems to take place exclusively toward the tip of the head and not to any lateral direction, as judged by the absence of plasmodesmata at the anticlinal walls and by the presence of osmiophilic residues only at the apical periclinal walls (due to their impregnation by the oil) (Figure 2.59). These features favour the suggestion that head cells secrete independently from each other (presumably producing different components of the essential oil) and that the final mixture of the partial components is performed within the subcuticular space.

The above pattern of eccrine secretion presupposes a specific structure of the plasma membrane lining the apical walls of the head cells and a proper enzymic equipment, so that the exudate can pass through this membrane without causing any damage to it. On the other hand, in the apical plasma membrane a "pump mechanism" of high potential should exist, which, within the scope of cell polarity, draws the intracellular exudate toward the head apex. The important roles ascribed to the apical membrane become ultrastructurally evident by its greater thickness and electron density. The accumulation of many energy providing mitochondria along the apical plasma membrane (Figures 2.59, 2.60), further supports the view of a specific function of this membrane.

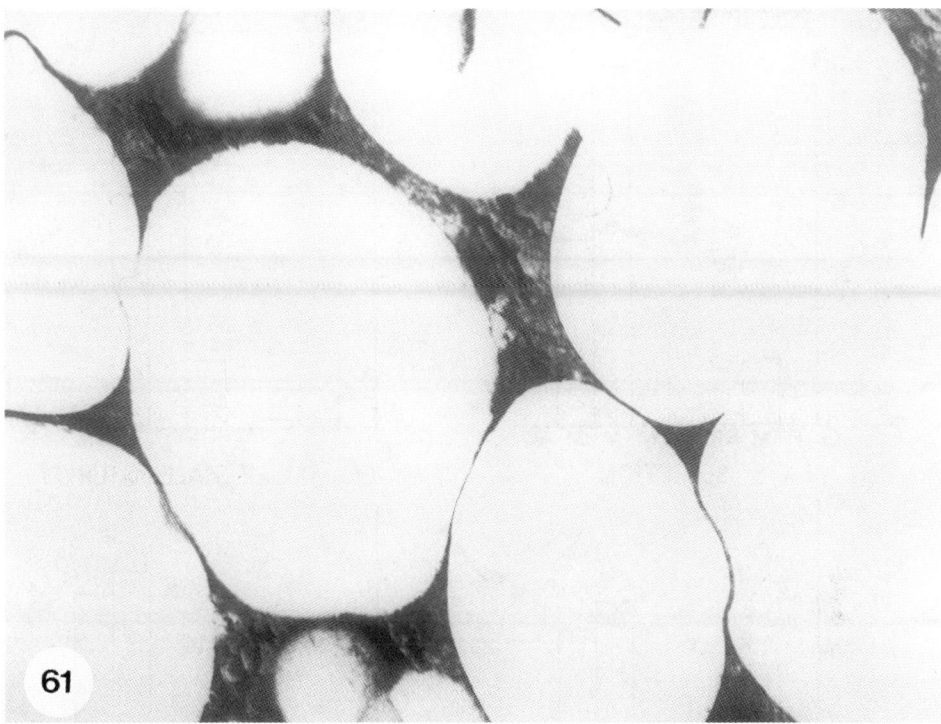

Figure 2.61 Large droplets with an electron translucent content embedded in an osmiophilic matrix, within the subcuticular space of peltate glandular hairs (×25 000).

The continuous discharge of the essential oil into the subcuticular space results in a progressive raising of the cuticle to finally create a bladder-like chamber filled with oil (Figure 2.60). In most of the peltate hairs on the leaves of *Origanum*, the content of the subcuticular space appears osmiophilic homogenous, but in many hairs it appears bearing large globular compartments of low electron density (Figure 2.61). These compartments were found to correspond to flavone aglycones (Bosabalidis et al., 1998a). Subcuticular spaces with a homogenous content have been reported in the glandular hairs of *Mentha* (Amelunxen, 1965), *Arctium* (Schnepf, 1969a), *Artemisia* (Duke and Paul, 1993), etc., while those with a heterogenous content of two chemical phases in the glandular hairs of *Monarda* (Heinrich, 1973), *Cannabis* (Mahlberg et al., 1984), *Salvia* (Serrato-Valenti et al., 1997; Bisio et al., 1999), etc.

A point of consideration is whether the secreted essential oil remains within the subcuticular space throughout glandular hair lifetime or it becomes released to the outside. In some species, like those of *Inula* (Werker and Fahn, 1982), *Tamus* (Bruni et al., 1987), *Chrysanthemum* (Vermeer and Peterson, 1979b), etc., the raised cuticle covering the subcuticular space has been observed to possess pores, through which the entire secretory product or at least a part of it is released.

Discharge of the essential oil from the subcuticular space may also be performed by rupture of the cuticle. This manner of excretion has been reported to occur in the glandular hairs of *Viscaria* (Tsekos and Schnepf, 1974), *Artemisia* (Corsi and Nencioni,

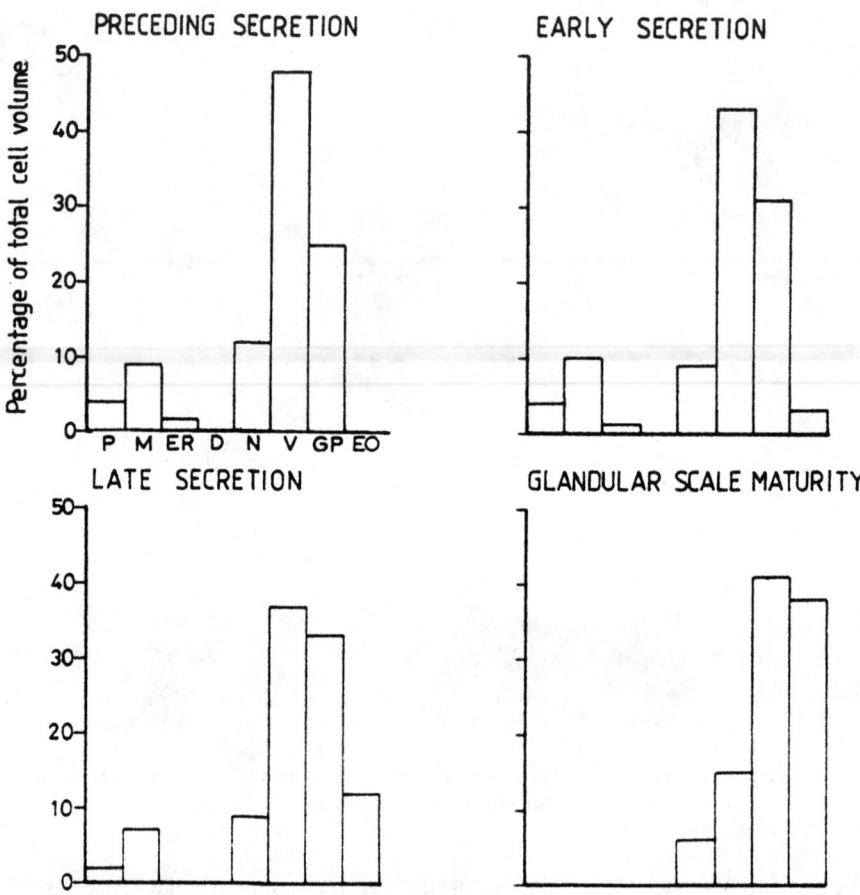

Figure 2.62 Average volume percentages of secretory cell (head cell) components at four stages of peltate hair development. P, plastids; M, mitochondria; ER, endoplasmic reticulum; D, dictyosomes; N, nucleus; V, vacuoles; GP, ground plasm; EO, cytoplasmic essential oil.

1995), *Newcastelia* (Dell and McComb, 1975), *Medicago* (Kreitner and Sorensen, 1979) etc. In the glandular hairs of *Thymus* (Bruni and Modenesi, 1983), *Salvia* (Serrato-Valenti *et al.*, 1997; Bisio *et al.*, 1999), *Plectranthus* (Ascensao *et al.*, 1999) and *Leonotis* (Ascensao *et al.*, 1995), the rupture of the cuticle seems to take place along a predetermined line of apparent weakness at the equatorial plane of the ballooned subcuticular space. In *Pelargonium* (Oosthuizen and Coetzee, 1983), cuticular rupture does not proceed horizontally, but vertically. To the above views supporting a process of release of the secretory product from the subcuticular space through cuticular pores or gaps, other contradictory views have been expressed in which glandular hairs were observed to have a continuous cuticle enclosing the secretion (Schnepf, 1969b; Henderson, *et al.*, 1970; Bory and Clair-Maczulajtys, 1981; Bosabalidis and Tsekos, 1982a; Oosthuizen and Coetzee, 1983; Bell and Curtis, 1985; Werker *et al.*, 1985a; Brun *et al.*, 1991).

In order to have an integrated idea about the alterations taking place in the various head cell components during essential oil secretion, morphometric assessments have been performed in the peltate hairs of *O. dictamnus* (Bosabalidis, 1990a). Four stages (preceding secretion, early secretion, late secretion and maturity) have been selected by using as criteria the relative volumes of the intracellular components, the time essential oil droplets first appear in the ground plasm, the increase in the number and size of the oil droplets, the presence of extensive cytoplasmic accumulations of essential oil and, finally, the degree of detachment of the cuticle from the head apical walls to form a subcuticular space. The changes in volume of the secretory cell components were determined with respect to the total cell volume (Figure 2.62). During all stages, the relative volumes of the plastids, endoplasmic reticulum elements, mitochondria and dictyosomes, do not exceed 10 per cent of the total secretory cell volume. At the stage preceding secretion, the volume percentages of the ground plasm and the nucleus are 25 per cent and 12 per cent, respectively. The volume of the cytoplasmic oil droplets at the stage of early secretion was estimated to be 3 per cent of the total cell volume, while at the stage of late secretion, this volume increases to 12 per cent. At glandular hair maturity, the cytoplasmic oil accumulations were found to occupy a relative cell volume of 38 per cent.

An attempt has been further made to morphometrically estimate the essential oil yield of *O. dictamnus* dry leaves by using only micrographs and not any kind of distillatory apparatuses and large amounts of material (Bosabalidis, 1990a). The total volume of the essential oil in a peltate hair is the sum of the oil volume within the subcuticular space plus the oil volume within the secretory cells. The geometrical shape of the subcuticular space is roughly an ellipsoid, in which the two crossed horizontal axes are of the same length (peltate hair surface view). The third vertical axis of the ellipsoid corresponds to the vertical distance between the highest point of the raised cuticle and the lowest point of the mean surface of the apical walls of the head cells. The volume of the oil in the subcuticular space ($V = 4/3 \pi R_{vtc} R^2$ hzl) can be calculated on light micrographs, while that of the cell interior on electron micrographs. Multiplication of the oil content of a single peltate hair by the number of hairs occuring on the upper and lower leaf surfaces (stereo micrographs) results in the oil content of a leaf. Considering the number of leaves held in 100 g dry leaf material, a theoretical essential oil yield of 1.4 per cent is concluded. This value is within the limits of acceptance, as deduced from studies in which a distillatory procedure was applied (Katsiotis and Oikonomou, 1986).

Phase of senescence   When the process of secretion is completed and the subcuticular space is filled with essential oil, the peltate glandular hairs in the leaves of *Origanum* start to disintegrate (Bosabalidis and Tsekos, 1982a,b). In the head cells, the nucleus acquires a flake-like appearance and the nuclear envelope is not visible any longer. The electron density of the plastid matrix is very much reduced. In the mitochondria, the sacculi are not easily distinguishable and the outer envelope is locally dilated or its continuity is interrupted by small accumulations of a substance probably derived from membrane disorganization.

At more advanced senescence of the secretory cells, the electron density of the cytoplasm highly increases and the whole protoplast becomes greatly detached from the cell walls, so that cells appear plasmolyzed (Figure 2.63). The cell walls are not stretched, but they get strongly folded. Features of disintegration are also met in the stalk cell of the peltate hair, whereas the basal cell seems to retain its vital characteristics (Figure 2.64). Disorganization of the peltate glandular hairs after integration of secretion has been also mentioned to occur in other members of Lamiaceae (Amelunxen, 1965;

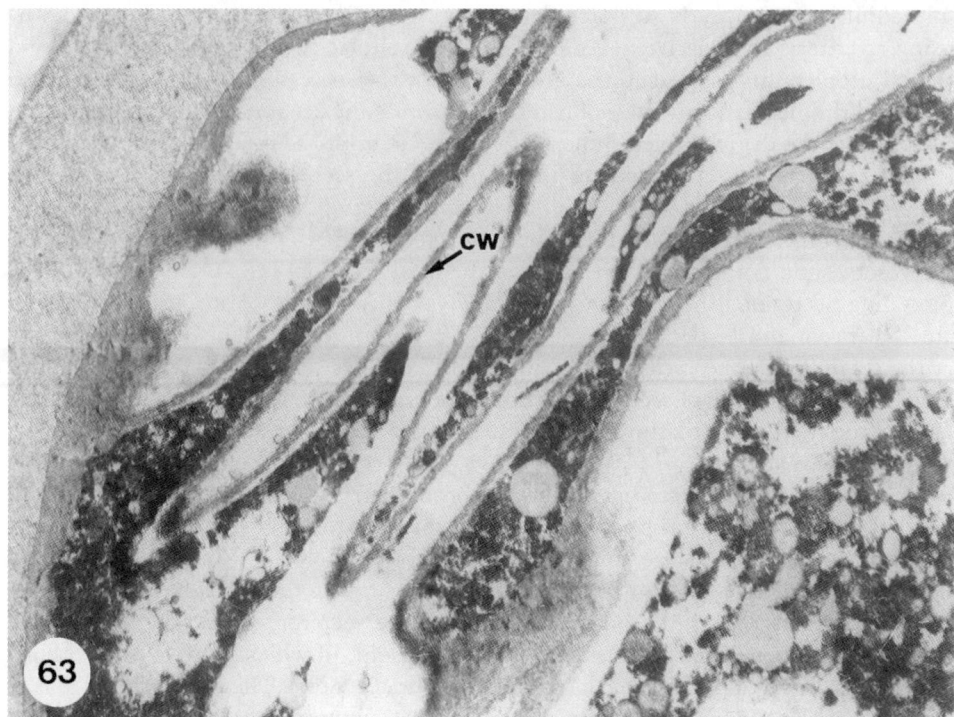

Figure 2.63 Disintegrated head cells of a senescent peltate hair. The cell walls (cw) are highly refolded, while the cell contents are disorganized and detached from the cell walls (×17 500).

Fischer and Hecht-Buchholz, 1985; Werker *et al.*, 1993), as well as in members of other families, like Myricaceae (Bell and Curtis, 1985), Dicrastylidaceae (Dell and McComb, 1975), Simarubaceae (Bory and Clair-Maczulajtys, 1981), Asteraceae (Carlquist, 1958), Cannabaceae (Hammond and Mahlberg, 1973), etc.

The mesophyll

The structure of the mesophyll in *Origanum* species follows the typical pattern of the dicot leaf. The chlorenchymatic cells of the palisade parenchyma appear in longitudinal section (leaf cross section) elongated (see Figures 2.3–2.5), while in cross section (leaf paradermal section) rounded (Figure 2.65). These profiles disclose a cylindrical 3D shape. The palisade parenchyma cells leave between them small intercellular spaces. The chloroplasts and the other cytoplasmic organelles, as well as the nucleus, are peripherally arranged. The chlorenchymatic cells of the spongy parenchyma have an irregular shape and among them large intercellular spaces are formed (Figure 2.66).

At the submicroscopic level, the mesophyll cells appear composed of a large central vacuole, which limits the cytoplasm to a thin parietal layer (Figure 2.67). Within this layer, various organelles, such as nucleus, chloroplasts, mitochondria, dictyosomes, endoplasmic reticulum elements, peroxisomes, etc. are located. The chloroplasts display a lens-like profile and contain large starch grains and a well-developed grana/fret

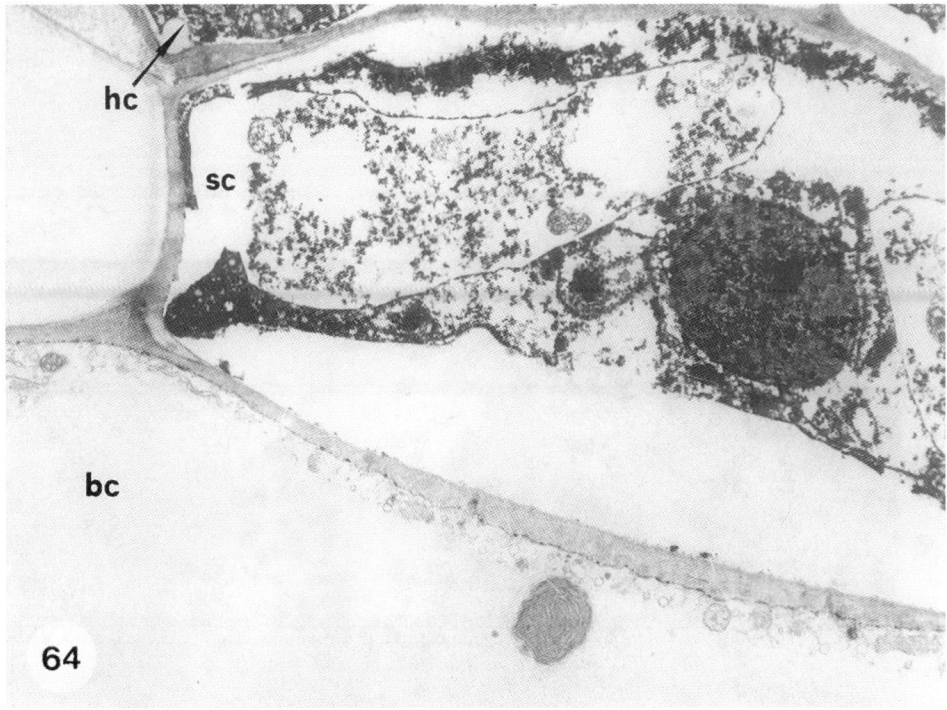

Figure 2.64 Disintegrated stalk cell (sc) of a senescent peltate hair. The basal cell (bc) seems to retain its vitality, as judged by the presence of a thin parietal cytoplasmic layer in which organelles are embedded (×9000).

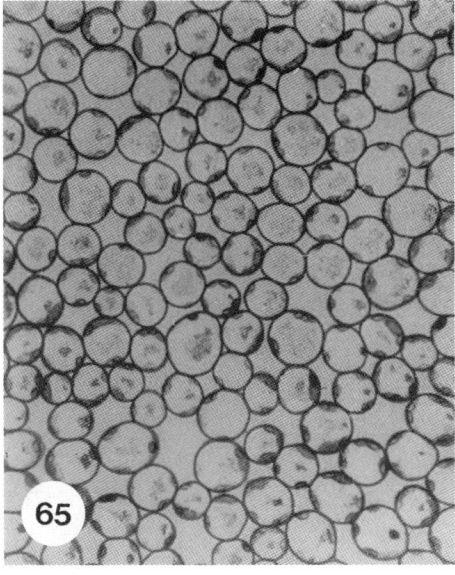

Figure 2.65 Cross section of palisade parenchyma cells (leaf paradermal section) in the mesophyll of *Origanum vulgare*. The cells exhibit a round profile and form small intercellular spaces (×500).

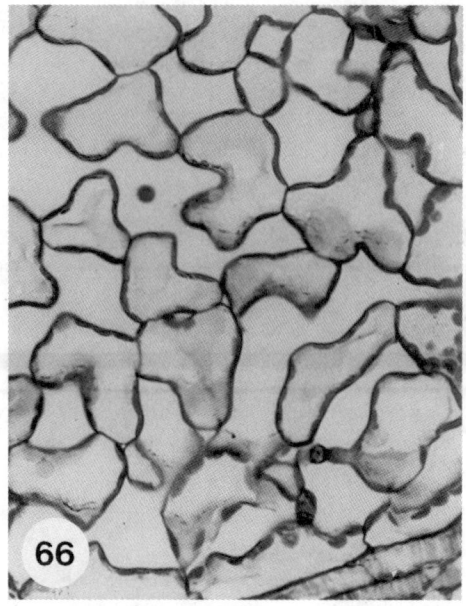

Figure 2.66 Spongy parenchyma cells, as they appear in a paradermal section of the leaf of *Origanum vulgare*. The cells have an irregular shape and form large intercellular spaces (×500).

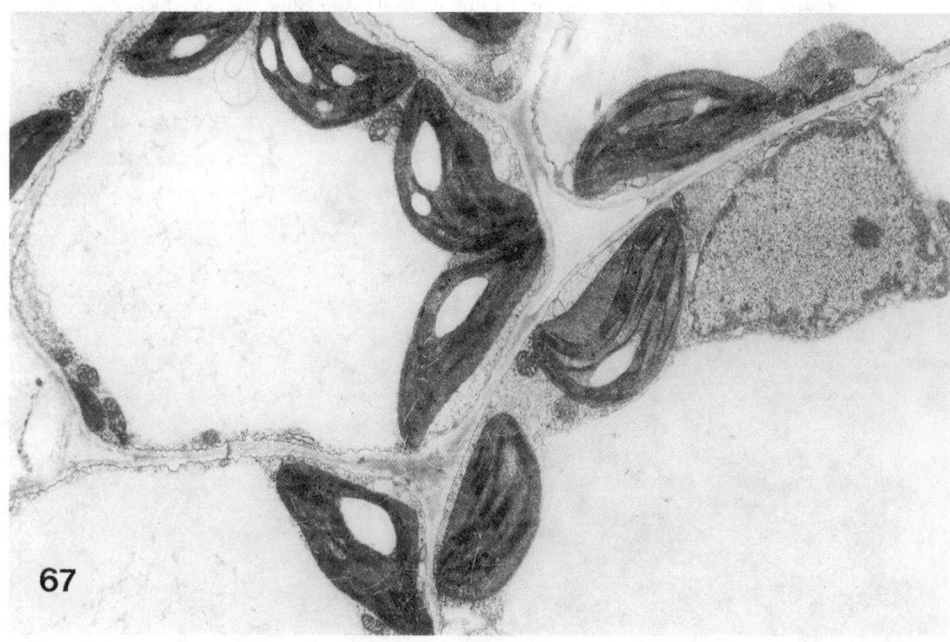

Figure 2.67 Ultrathin section of the chlorenchymatous mesophyll of *Origanum vulgare*. The chloroplasts along with other organelles are peripherally arranged (×8500).

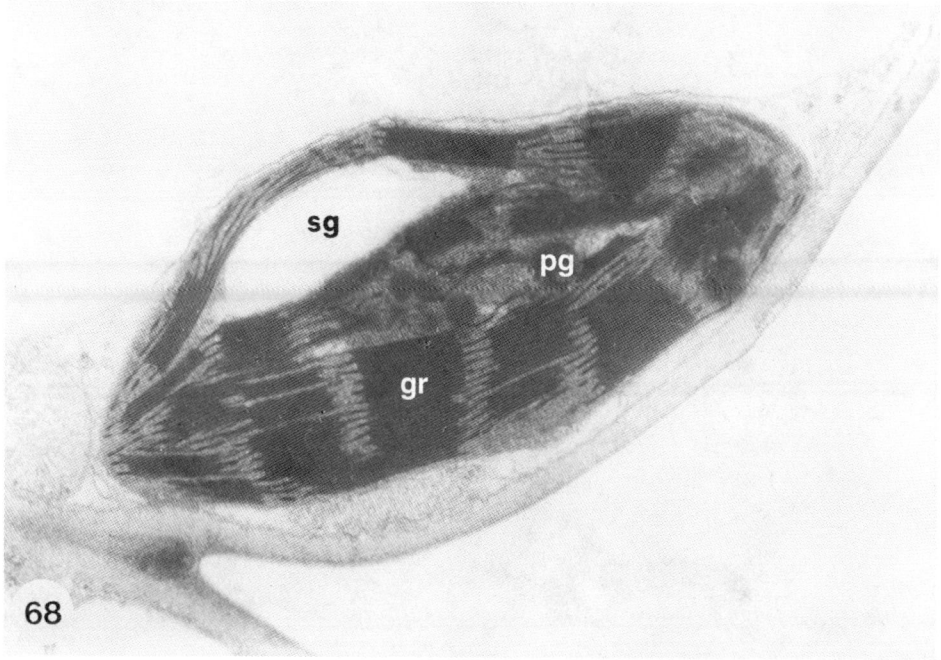

Figure 2.68 A mesophyll chloroplast in *Origanum vulgare* leaf. The chloroplast contains variously-sized grana (gr), plastoglobuli (pg) and starch grains (sg) (×28 500).

system (Figure 2.68). They also contain plastoglobuli of high electron density. The results of morphometric assessments conducted on mesophyll chloroplasts of *O. vulgare* subsp. *hirtum* are presented in Table 2.9.

A cytoplasmic component often observed in the mesophyll cells of *Origanum* species is a crystalloid inclusion (Bosabalidis and Papadopoulos, 1983). This inclusion exhibits various profiles (trapezoidal, rectangular, triangular, etc.) and its internal substructure may be amorphous, striated, or a square lattice (Figures 2.69, 2.70, 2.72–2.75). The chemical nature of the inclusion was cytochemically demonstrated to be proteinaceous. Thus, after treatment of ultrathin sections with the proteolytic enzyme pepsin, the crystalloid becomes entirely digested (Figures 2.70, 2.71).

To interpret the substructure of the crystalloid and determine the interspacings, the specimens in the electron microscope were rotated/tilted with respect to the electron

Table 2.9 Morphometric assessments of chloroplasts in leaves of *Origanum vulgare* subsp. *hirtum*

Parameter	n	(Mean ± SD)
Number of chloroplasts/palisade cell section	52	13.2 ± 1.6
Number of chloroplasts/spongy cell section	52	5.9 ± 0.8
Chloroplast surface area ($\mu m^2/\mu m^3$ of cell)	19	0.48 ± 0.01
Volume fraction of grana and frets/chloroplast section	24	34.9 ± 5.5
Volume fraction of starch grains/chloroplast section	24	48.8 ± 7.9
Volume fraction of plastoglobuli/chloroplast section	24	1.8 ± 0.6

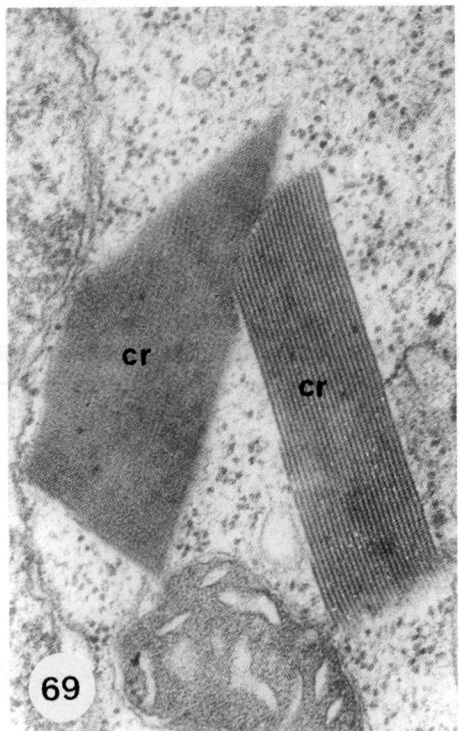

Figure 2.69 Two crystalloids in the cytoplasm of a leaf chlorenchyma cell of *Origanum vulgare* (×62 000).

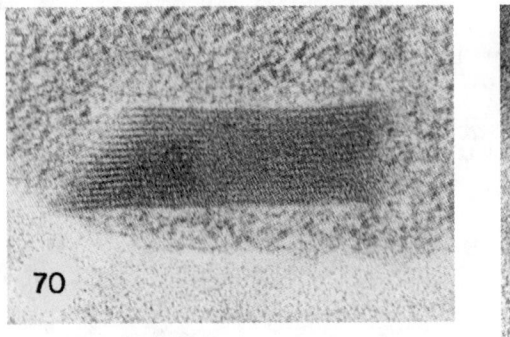 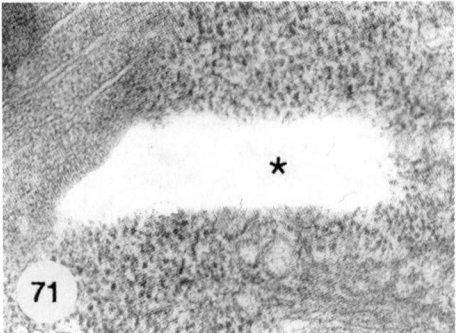

Figures 2.70 and 2.71 Trapezoidal cytoplasmic crystalloids before (Figure 2.70) and after (Figure 2.71) pepsin treatment. The electron transparent region (asterisk) in Figure 2.71 corresponds to a digested crystalloid, disclosing thus its proteinaceous constitution (×58 000).

beam and the obtained diffraction patterns were recorded at various angles. Thus, when the crystalloid is oriented in the specimen, so that the electron beam becomes parallel to the crystallographic plane along its long axis, the image shows a sequence of parallel striations which correspond to the parallel planes (Figure 2.72). The spacings between the striations are estimated to be 100 A. Tilting and rotating of the specimen until it is

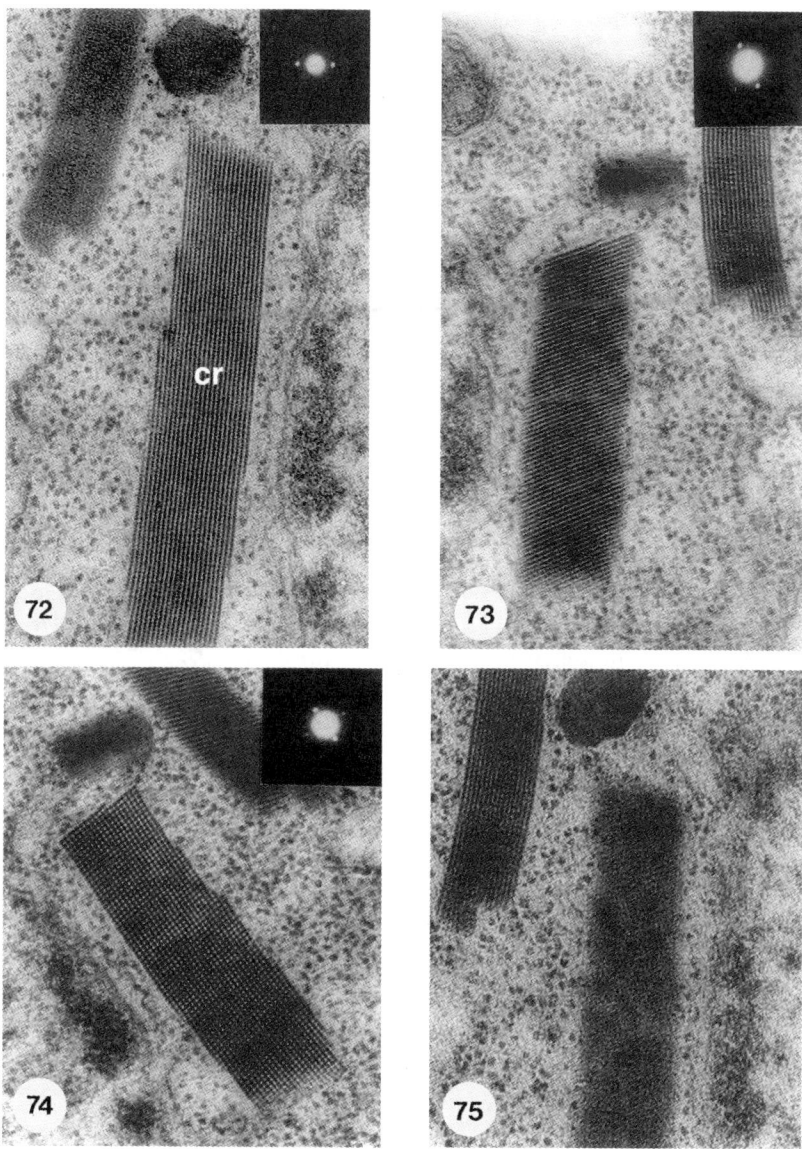

Figures 2.72–2.75 A cytoplasmic crystalloid (in a leaf chlorenchyma cell of *Origanum dictamnus*) tilted at various angles to the TEM electron beam. Its internal structure appears to consist of striations running parallel to the long axis (Figure 2.72) or the short axis (Figure 2.73). It also appears as a square lattice (Figure 2.74) or amorphous (Figure 2.75). The diffraction patterns are shown in the corresponding insets (×70 000).

observed parallel to the crystallographic planes along the short axis of the crystalloid, cause the orientation of the striations to change, but the spacings still remain the same (Figure 2.73). When the electron beam becomes simultaneously parallel to both of these planes, the internal structure of the crystalloid appears as a square lattice of electron dense dots 70–80 Å in diameter, spaced at 100 Å centre-to-centre (Figure 2.74).

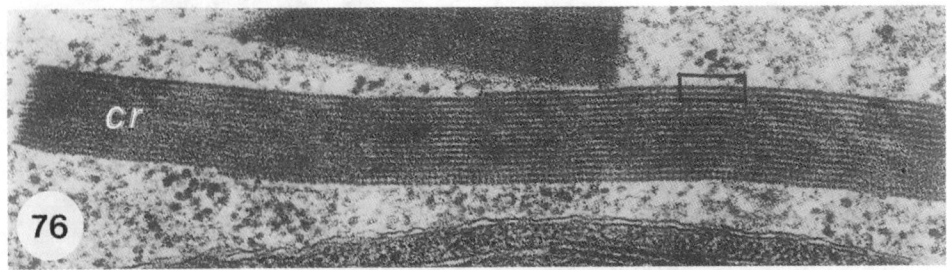

Figure 2.76 A cytoplasmic crystalloid in which the boxed portion is highly magnified (Figures 2.77, 2.78) to demonstrate the substructure of the striations (×56 000).

Crystalloids oriented randomly with respect to the electron beam appear amorphous (Figure 2.75). This is due to the disappearance of the crystallographic planes, which are not in a Bragg reflection position. The diffraction patterns of the above images are shown in the corresponding insets to the figures.

High magnification of the striations discloses that they consist of double helices bridged together via short fibrils (Figures 2.76–2.78). Our selection of right and left-handed helices illustrated in Figure 2.78, is arbitrary. The double-helical composition of the striations is further supported by the form of the diffraction patterns recorded when a crystalloid was tilted at various angles (Figures 2.72–2.75). The existence of three spots, i.e. a central one and a ±1 order, indicates that the striations comprise a sinusoidal grating. This implies that the density of the two groups of striations follows the sinusoidal distribution, which occurs if each striation can be projected as an ellipsoid of revolution (Moore, 1972). Since the crystalloids have been cytochemically proved to be proteinaceous, the above mentioned helices most probably represent protein macromolecules. Lawton (1978), studying the *p*-protein crystalloids in the sieve elements of pea hypocotyl, has proposed a model according to which striations consist of helically wound filaments. A similar twisted form of striations has been described by Renaudin and Cheguillaume (1977) in the cytoplasmic crystalloid inclusions of *Thesium humifusum* haustoria. Observations made by Arsanto (1982) on the *p*-protein crystalloids of some dicots, revealed that striations are composed of double helices 100–150 Å in diameter.

The stem

Anatomical and morphometric aspects

The stem in *Origanum* species displays in cross section a square profile, like in all members of Lamiaceae (Figure 2.79). A group of collenchyma cells occupies each corner and often extends as a single layer at the subepidermal region of the stem sides. The cortex is relatively thin and is composed of small parenchyma cells with a round or ellipsoid contour. An endodermis of flattened cells also exists in the cortical region (Parry, 1969). In very young stems, the conductive tissue consists of variously-sized vascular bundles arranged in a circle between the cortex and the pith. In older stems, the phloem poles of the bundles fuse with each other to create a continuous band of phloem. The same process also takes place in the xylem poles. The width of the xylem band, however, is not uniform,

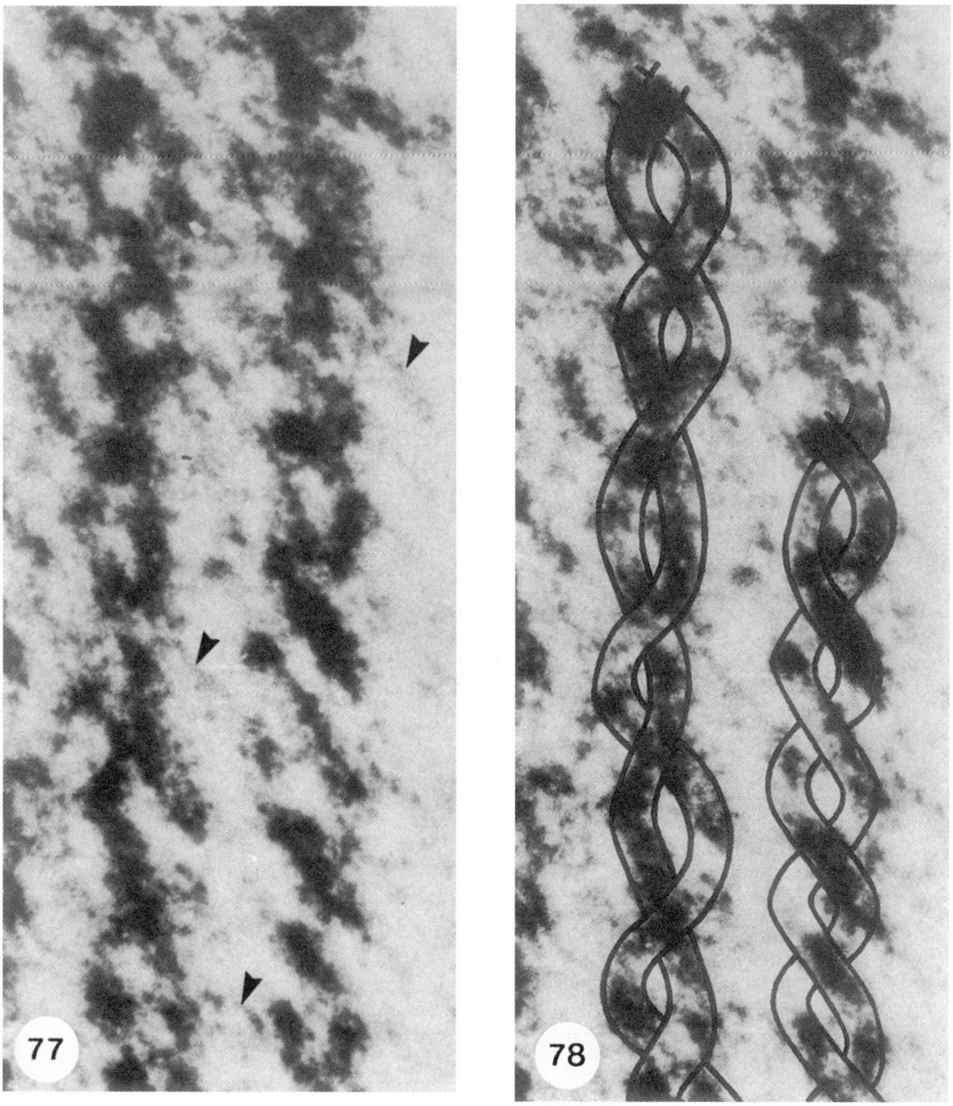

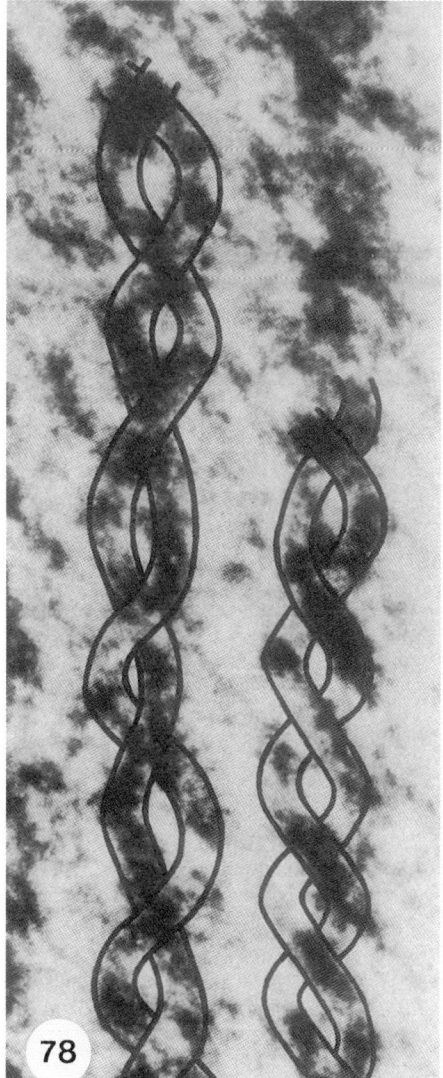

Figures 2.77 and 2.78 The same highly magnified image in which crystalloid striations appear as double helices. These helices are locally cross-bridged via short fibrils (arrowheads in Figure 2.77) (×1 800 000).

but it locally projects to the pith (more numerous vessels) at the regions facing the corners of the stem. The pith is composed of storage parenchyma with small intercellular spaces. The size of the pith cells decreases from the centre to the periphery. Morphometric estimations on the relative volumes of the histological components of the stem in *O. vulgare* subsp. *hirtum* (Table 2.10) have revealed that the xylem is about twice as much voluminous than the phloem, while at the same time a great amount of the total stem volume is occupied by the cortex and the pith (35.6 per cent and 25.4 per cent, respectively).

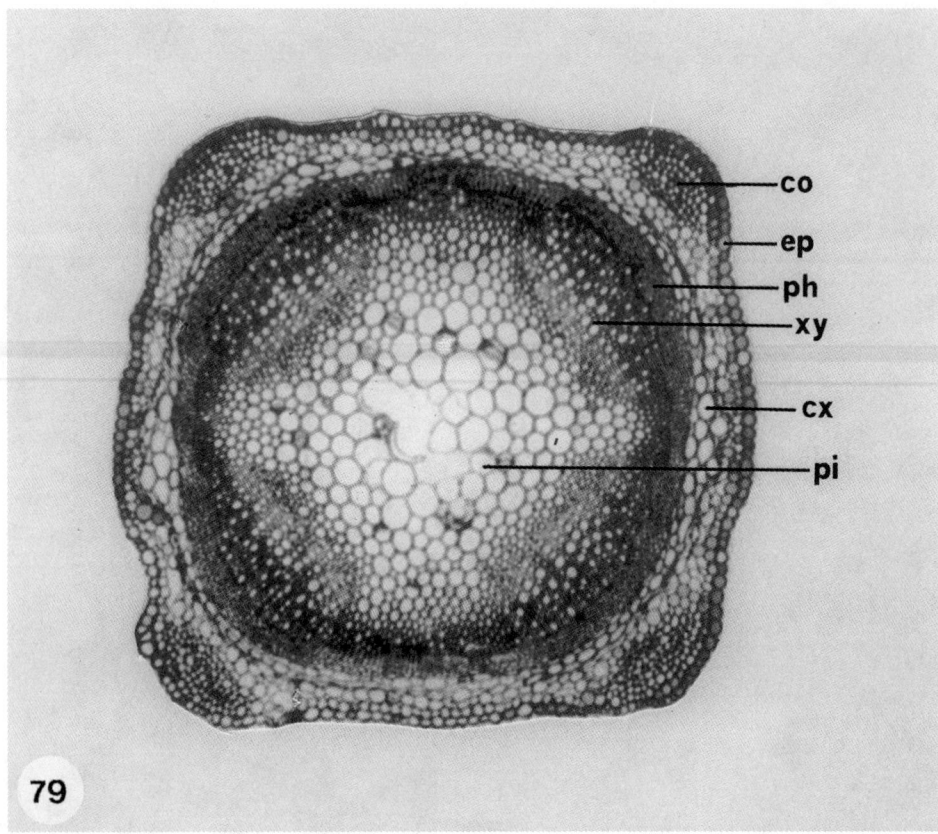

Figure 2.79 Cross section of the stem of *Origanum vulgare* to show the arrangement of the tissues. co, collenchyma; cx, cortex; ep, epidermis; ph, phloem; pi, pith; xy, xylem (×66).

Table 2.10 Relative volumes (±standard deviation) of the stem histological components in *Origanum vulgare* subsp. *hirtum*

Tissue	n	RV (%)
Epidermis	12	3.43 ± 0.6
Cortex	12	35.6 ± 2.7
Phloem	12	14.4 ± 3.9
Xylem	12	28.9 ± 4.1
Pith	12	25.4 ± 2.1

The root

Anatomical and morphometric aspects

In a cross section at a level above the root hair zone, the primary root of *Origanum* appears typically composed of an epidermis, a cortex and a vascular cylinder (Figure

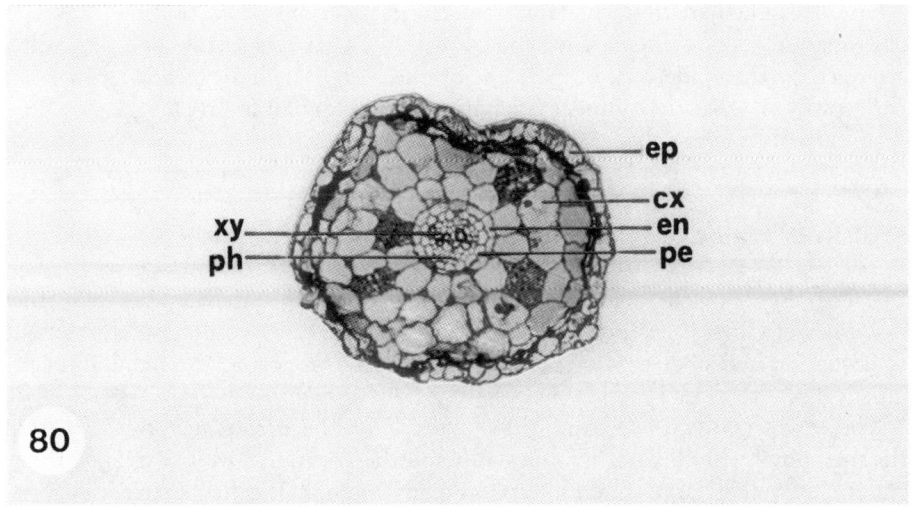

Figure 2.80 Cross section of the primary root of *Origanum vulgare*. cx, cortex; en, endodermis; ep, epidermis; per, pericycle; ph, phloem; xy, xylem (×40).

2.80). The epidermal cells have thin walls and are devoid of a cuticle. They are in close contact with each other forming a single-layered epidermis.

The cortex occupies the greatest part of the root and it consists of large parenchyma cells. The content of some of them is dark, because of the presence within the vacuoles of secondary metabolites (phenolics, tannins, etc.). The arrangement of the cells in the cortex seems to follow a radial pattern due to the successive periclinal divisions of the initial cortical cells. Intercellular spaces are not prominent. The cortical cell plastids lack chlorophyll and they also do not contain starch grains. The endodermis, which constitutes the inner boundary of the cortex, is one-layer thick and it is composed of small cells with no intercellular spaces. Between some endodermal cells, passage cells exist.

The vascular cylinder occupies the central portion of the root. Its outermost region, the pericycle, is composed of small parenchyma cells unistratosely arranged. The primary xylem forms two opposite continuous rays, on either side of which one pole of primary phloem is seated (diarch root). The protoxylem elements are small and face the pericycle, while the metaxylem ones are large and centrally located. Because of the

Table 2.11 Relative volumes (±standard deviation) of the root histological components in *Origanum vulgare* subsp. *hirtum*

Tissue	n	RV (%)
Epidermis	12	19.79 ± 2.14
Cortex	12	55.24 ± 4.29
Endodermis	12	6.01 ± 0.28
Pericycle	12	5.35 ± 0.59
Phloem	12	6.1 ± 0.39
Xylem	12	7.51 ± 0.49

location of the metaxylem elements in the centre of the root, the latter does not possess a typical parenchymatic pith. Morphometric estimations conducted on cross sections of primary roots of *Origanum* showed that nearly 20 per cent of the total root volume is occupied by the epidermis, 60 per cent by the cortex (including endodermis), and 20 per cent by the vascular cylinder (including pericycle) (Table 2.11).

SUMMARY AND CONCLUSIONS

In the present article, information originating from research studies on structural dynamics of *Origanum* is presented, in order for the reader to get a picture about the construction and operation of the specific tissues composing the plant fundamental organs (leaf, stem, root).

The leaves of *Origanum* exhibit the typical for the dicots anatomical pattern. In the mesophyll, chloroplasts are numerous and they contain a well-developed thylakoid system, as well as large starch grains. In many mesophyll cells, proteinaceous crystalloids constructed of parallel striations (protein double helices) are encountered free in the cytoplasm. The leaf epidermis bears glandular hairs and nonglandular hairs, as well. Glandular hairs are of three distinct types, i.e. peltate hairs, capitate hairs Type I, and capitate hairs Type II. The peltate hairs are composed of a 12-celled head, a unicellular stalk and an also unicellular base. These hairs secrete the bulk of the essential oil produced by the *Origanum* plants. The capitate hairs are composed of one head cell, one stalk cell, one intermediate cell (only in Type II), and one basal cell.

Analyses of the ontogenetic patterns of the glandular and nonglandular hairs showed that each hair originates from a protodermal initial cell which further undergoes a series of periclinal and anticlinal divisions. In the glandular hairs, the essential oil appears to be biosynthesized in the ground plasm of the head cells, from where it becomes released into a space formed by separation of the cuticle from the apical cell walls. After the head cells have concluded their secretory function, they progressively get disorganized.

The stems of *Origanum* plants exhibit in cross section a square profile. In the young stems, the cortex contains chlorenchyma cells and at each of the four stem corners, a pole of angular collenchyma develops. In front of each collenchyma pole, a vascular bundle exists, while in the centre of the stem the pith is localized. The roots are diarch (two opposite rays of primary xylem crossed with two groups of primary phloem) and their major volume is occupied by large cortical cells.

The above structural features may constitute a pool of data which can be utilized for other fields of *Origanum* research, such as physiology, biochemistry, phytochemistry, biotechnology, etc. Considering that "structure" and "function" are dependent on each other, the importance and necessity of knowing the structural components participating in a certain metabolic process, as well as their temporal alterations and modifications become obvious. Important is also the determination of the ontogenetic patterns of the histological structures, so that one is able to follow step by step sum of the developmental changes taking place from the initial stage until the stages of full growth and subsequent senescence. Recently developed in plant sciences techniques and sophisticated devices can be used, in order specific subjects to be faced from many sides and derived studies to be well-documented.

REFERENCES

Amelunxen, F. (1964) Elektronenmikroskopische Untersuchungen an den Drüsenhaaren von *Mentha piperita* L. *Planta Med.* 12, 121–139.

Amelunxen, F. (1965) Elektronenmikroskopische Untersuchungen an den Drüsenschuppen von *Mentha piperita* L. *Planta Med.* 13, 457–473.

Amelunxen, F., Wahlig, T. and Arbeiter, H. (1969) Über den Nachweis des ätherischen Öls in isolierten Drüsenhaaren und Drüsenschuppen von *Mentha piperita* L. *Z. Pflanzenphysiol.* 61, 68–72.

Ames, I.H. and Pivorun, J.P. (1974) A cytochemical investigation of a chloroplast inclusion. *Amer. J. Bot.* 61, 794–797.

Arsanto, J.P. (1982) Observations on *p*-protein in dicotyledons. Substructural and developmental features. *Amer. J. Bot.* 69, 1200–1212.

Ascensao, L., Marques, N. and Pais, M.S. (1995) Glandular trichomes on vegetative and reproductive organs of *Leonotis leonurus* (Lamiaceae). *Ann. Bot.* 75, 619–626.

Ascensao, L. and Pais, M.S. (1998) The leaf capitate trichomes of *Leonotis leonurus*: Histochemistry, ultrastructure and secretion. *Ann. Bot.* 81, 263–271.

Ascensao, L., Figueiredo, A.C., Barroso, J.G., Pedro, L.G., Schripsema, J., Deans, S.G. and Scheffer, J.J.C. (1998) *Plectranthus madagascariensis*: Morphology of the glandular trichomes, essential oil composition and its biological activity. *Int. J. Plant Sci.* 159, 31–38.

Ascensao, L., Mota, L. and Castro, M.D.M. (1999) Glandular trichomes on the leaves and flowers of *Plectanthrus ornatus*: Morphology, distribution and histochemistry. *Ann. Bot.* 84, 437–447.

Beerling, D.J. and Kelly, C.K. (1997) Stomatal density responses of temperate woodland plants over the past seven decades of CO_2 increase: A comparison of Salisbury (1927) with contemporary data. *Amer. J. Bot.* 84, 1572–1583.

Bell, J.M. and Curtis, J.D. (1985) Development and ultrastructure of foliar glands of *Comptonia peregrina* (Myricaceae). *Bot. Gaz.* 146, 288–292.

Berta, G., Dela Pierre, M. and Maffei, M. (1993) Nuclear morphology and DNA content in the glandular trichomes of peppermint (*Mentha* × *piperita* L.). *Protoplasma* 175, 85–92.

Bestmann, H.J., Classen, B., Kobold, U., Vostrowsky, O. and Klingauf, F. (1987) Botanical insecticides. IV. The insecticidal effect of the essential oil from the costmary chrysanthemum *Chrysanthemum balsamita* L. *Anz. Schädl. Pflanz. Umweltsch.* 60, 31–34.

Bini-Maleci, L., Corsi, G. and Pagni, A.M. (1983) Trichomes tecteurs et secreteurs dans la sauge (*Salvia officinalis* L.). *Pl. Med. Phytother.* 17, 4–17.

Bini-Maleci, L. and Servettaz, O. (1991) Morphology and distribution of trichomes in Italian species of *Teucrium* sect. *Chamaedrys* (Labiatae) – a taxonomical evaluation. *Pl. Syst. Evol.* 174, 83–91.

Bisio, A., Corallo, A., Gastaldo, P., Romussi, G., Ciarallo, G., Fontana, N., De Tommas, N. and Profumo, P. (1999) Glandular hairs and secreted material in *Salvia blepharophylla* Brandegee ex Epling grown in Italy. *Ann. Bot.* 83, 441–452.

Bory, G. and Clair-Maczulajtys, D. (1981) Morphology, ontogeny and cytology of trichomes of *Ailanthus altissima*. *Phytomorphology* 30, 67–78.

Bosabalidis, A.M. and Tsekos, I. (1982a) Glandular scale development and essential oil secretion in *Origanum dictamnus* L. *Planta* 156, 496–504.

Bosabalidis, A.M. and Tsekos, I. (1982b) Ultrastructure of the essential oil secretion in glandular scales of *Origanum dictamnus* L. leaves. In N. Margaris, A. Koedam and D. Vokou (eds), *Aromatic Plants: Basic and Applied Aspects*, Martinus Nijhoff Publ., The Hague, pp. 3–12.

Bosabalidis, A.M. and Tsekos, I. (1982c) Ultrastructural studies on the secretory cavities of *Citrus deliciosa* Ten. II. Development of the essential oil-accumulating central space of the gland and process of active secretion. *Protoplasma* 112, 63–70.

Bosabalidis, A.M. and Papadopoulos, D. (1983) Ultrastructure, organization and cytochemistry of cytoplasmic crystalline inclusions in *Origanum dictamnus* L. leaf chlorenchyma cells. *J. Cell Sci.* 64, 231–244.

Bosabalidis, A.M. and Tsekos, I. (1984) Glandular hair formation in *Origanum* species. *Ann. Bot.* 53, 559–563.

Bosabalidis, A.M. (1987a) Origin, differentiation and cytochemistry of membrane-limited inclusion bodies in leucoplasts of leaf epidermal cells of *Origanum dictamnus* L. *Cytobios* 50, 77–88.

Bosabalidis, A.M. (1987b) Morphometric evaluation of inclusion body-containing leucoplasts in leaf epidermal cells of *Origanum dictamnus* L. *Bot. Helv.* 97, 315–321.

Bosabalidis, A.M. (1990a) Quantitative aspects of *Origanum dictamnus* L. glandular scales. *Bot. Helv.* 100, 199–206.

Bosabalidis, A.M. (1990b) Glandular trichomes in *Satureja thymbra* leaves. *Ann. Bot.* 65, 71–78.

Bosabalidis, A.M. and Exarchou, F. (1995) Effect of NAA and GA3 on leaves and glandular trichomes of *Origanum* × *intercedens* Rech.: Morphological and anatomical features. *Int. J. Plant Sci.* 156, 488–495.

Bosabalidis, A.M. (1996) Ontogenesis, ultrastructure and morphometry of the petiole oil ducts of celery (*Apium graveolens* L.). *Flav. Fragr. J.* 11, 269–274.

Bosabalidis, A.M. and Kokkini, S. (1997) Infraspecific variation of leaf anatomy in *Origanum vulgare* grown wild in Greece. *Bot. J. Linn. Soc.* 123, 353–362.

Bosabalidis, A.M., Gabrieli, C. and Niopas, I. (1998a) Flavone aglycones in glandular hairs of *Origanum* × *intercedens*. *Phytochemistry* 49, 1549–1553.

Bosabalidis, A.M. and Skoula, M. (1998b) A comparative study of the glandular trichomes on the upper and lower leaf surfaces of *Origanum* × *intercedens* Rech. *J. Essent. Oil Res.* 10, 277–286.

Bourett, T.M., Howard, R.J., O'Keefe, D.P. and Hallahan, D.L. (1994) Gland development on leaf surfaces of *Nepeta racemosa*. *Int. J. Plant Sci.* 155, 623–632.

Brun, N., Colson, M., Perrin, A. and Voirin, B. (1991) Chemical and morphological studies of the effects of ageing on monoterpene composition in *Mentha* × *piperita* leaves. *Can. J. Bot.* 69, 2271–2278.

Bruni, A. and Modenesi, P. (1983) Development, oil storage and dehiscence of peltate trichomes in *Thymus vulgaris* (Lamiaceae). *Nord. J. Bot.* 3, 245–251.

Bruni, A., Tosi, B. and Modenesi, P. (1987) Morphology and secretion of glandular trichomes in *Tamus communis*. *Nord. J. Bot.* 7, 79–84.

Cantino, P. and Sanders, R. (1986) Subfamilial classification of Labiatae. *Syst. Bot.* 11, 163–185.

Cantino, P.D. (1990) The phylogenetic significance of stomata and trichomes in the Labiatae and Verbenaceae. *J. Arnold Arbor.* 71, 323–370.

Carlquist, S. (1958) Structure and ontogeny of glandular trichomes of Madinae (Compositae). *Amer. J. Bot.* 45, 675–682.

Circella, G. and D'Andrea, L. (1993) Comparative study on biology, growth and productivity of different taxa and ecotypes of genus *Origanum*. *Acta Horticult.* 330, 115–121.

Corsi, G. and Pagni, A.M. (1990) The glandular hairs of *Valeriana officinalis* subsp. *collina*. I. Some unusual features in their development and differentiation. *Bot. J. Linn. Soc.* 104, 381–388.

Corsi, G. and Nencioni, S. (1995) Secretory structures in *Artemisia nitida* Bertol. (Asteraceae). *Isr. J. Plant Sci.* 43, 359–365.

Corsi, G. and Bottega, S. (1999) Glandular hairs of *Salvia officinalis*: New data on morphology, localization and histochemistry in relation to function. *Ann. Bot.* 84, 657–664.

Danilova, M.F. and Kashina, T.K. (1989) Ultrastructure of glandular hairs in *Perilla ocymoides* (Lamiaceae) in connection with their possible involvement in photoperiodic induction of flowering. *Phytomorphology* 39, 265–275.

de Bary, H. (1877) *Vergleichende Anatomie der Vegetationsorgane der Phanerogamen und Farne*. Leipzig.

Dell, B. and McComb, A.J. (1975) Glandular hairs, resin production, and habitat of *Newcastelia viscida* E. Pritzel (Dicrastylidaceae). *Aust. J. Bot.* 23, 373–390.

Dell, B. and McComb, A.J. (1977) Glandular hair formation and resin secretion in *Eremophila fraserii* F. Menll (Myoporaceae). *Protoplasma* 92, 71–86.

Dolzmann, P. (1964) Elektronenmikroskopische Untersuchungen an den Saughaaren von *Tillandsia usneoides* (Bromeliaceae). I. Feinstruktur der Kuppelzelle. *Planta* 60, 461–472.

Duke, S.O. and Paul, R.N. (1993) Development and fine structure of the glandular trichomes of *Artemisia annua* L. *Int. J. Plant Sci.* 154, 107–118.

Economou-Amilli, A., Vokou, D., Anagnostidis, K. and Margaris, N.S. (1982) Leaf morphology of *Thymus capitatus* (Labiatae) by scanning electron microscopy. In N. Margaris, A. Koedam and D. Vokou (eds), *Aromatic Plants: Basic and Applied Aspects*. Martinus Nijhoff Publ. The Hague, pp. 13–24.

Fahn, A. (1979) Ultrastructure of nectaries in relation to nectar secretion. *Amer. J. Bot.* 66, 977–985.

Fahn, A. and Shimony, C. (1996) Glandular trichomes of *Fagonia* L. (Zygophyllaceae) species: Structure, development and secreted materials. *Ann. Bot.* 77, 25–34.

Fahn, A. and Shimony, C. (1998) Ultrastructure and secretion of the secretory cells of two species of *Fagonia* L. (Zygophyllaceae). *Ann. Bot.* 81, 557–565.

Favali, M.A. and Patrignani, G. (1973) Ultrastructure of the stem of the aquatic plant *Callitriche stagnalis* L. *Österr. Bot. Z.* 122, 167–175.

Fischer, G. and Hecht-Buchholz, C. (1985) The influence of boron deficiency on glandular scale development and structure in *Mentha piperita*. *Planta Med.* 51, 371–377.

Franceschi, V.R. and Gianquinta, R.T. (1983) Glandular trichomes of soybean leaves: Cytological differentiation from initiation through senescence. *Bot. Gaz.* 144, 175–184.

Frey-Wyssling, A. and Mühlethaler, K. (1959) Über das submikroskopische Geschehen bei der Kutinisierung pflanzenlicher Zellwände. *Viertj. Naturf. Ges.* 104, 294–299.

Gailhofer, M. and Thaler, I. (1974) Eiweisskristalle in den Plastiden der keimpflanzen einiger Palmen. *Phyton* 15, 251–258.

Gavalas, N., Bosabalidis, A.M. and Kokkini, S. (1998) Comparative study of leaf anatomy and essential oils of the hybrid *Mentha* × *villoso-nervata* and its parental species *M. longifolia* and *M. spicata*. *Isr. J. Plant Sci.* 46, 27–33.

Gershenzon, J., Maffei, M. and Croteau, R. (1989) Biochemical and histochemical localization of monoterpene biosynthesis in the glandular trichomes of spearmint (*Mentha spicata*). *Plant Physiol.* 89, 1351–1357.

Gleizes, M., Pauly, G., Carde, J.P., Marpeau, A. and Bernard-Dagan, C. (1983) Monoterpene hydrocarbon biosynthesis by isolated leucoplasts of *Citrofortunella mitis*. *Planta* 159, 373–381.

Goertzen, L.R. and Small, E. (1993) The defensive role of trichomes in black medick (*Medicago lupulina*, Fabaceae). *Pl. Syst. Evol.* 184, 101–111.

Gradmann, H. (1923) Die Windschutzeinrichtungen an den Spaltöffnungen der Pflanzen. *Jahr. Wiss. Bot.* 62, 449.

Gray, J.C. (1987) Control of isoprenoid biosynthesis in higher plants. *Adv. Bot. Res.* 14, 25–91.

Gupta, M.L. and Bhambie, S. (1978) Studies in Lamiaceae. IV. Foliar appendages in *Ocimum* L. and their taxonomic significance. *Proc. Indian Natn. Sci. Acad.* 44, 154–160.

Hammond, C.T. and Mahlberg, P.G. (1973) Morphology of glandular hairs of *Cannabis sativa* from scanning electron microscopy. *Amer. J. Bot.* 60, 524–528.

Hanlidou, E., Kokkini, S., Bosabalidis, A.M. and Bessiere, J.M. (1991) Glandular trichomes and essential oil constituents of *Calamintha menthifolia* (Lamiaceae). *Pl. Syst. Evol.* 177, 17–26.

Harborne, J.B., Tomas-Barberan, F.A., Williams, C.A. and Gil, M.I. (1986) A chemotaxonomic study of flavonoids from european *Teucrium* species. *Phytochemistry* 25, 2811–2816.

Heinrich, G. (1969) Elektronenmikroskopische Beobachtungen zur Entstehungsweise der Exkretbehälter von *Ruta graveolens*, *Citrus limon* und *Poncirus trifoliata*. *Österr. Bot. Z.* 117, 397–403.

Heinrich, G. (1973) Entwicklung, Feinbau und Ölgehalt der Drüsenschuppen von *Monarda fistulosa*. *Plant Med.* 23, 154–166.

Henderson, W., Hart, J.W., How, P. and Judge, J. (1970) Chemical and morphological studies on sites of sesquiterpene accumulation in *Pogostemon cablin* (patchouli). *Phytochemistry* 9, 1219–1228.

Hurkman, W.J. and Kennedy, G.S. (1976) Fine structure and development of proteoplasts in primary leaves of mungbean. *Protoplasma* 89, 171–184.

Ietswaart, J.H. (1980) *A taxonomic revision of the genus Origanum* (Labiatae). Leiden Univ. Press, The Hague.

Inamdar, J.A. and Bhatt, D.C. (1972) Structure and development of stomata in some Labiatae. *Ann. Bot.* 36, 335–344.

Israel, H.W. and Steward, F.C. (1967) The fine structure and development of plastids in cultured cells of *Daucus carota*. *Ann. Bot.* 31, 1–18.

Karousou, R., Bosabalidis, A.M. and Kokkini, S. (1992) *Sideritis syriaca* ssp. *syriaca*: Glandular trichome structure and development in relation to systematics. *Nord. J. Bot.* 12, 31–37.

Karpouhtsis, I., Pardali, E., Feggou, E., Kokkini, S., Scouras, Z. and Mavragani-Tsipidou, P. (1998) Insecticidal and genotoxic activities of oregano essential oils. *J. Agric. Food Chem.* 46, 1111–1115.

Katsiotis, S. and Oikonomou, G.N. (1986) Vergleichende Untersuchung verschiedener wildwachsender und in Kreta angebauter Muster von *Origanum dictamnus* L. *Sci. Pharm.* 54, 49–52.

Klingauf, F., Bestmann, H.J., Vostrowsky, O. and Michaelis, K. (1983) The effect of essential oils on insect pests. *Mitt. Deutsch. Ges. Allg. Angew. Entomol.* 4, 123–126.

Kokkini, S., Vokou, D. and Karoussou, R. (1991) Morphological and chemical variation of *Origanum vulgare* L. in Greece. *Bot. Chron.* 10, 337–346.

Konstantonopoulou, I., Vassilopoulou, L., Mavragani-Tsipidou, P. and Scouras, Z.G. (1992) Insecticidal effects of essential oils. A study of the effects of essential oils extracted from eleven Greek aromatic plants on *Drosophila auraria*. *Experientia* 48, 616–619.

Kowalski, S.P., Eanneta, N.T. and Steffens, J.C. (1988) Insect resistance in potato: purification and characterization of a polyphenol oxidase in glandular trichomes of wild potato. *Plant Physiol.* 86, 107.

Kreitner, G.L. and Sorensen, E.L. (1979) Glandular secretory system of alfalfa species. *Crop. Sci.* 19, 499–502.

Lawton, D.M. (1978) Ultrastructural comparison of the tailed and tailless *p*-protein crystals respectively of runner bean (*Phaseolus multiflorus*) and garden pea (*Pisum sativum*) with tilting stage electron microscopy. *Protoplasma* 97, 1–11.

Letchamo, W. and Gosselin, A. (1996) Transpiration, essential oil glands, epicuticular wax and morphology of *Thymus vulgaris* are influenced by light intensity and water supply. *J. Hort. Sci.* 71, 123–134.

Levin, D.A. (1973) The role of trichomes in plant defense. *Q. Rev. Biol.* 48, 3–15.

Lovett, J.V. and Speak, M.D. (1979) Studies of *Salvia reflexa* Hornem. II. Examination of specialized leaf surface structures. *Weed Res.* 19, 359–362.

Maffei, M., Gallino, M. and Sacco, T. (1986) Glandular trichomes and essential oils of developing leaves in *Mentha viridis lavanduliodora*. *Planta Med.* 31, 187–193.

Maffei, M., Chialva, F. and Sacco, T. (1989) Glandular trichomes and essential oils in developing peppermint leaves. I. Variation of peltate trichome number and terpene distribution within leaves. *New Phytol.* 111, 707–716.

Maffei, M., Mucciarelli, M. and Scannerini, S. (1993) Environmental factors affecting the lipid metabolism in *Rosmarinus officinalis* L. *Bioch. Syst. Ecol.* 21, 765–784.

Mahlberg, P.G., Hammond, C.T., Turner, J.C. and Hemphill, J.K. (1984) Structure, development and composition of glandular trichomes of *Cannabis sativa* L. In E. Rodriguez, P.L. Healey and I. Mehta (eds), *Biology and Chemistry of Plant Trichomes*, Plenum Press, New York, pp. 23–51.

Manetas, Y. (1999) Is enhanced UV-B radiation really damaging for plants? Some case studies with European mediterranean species. In J. Rozema (ed.), *Stratospheric Ozone Depletion: The Effects of Enhanced UV-B Ratiation on Terrestrial Ecosystems*. Backhyus Publ. Leiden, pp. 251–263.

Mc Caskill, D., Gerschenzon, J. and Croteau, R. (1992) Morphology and monoterpene biosynthetic capabilities of secretory cell clusters isolated from glandular trichomes of peppermint (*Mentha piperita* L.). *Planta* 187, 445–454.

Mc Caskill, D. and Croteau, R. (1995) Monoterpene and sesquiterpene biosythesis in glandular trichomes of peppermint (*Mentha* × *piperita*) rely exclusively on plastid-derived isopentenyl diphosphate. *Planta* 197, 49–56.

Metcalfe, C.R. and Chalk, L. (1979) *Anatomy of the Dicotyledons*, Vol. I, 2nd edn, Oxford: Clarendon Press.

Modenesi, P., Serrato-Valenti, G. and Bruni, A. (1984) Development and secretion of clubbed trichomes in *Thymus vulgaris* L. *Flora* 175, 211–219.

Moore, W.J. (1972) *Physical Chemistry*. pp. 928–959. London: Longman.

Nessler, C.L. and Mahlberg, P.G. (1979) Plastids in laticifers of *Papaver*. I. Development and cytochemistry of laticifer platids in *P. somniferum* L. (Papaveraceae). *Amer. J. Bot.* 66, 226–273.

Netolitzky, F. (1932) Die Pflanzenhaare. In K. Linsbauer (ed.), *Handbuch der Pflanzenanatomie*. Gebr. Bornträger Vlg. Berlin.

Nielsen, M.T., Jones, G.A. and Collins, G.B. (1982) Inheritance pattern for secreting and non-secreting glandular trichomes in tobacco. *Crop Sci.* 22, 1051–1053.

Nielsen, M.T., Akers, C.P., Järlfors, U.E., Wagner, G.J. and Berger, S. (1991) Comparative ulstructural features of secreting and nonsecreting glandular trichomes of two genotypes of *Nicotiana tabacum* L. *Bot. Gaz.* 152, 13–22.

Oosthuizen, L.M. and Coetzee, J. (1983) Morphogenesis of trichomes of *Pelargonium scabrum*. *S. Afr. J. Bot.* 2, 305–310.

Oosthuizen, L.M. and Coetzee, J. (1984) Trichome initiation during leaf growth in *Pelargonium scabrum*. *S. Afr. Tydskr. Plankt.* 3, 50–54.

Parry, J.W. (1969) *Spices. Vol. II. Morphology, Histology, Chemistry*. New York: Chemical Publ. Co.

Renaudin, S. and Cheguillaume, N. (1977) On the localization, the fine structure, and the chemical composition of crystalline inclusions in the *Thesium humifusum* haustoria. *Protoplasma* 93, 223–229.

Rosenthal, G.A. and Berenbaum, M.R. (1991) *Herbivores. Their Interactions with Secondary Plant Metabolites*. San Diego, New York, Academic Press.

Sawidis, T., Eleftheriou, E.P. and Tsekos, I. (1989) The floral nectaries of *Hibiscus rosa-sinensis*. III. A morphometric and ultrastructural approach. *Nord. J. Bot.* 9, 63–71.

Scheffer, J.J.C., Looman, A. and Baerheim-Svendsen, A. (1986) The essential oils of three *Origanum* species grown in Turkey. In E.-J. Brunke (ed.), *Progress in Essential Oil Research*, Walter de Gruyte & Co., Berlin.

Schnepf, E. (1969a) Über den Feinbau von Öldrüsen. I. Die Drüsenhaare von *Arctium lappa*. *Protoplasma* 67, 185–194.

Schnepf, E. (1969b) Über den Feinbau von Öldrüsen. II. Die Drüsenhaare in *Calceolaria* – Blüten. *Protoplasma* 67, 195–203.

Serrato-Valenti, G., Bisio, A., Cornara, L. and Ciarallo, G. (1997) Structural and histochemical investigation of the glandular trichomes of *Salvia aurea* L. leaves and chemical analysis of the essential oil. *Ann. Bot.* 79, 329–336.

Servettaz, O., Pinetti, A., Bellesia, F. and Bini-Maleci, L. (1994) Micromorphological and phytochemical research on *Teucrium scorodonia* and *Teucrium siculum* from the Italian flora. *Bot. Acta* 107, 416–421.

Sharaby, A. (1988) Anti-insect properties of the essential oil of lemongrass, *Cymbopogon citratus* against the cotton leafworm *Spodoptera exigua* (Itbn). *Ins. Sci. Appl.* 9, 77–80.

Sivropoulou, A., Papanikolaou, E., Nikolaou, C., Kokkini, S., Lanaras, T. and Arsenakis, M. (1996) Antimicrobial and cytotoxic activities of *Origanum* essential oils. *J. Agric. Food Chem.* 44, 1202–1205.

Spring, O. and Bienert, U. (1987) Capitate glandular hairs from sunflower leaves: Development, distribution and sesquiterpene lactone content. *J. Plant Physiol.* 130, 441–448.

Srivastava, L.M. (1966) On the fine structure of the cambium of *Fraxinus americana* L. *J. Cell Biol.* 31, 79–93.

Staudermann, W. (1924) Die Haare der Monokotylen. *Bot. Arch.* 8, 105–184.

Stebbins, L.G. and Shah, S.S. (1960) Developmental studies of cell differentiation in the epidermis of monocotyledons. II. Cytological features of stomatal development in the Graminae. *Devel. Biol.* 2, 477–500.

Tsekos, I. (1974) Zur Feinstruktur der Drüsen von *Ribes sanguineum* Pursch. *Sci. Ann. Fac. Phys. Math. Univ. Thes.* 14, 25–38.

Tsekos, I. and Schnepf, E. (1974) Der Feinbau der Drüsen der Pechnelke *Viscaria vulgaris. Bioch. Physiol. Pflanzen* 165, 265–270.

Venkatachalam, K.V., Kjonaas, R. and Croteau, R. (1984) Development and essential oil content of secretory glands of sage (*Salvia officinalis*). *Plant Physiol.* 76, 148–150.

Vermeer, J. and Peterson, R.L. (1979a) Glandular trichomes on the inflorescence of *Chrysanthemum morifolium* cv. Dramatic (Compositae) I. Development and morphology. *Can. J. Bot.* 57, 705–713.

Vermeer, J. and Peterson, R.L. (1979b) Glandular trichomes on the inflorescence of *Chrysanthemum morifolium* cv. Dramatic (Compositae) II. Ultrastructure and histochemistry. *Can. J. Bot.* 57, 714–729.

Vokou, D. and Margaris, N.S. (1986) Autoallelopathy of *Thymus capitatus. Acta Ecol.* 7, 157–163.

Vokou, D., Kokkini, S. and Bessiere, J.-M. (1993) Geographic variation of Greek oregano (*Origanum vulgare* ssp. *hirtum*) essential oils. *Bioch. System. Ecol.* 21, 287–295.

Walt van der, J.J.A. and Demarne, F. (1988) *Pelargonium graveolens* and *P. radens*: A comparison of their morphology and essential oils. *S. Afr. J. Bot.* 54, 617–622.

Walters, D.S., Grossman, H., Craig, R. and Mumma, R.O. (1989) *Geranium* defensive agents. IV. Chemical and morphological bases of resistance. *J. Chem. Ecol.* 15, 357–372.

Weiss, A. (1867) *Die Pflanzenhaare*, pp. 369–677. Karsten Bot. Unters.

Werker, E. and Fahn, A. (1982) *Inula* hairs. Structure, ultrastructure and secretion. In N. Margaris, A. Koedam and D. Vokou (eds), *Aromatic Plants: Basic and Applied Aspects*. Martinus Nijhoff Publ. The Hague, pp. 25–37.

Werker, E., Ravid, U. and Putievsky, E. (1985a) Structure of glandular hairs and identification of the main components of their secreted material in same species of the Labiatae. *Isr. J. Bot.* 34, 31–45.

Werker, E., Putievsky, E. and Ravid, U. (1985b) The essential oils and glandular hairs in different chemotypes of *Origanum vulgare* L. *Ann. Bot.* 55, 793–801.

Werker, E. (1993) Function of essential oil-secreting glandular hairs in aromatic plants of the Lamiaceae. A review. *Flav. Fragr. J.* 8, 249–255.

Werker, E., Putievsky, E., Ravid, U., Dudai, N. and Katzir, I. (1993) Glandular hairs and essential oil in developing leaves of *Ocimum basilicum* L. (Lamiaceae). *Ann. Bot.* 71, 43–50.

Wollenweber, E. and Schnepf, E. (1970) Vergleichende Untersuchungen über die flavonoiden Exkrete von "Mehl"- und "Öl"-Drüsen bei Primeln und die Feinstruktur der Drüsenzellen. *Z. Pflanzenphysiol.* 62, 216–227.

Wollenweber, E. (1984) The systematic implication of flavonoids secreted by plants. In E. Rodriguez, P.L. Healey and I. Mehta (eds), *Biology and Chemistry of Plant Trichomes*, Plenum Press, New York and London, pp. 53–69.

Yamaura, T., Tanaka, S. and Tabata, M. (1992) Localization of the biosynthesis and accumulation of monoterpenoids in glandular trichomes of thyme. *Planta Med.* 58, 153–158.

Part 3

Taxonomy and chemistry

# 3 The taxonomy and chemistry of *Origanum*

Melpomeni Skoula and Jeffrey B. Harborne

"WHAT IS ORIGANUM?"

The Linnaean concept of the genus *Origanum* (1754) can be summarized as following:

> Labiatae with flowers in more or less dense spikes, bracts conspicuous and often coloured, calyces with five equal teeth or 2-lipped or lower lip reduced, corollas 2-lipped

Bentham (1834) recognized three genera: *Amaracus*, *Majorana* and *Origanum* but later (1848) he returned to the Linnaean concept and described within *Origanum* four sections: *Amaracus*, *Majorana*, *Origanum* and *Anatolicon*. Briquet (1895) accepted three separate genera (*Amaracus*, *Majorana* and *Origanum*) and described some more sections in the genus *Majorana* (*Schizocalyx*, *Holocalyx* and *Chilocalyx*). Until recently, both the three genera concept and the one genus concept have been used by taxonomists.

Ietswaart (1980), in his taxonomic revision of *Origanum* which is widely recognized today, accepted the original broad concept of Linnaeus. Following morphological criteria he describes *Origanum* as follows:

> Labiatae with several erect, medium sized stems per specimen, and with glandular punctate ±ovate leaves. Few (sub)sessile flowers in a verticillaster. Verticillasters arranged in (dense) spikes with differentiated (coloured) bracts. Inflorescences more or less paniculate. Calyces straight, very variable: regularly 5-toothed, 2-or 1-lipped for 9/10 to 1/5; tubular campanulate or flattened; throats pillose or not. Corollas variable: 2-lipped for 2/5 to 1/7, gibbous (saccate) or not, sometimes flattened; tube straight or slightly curved. Stamen subequal to very unequal in length; accenting, straight or divergent, extensively protruding to included.

INFRAFAMILIAR CLASSIFICATION

The classification of the Labiatae above the genus level has been controversial. The delimitations and definition of tribes and subfamilies has been, and is particularly difficult and nobody with wide experience of the family can have much confidence in the reality of the majority of the existing supra-generic taxa (Hedge, 1992). However, according to the current status and classification of the genera within the Labiatae family, eight subfamilies are recognized: Ajugoideae, Chloanthoideae, Lamioideae, Nepetoideae, Pogostemonoideae, Scutellarioideae, Teucrioideae and Viticoideae

(Cantino *et al.*, 1992); this infrafamiliar classification has taken into consideration phylogenetic analyses based mainly on floral, fruit or vegetative morphology but also on embryology palynology, phytochemistry and leaf epidermal anatomy (Cantino, 1992), as well as on molecular evidence from cpDNA restriction site variation (Wagstaff *et al.*, 1995). Only the Nepetoideae, the largest subfamily, is further divided into four tribes: Escholtzieae, Lavanduleae, Mentheae and Ocimeae. The above infrafamiliar classification encompasses *Amaracus* and *Majorana* in the genus *Origanum*, which is within the tribe Mentheae.

In 1994, Kaufmann and Wink, attempted a phylogenetic classification of the subfamily Nepetoideae using *rbc*L gene sequences. Their analysis agreed with the previously mentioned classification, regarding the three tribes of Nepetoideae, namely: Escholtzieae, Lavanduleae, and Ocimeae, while they proposed a further subdivision of the Mentheae tribe into at least three main groups. One group includes *Hyssopus, Dracocephalum, Satureja, Glechoma, Agastache, Nepeta, Horminum,* and *Prunella*, another includes *Melissa, Majorana,* and *Ocimum*, and the third includes *Mentha, Monandra, Origanum, Thymus, Salvia, Rosmarinus* and *Perovskia*. All these genera belong to the tribe Mentheae according to Wagstaff *et al.* (1995) except that *Ocimum* is included in the tribe Ocimeae. Even though the *Melissa–Majorana–Ocimum* relation was not strongly supported by the bootstrap analysis, Kaufmann and Wink (1994) proposed that *Majorana* should be treated as genus on its own as it differed quite significantly from *Origanum*, based on relationships derived from the *rbc*L sequences.

Nevertheless, in the present work the classification of Ietswaart is followed, thus *Amaracus, Majorana,* and *Origanum* are considered as one genus.

INFRAGENERIC CLASSIFICATION

Within the genus *Origanum*, Ietswaart (1980), recognized three groups, 10 sections, 38 species, six subspecies and 17 hybrids. Ietswaart's classification was based on the following morphological characters: length of stems, indumentum of stems and leaves, arrangements, number and length of branches, shape of leaves, length of petioles, number of sessile glands on leaves, arrangement of verticillasters, shape of spikes, number of flowers in a verticillaster, shape, size, texture and colour of bracts, general shape and length of calyces, shape of lips and teeth in calyces, general shape and colour of corollas, shapes of lips and lobes of corollas, arrangement of stamens, length of staminal filament and length of styles. Since Ietswaart's publication, five more species (Carlström, 1984; Danin, 1990; Duman *et al.*, 1995; Danin and Künne, 1996) and one more hybrid (Duman *et al.*, 1998) have been recognized, raising the number of species to 43 and the number of hybrids to 18.

The recognition of the three groups of Ietswaart's classification is the following:

1. Group A has 2 or 1-lipped, rather large calyces, 4–12 mm long. Bracts are rather large, 4–25 mm long, membranous, usually purple, sometimes yellowish green, more or less glabrous.
2. Group B has 2 or 1-lipped rather small calyces, 1.3–3.5 mm long. Bracts are rather small 1–5 mm leaf-like in texture and colour, more or less hairy.
3. Group C has calyces with 5 (sub)equal teeth.

Table 3.1 shows the classification of *Origanum* species within the three groups and the 10 sections defined by Ietswaart as well as their distribution. The members of the

Table 3.1 List of accepted taxa within the genus Origanum

No. of species	No. of taxon	Group	Section	Species / Subspecies / Varieties	Distribution[a]
		A	Amaracus Bentham		
1	1			O. boissieri Ietswaart	Tu
2	2			O. calcaratum Jussieu	Gr-AE, Gr-Cr
3	3			O. cordifolium Vogel	Cy, Sy
4	4			O. dictamnus L.	Gr-Cr
5	5			O. saccatum Davis	Tu
6	6			O. solymicum Davis	Tu
7	7			O. symes Carlström[b]	Gr-AE
			Anatolicon Bentham		
8	8			O. akhdarense Ietswaart et Boulos	Li
9	9			O. cyrenaicum Beguinot et Vaccari	Li
10	10			O. hypericifolium Schwarz et Davis	Tu
11	11			O. libanoticum Boissier	Le
12	12			O. pampaninii Ietswaart	Li
13	13			O. scabrum Boissier et Heldreich	Gr, Gr-Cr
14	14			O. sipyleum L.	Tu, Gr-AE
15	15			O. vetteri Briquet et Barbey	Gr-AE
			Brevifilamentum Ietswaart		
16	16			O. acutidens Ietswaart	Tu
17	17			O. bargyli Mouterde	Tu, Sy
18	18			O. brevidens Dinsmore	Tu
19	19			O. haussknechtii Boissier	Tu
20	20			O. husnucan-baseri H. Duman, Z. Aytaç et A. Duran[c]	Tu
21	21			O. leptocladum Boissier	Tu
22	22			O. rotundifolium Boissier	Tu, Georgia

Table 3.1 (Continued)

No. of species	No. of taxon	Group	Section	Species / Subspecies / Varieties	Distribution[a]
			Longitubus Ietswaart		
23	23			O. amanum Post	Tu
		B	Chilocalyx Ietswaart		
24	24			O. bilgeri Davis	Tu
25	25			O. micranthum Vogel	Tu
26	26			O. microphyllum Vogel	Gr-Cr
27	27			O. minutiflorum Schwarz et Davis	Tu
			Majorana Bentham		
28	28			O. majorana L.	Gr-AE, Cy, Tu
29	29			O. onites L.	Gr, Gr-AE, Gr-Cr, Tu, Eg-Sn
30	30			O. syriacum L.	Tu, Cy, Sy, Le, Is, Jd, Eg-Sn
				O. syriacum L. var. syriacum	Is, Jd, Sy
				O. syriacum L. var. bevanii Ietswaart	Tu, Cy, Sy, Le
				O. syriacum L. var. sinaicum Ietswaart	Eg-Sn
		C	Campanulaticalyx Ietswaart		
31	31			O. dayi Post	Is
32	32			O. isthmicum Danin	Sn
33	33			O. jordanicum Danin and Künne[d]	Jd
34	34			O. petraeum Danin[e]	Jd
35	35			O. punonense Danin[e]	Jd
36	36			O. ramonense Danin	Is-Negev
			Elongatispica Ietswaart		
37	39			O. elongatum Emberger ex Maire	Mo
38	40			O. floribundum Munby	Ag
39	41			O. grosii Pau et Font Quer ex Ietswaart	Mo

	Origanum		
40		*O. vulgare* L.	From Az. to Taiwan and from N. Africa to Scandinavia
42		*O. vulgare* L. ssp. *vulgare*	From Britain and Scandinavia to Taiwan
43		*O. vulgare* L. ssp. *glandulosum* Ietswaart	Ag, Tn
44		*O. vulgare* L. ssp. *gracile* Ietswaart	From E. Turkey to Afganistan and S. Siberia
45		*O. vulgare* L. ssp. *hirtum* Ietswaart	Balkan, Tu
46		*O. vulgare* L. ssp. *virens* Ietswaart	Az., Ma, Balearic Is, Po, Sp, Mo
47		*O. vulgare* L. ssp. *viride* Hayek	Corse to C. China
	Prolaticorolla Ietswaart		
41		*O. compactum* Bentham	Sp, Mo
42		*O. ehrenbergii* Boissier	Le
43		*O. laevigatum* Boissier	Tu, Cy

Notes

a Ag, Algeria; Az, Azores Is.; Cy, Cyprus; Eg-Si, Egypt-Sinai; Gr, Mainland of Greece; Gr-AE, Greece – Aegean Islands; Gr-Cr, Greece-Crete; Is, Israel; Jd, Jordan; Le, Lebanon; Li, Libya; Ma, Madeira; Mo, Morroco; Po, Portugal; Sp, Spain; Sy, Syria; Tn, Tunisia; Tu, Turkey.
b Identified after Ietswaart's revision of the genus by Carlström, 1984.
c Identified after Ietswaart's revision by Duman *et al.*, 1995.
d Identified after Ietswaart's revision by Danin and Künne, 1996.
e Identified after Ietswaart's revision by Danin, 1990.

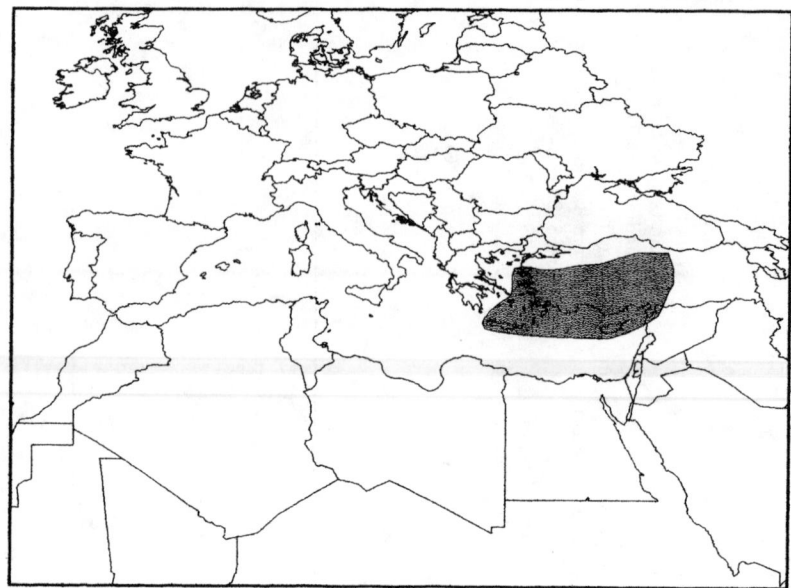

Figure 3.1 Distribution of the section *Amaracus*.

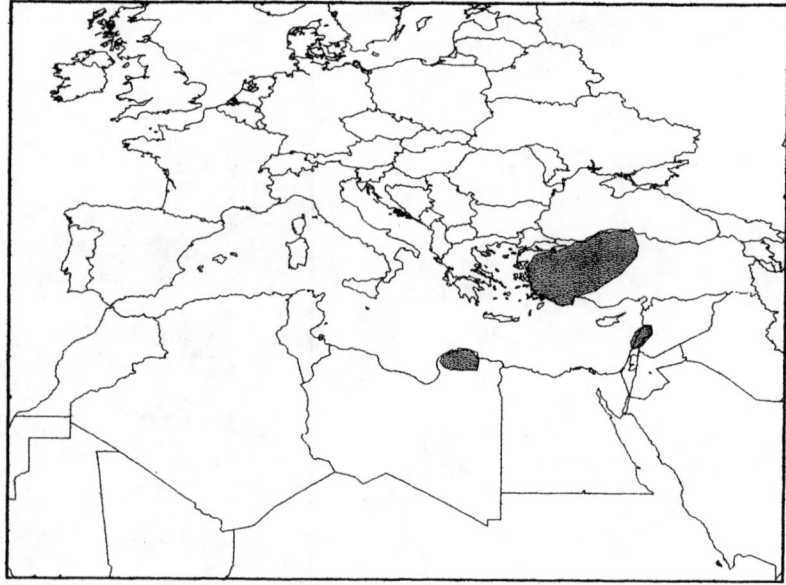

Figure 3.2 Distribution of the section *Anatolicon*.

genus are mainly distributed around the Mediterranean region while more than 81 per cent (35 out of 43) occur exclusively in the East Mediterranean, (Greuter *et al.*, 1986); four species are restricted in the West Mediterranean, while three are endemic to Libya. Figures 3.1–3.8 show the distribution of each *Origanum* section: from group A, the

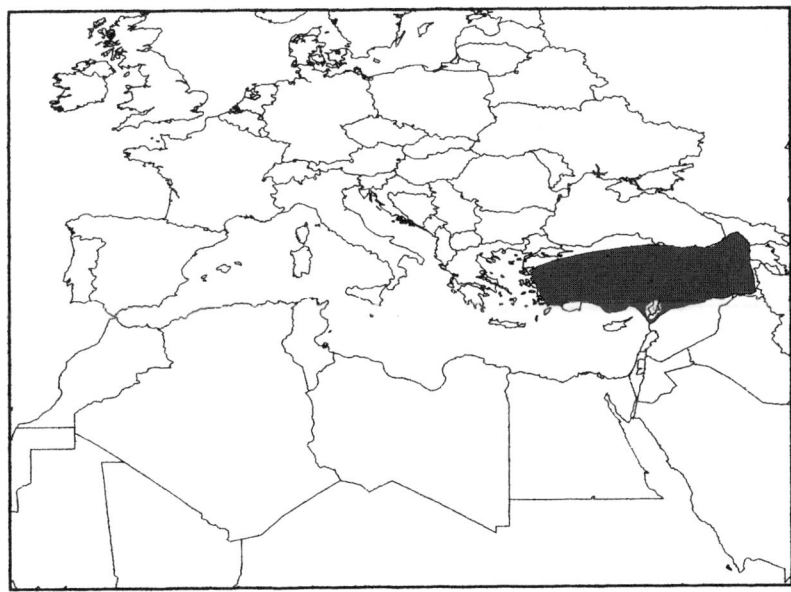

Figure 3.3 Distribution of the sections *Brevifilamentum* (covering most Anatolia) and *Longitubus*.

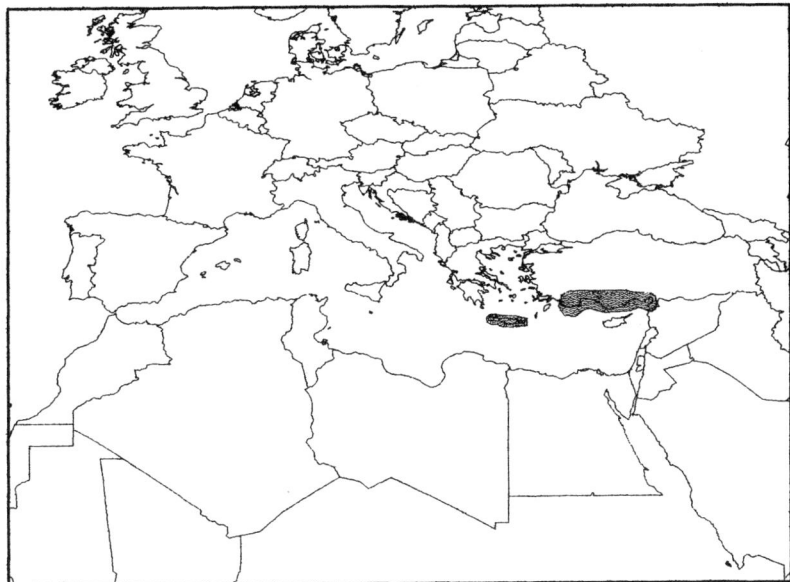

Figure 3.4 Distribution of the section *Chilocalyx*.

sections *Amaracus*, *Brevifilamentum* and *Longitubus* are restricted to the East Mediterranean, while *Anatolicon* extends up to Libya; both sections of group B, *Chilocalyx* and *Majorana* are confined to the East Mediterranean; in group C, *Campanulaticalyx* is found in a small part of the East Mediterranean, *Elongataspica* is

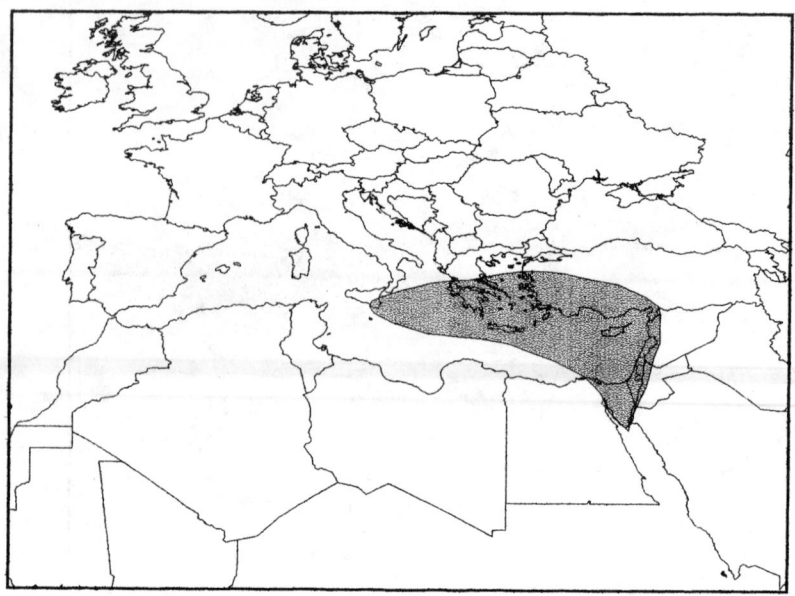

Figure 3.5 Distribution of the section *Majorana*.

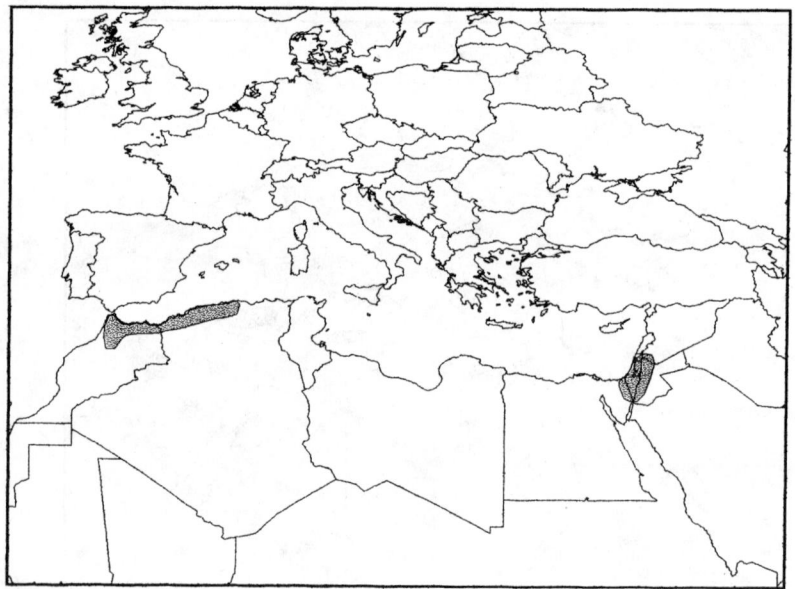

Figure 3.6 Distribution of the section *Campanuaricalyx* (east) and *Elongataspica* (west).

restricted to a small area of the south-west Mediterranean, and section *Prolaticorolla* is found in two spots at the East and West extremes of the Mediterranean, whereas section *Origanum* (*O. vulgare*) extends from the Azores and Canary Islands to Britain and Scandinavia, and then up to China and Taiwan.

Taxonomy and chemistry of Origanum 75

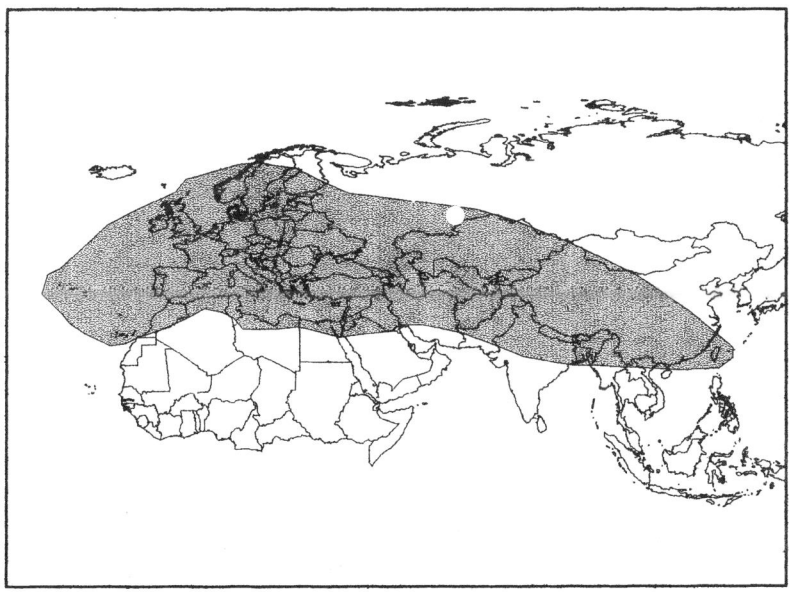

Figure 3.7 Distribution of the section *Origanum*.

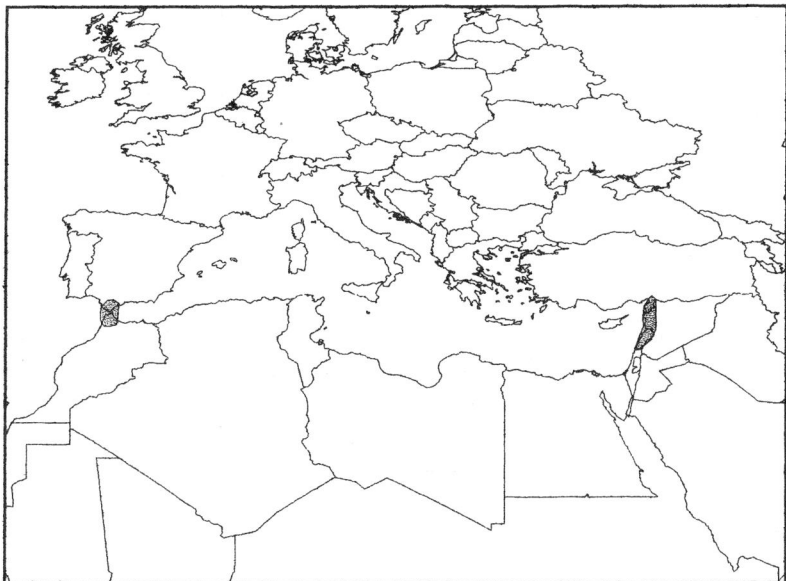

Figure 3.8 Distribution of the section *Prolaticorolla*.

Table 3.2 shows the hybrids that have been found when *Origanum* species co-occur, either in natural or in artificial (Botanical Gardens) conditions. Often hybrids have been considered initially as species, as in the case of *Majorana leptoclados* (*O.* × *minoanum*), *O. paniculatum* Koch. (*O.* × *aplii* Boros), *O. pulchellum* Boissier (*O.* × *hybridinum* Miller),

Table 3.2 List of recognized hybrids within the genus Origanum

No. of hybrid	No. of taxon	Hybrid	Parents	Distribution
h1	51	O. × adamense Baser et Duman[a]	O. bargyli × O. laevigatum	Tu
h2	52	O. × adonidis Mouterde	O. libanoticum × O. syriacum var. bevanii	Le
h3	53	O. × aplii Boros	O. majorana × O. vulgare ssp. vulgare	Cultivated
h4	54	O. × barbarae Bornmüller	O. ehrenbergii × O. syriacum var. bevanii	Le
h5	55	O. × dolichosiphon Davis	O. amanum × O. laevigatum	Tu
h6	56	O. × font-queri Pau[b]	O. grosii × O. compactum	Mo
h7	57	O. × hybridinum Miller	O. dictamnus × O. sipyleum	Cultivated
h8	58	O. × intercedens Rechinger	O. onites × O. vulgare ssp. hirtum	Gr, Gr-Cr, Gr-AE, Tu
h9	59	O. × intermedium Davis	O. onites × O. sipyleum	Tu
h10	60	O. × lirium Heldreich ex Halacsy	O. scabrum × O. vulgare ssp. hirtum	Gr
h11	61	O. × majoricum Cambessedes	O. majorana × O. vulgare ssp. virens	Sp, Po
h12	62	O. × minoanum Davis	O. microphyllum × O. vulgare ssp. hirtum	Gr-Cr
h13	63	O. × pabotii Mouterde	O. bargyli × O. syriacum var. bevanii	Sy
h14	64	O. × symeonis Mouterde	O. laevigatum × O. syriacum var. bevanii	Tu
h15	65	O. amanum × O. dictamnus	O. amanum × O. dictamnus	Cultivated
h16	66	O. calcaratum × O. dictamnus	O. calcaratum × O. dictamnus	Cultivated
h17	67	O. micranthum × O. vulgare ssp. hirtum	O. micranthum × O. vulgare ssp. hirtum	Tu
h18	68	O. sipyleum × O. vulgare ssp. hirtum	O. sipyleum × O. vulgare ssp. hirtum	Tu

Notes

a Recognized after Ietswaart's revision of the genus, by Duman et al., 1998.
b Ietswaart (1980), himself is uncertain as to whether O. × font-queri is really a hybrid between O. grosii and O. elongatum, or a synonym to O. grosii and O. × font-queri are products of hybridisation between O. elongatum and O. compactum.

Amaracus lirius Hayek (*O.* × *lirium* Heldreich ex Halacsy) and *O. haradjianii* Rechinger (*O.* × *symeonis* Mouterde) (Table 3.3). More hybrids are likely to be found as the genus *Origanum* is further studied in the field.

More than 300 scientific names have been given, during the last 150 years, to not more than 70 presently recognized *Origanum* species, subspecies, varieties and hybrids (Table 3.3). This plethora of different names reflects the extent of morphological variation the genus exhibits in nature. Ietswaart (1980), having comprehensively examined many specimens came to the conclusion that many species, subspecies and varieties can be discerned in their typical form; however nearly all of them gradually merge into at least one other form.

Table 3.3 List of accepted names (bold) and synonyms of members of *Origanum* with reference to Tables 3.1 and 3.2

Plant name	No. of species	No. of taxon
Amaracus akhdarensis Brullo et Furnari	8	8
A. amanus Bornmüller	23	23
A. brevidens Bornmüller	18	18
A. ciliatus Briquet	1	1
A. cordifolius Montbret et Aucher ex Bentham	3	3
A. cyrenaicus Rechinger	9	9
A. dictamnus Bentham	4	4
A. hausskenechtii Briquet	19	19
A. hausskenechtii Briquet var. *acutidens* Handel-Mazzetti	16	16
A. leptocladus Briquet	21	21
A. libanoticus Briquet	11	11
A. lirius Hayek	H10	60
A. majorana Schinz et Thellung	28	28
A. majorica Sampaio	H11	61
A. pampainii Brullo et Furnari	12	12
A. pulchellus Briquet	H7	57
A. pulcher Briquet	13	13
A. rotundifolius Briquet	22	22
A. scaber Briquet	13	13
A. sipyleus Rafanesque	14	14
A. syriacus Stokes	30	
A. tomentosus Moench	4	4
A. tournefortii Bentham	2	2
A. vetteri Briquet	15	15
A. vulgaris Hill	28	28
Dictamnus creticus Hill	4	4
Majorana aegyptiaca Kosteletzky	30	
M. crassa Moench	30	30
M. crassifolia Bentham	30	30
M. cretica Kosteletzky	28	28
M. cretica Miller	29	29
M. dictamnus Kosteletzky	4	4
M. dubia Briquet	28	28
M. fragrans Rafinesque	28	28
M. hortensis Moench	28	28
M. leptoclados Rechinger	H12	62
M. majorana Karsten	28	28
M. majorica Briquet	H11	61
M. majorica Briquet var. *lusitanica* Coutinho	H11	61

Table 3.3 (Continued)

Plant name	No. of species	No. of taxon
M. *maru* Briquet	30	30
M. *maru* Briquet var. *nervosa* Briquet	30	32
M. *maru* Hayek	26	26
M. *mexicana* Martius	28	28
M. *micrantha* Briquet	25	25
M. *microphylla* Bentham	26	26
M. *nervosa* Bentham	30	32
M. *onites* Bentham	29	29
M. *orega* Briquet	29	29
M. *orega* Walpers	29	29
M. *ovalifolia* Stokes	28	28
M. *scutellifolia* Stokes	30	
M. *sipylea* Kosteletzky	14	14
M. *smyrnaea* Kosteletzky	29	29
M. *suffruticosa* Rafinesque	28	28
M. *syriaca* Kosteletzky	30	
M. *tenuifolia* Gray	28	28
M. *tomentosa* Stokes	4	4
M. *vulgaris* Miller	28	28
Majoranamaracus zerniji Rechinger	H2	52
Onites tomentosa Rafinesque	29	29
Origanum acutidens Ietswaart	16	16
O. *aegyptiacum* Savi	30	
O. akhdarense Ietswaart et Boulos	8	8
O. *albiflorum* Koch var. *congestum* Koch	40	47
O. *album* Salisbury	29	29
O. amanum Post	23	23
O. *americanum* Rafinesque	40	42
O. *anglicum* Hill	40	42
O. *angustifolium* Koch	40	47
O. *balearicum* Pourret ex Lange	H11	61
O. *barcense* Simonkay	40	42
O. *barcense* Simonkay var. *microstachyum* Grecescu	40	42
O. bargyli Mouterde	17	17
O. *bevani* Holmes	30	31
O. bilgeri Davis	24	24
O. boissieri Ietswaart	1	1
O. brevidens Dinsmore	18	18
O. *brevidens* Dinsmore var. *pubescens* Thiebaut	17	17
O. *bucharicum* Bornmüller	40	44
O. calcaratum Jussieu	2	2
O. *capitatum* Wildenow ex Bentham	40	42
O. *ciliatum* Boissier et Kotschy	1	1
O. *cinereum* de Noë	38	40
O. compactum Bentham	41	48
O. *confertum* Savi	28	28
O. cordifolium Vogel	3	3
O. *crassa* Chevallier	30	30
O. *creticum* auct. non L.	40	45
O. *creticum* L.	40	42
O. *creticum* Schousbou ex Ball.	41	48
O. cyrenaicum Beguinot et Vaccari	9	9

O. dayi Post	31	33
O. decipiens Wallroth ex Bentham	40	42
O. dictamnifolium Saint-Lager	4	4
O. dictamnus L.	4	4
O. dubium Boissier	28	28
O. ehrenbergii Boisier var. *parviflorum* Bornmüller	42	49
O. ehrenbergii Boissier	42	49
O. elegans Sennen	40	42
O. elongatum Emberger ex Maire	37	39
O. floribundum Munby	38	40
O. floridum Salisbury	40	42
O. glandulosum Defontaines var. *elongatum* Bonnet	37	39
O. glandulosum Defontaines	40	43
O. glandulosum Salzmann ex Bentham	41	48
O. glaucum Rechinger et Edelberg	40	44
O. glaucum Rechinger et Edelberg var. *laxius* Rechinger et Edelberg	40	44
O. gracile Koch	40	44
O. grosii Pau et Font Quer ex Ietswaart	39	41
O. gussonei Tineo ex Lojacono Pojero	40	47
O. haradjianii Rechinger	H14	64
O. hausskenechtii Boissier	19	19
O. heracleoticum auct. non L.	40	45
O. heracleoticum auct. non L. f. *trichocalycinum* Rechinger	40	45
O. heracleoticum auct. non L. var. *albiflorum* Halácsy	40	45
O. heracleoticum auct. non L. var. *creticum* Halácsy	40	45
O. heracleoticum auct. non L. var. *creticum* Halácsy f. *glabra* Halácsy	40	45
O. heracleoticum auct. non L. var. *creticum* Halácsy f. *hirsuta* Halácsy	40	45
O. heracleoticum auct. non L. var. *rubriflorum* Halácsy	40	45
O. heracleoticum auct. non L. var. *trichocalycinum* Halácsy	40	45
O. heracleoticum hort. ex Koch	H3	53
O. heracleoticum L.	40	47
O. hirtum Link	40	45
O. hirtum Link f. *albiflorum* Hausknecht	40	45
O. hirtum Link f. *prismaticum* Hausknecht	40	45
O. hirtum Link f. *rubriflorum* Hausknecht	40	45
O. hirtum Link f. *trichocalycinum* Hausknecht	40	45
O. hirtum Link var. *corymbulosum* Candargy	40	45
O. hirtum Link var. *genuinum* Vogel	40	45
O. hirtum Link var. *humile* Bentham	40	45
O. hirtum Link var. *laxiflorum* Candargy	40	45
O. hirtum Link var. *macrostachyum* Candargy f. *macrostachyoides* Candargy	40	45
O. hirtum Link var. *oostachyum* Candargy	40	45
O. hirtum Link var. *prismaticum* Vogel	40	45
O. hirtum Link var. *subtypicum* Candargy	40	45
O. hirtum Link var. *typicum* Candargy	40	45
O. humile Miller	40	47
O. husnucan-baserii H. Duman, Z. Aytaç et A. Duran	20	20
O. hypercifolium Schwarz et Davis	10	10
O. hyrcanum Bornmüller	40	47
O. illiricum Scheele	40	45
O. isthmicum Danin	32	34

Table 3.3 (Continued)

Plant name	No. of species	No. of taxon
O. jordanicum Danin and Künne	33	35
O. laevigatum Boissier	43	50
O. laevigatum Boissier var. laxum Post	43	50
O. latifolium Miller	40	42
O. leptocladum Boissier	21	21
O. libanoticum Boissier	11	11
O. lusitanicum Rouy	H11	61
O. macrostachyum Hoffmannseg et Link var. genuinum Coutinho	40	46
O. majorana L.	28	28
O. majoranoides hort. ex Gams	40	42
O. majoranoides Willdenow	28	28
O. majoricum Cambessedes var. lusitanicum Rouy	H11	61
O. majus Garsault	40	42
O. maru L.	30	30
O. maru L. f. viridula Bornmüller	30	31
O. maru L. var. aegyptiacum Dinsmore	30	32
O. maru L. var. capitatum Post	30	30
O. maru L. var. sinaicum Boissier	30	32
O. maru sensu Sibthorp et Smith	26	26
O. megastachyum Link	40	45
O. micranthum Vogel	25	25
O. microphyllum Vogel	26	26
O. minus Garsault	40	47
O. minutiflorum Schwarz et Davis	27	27
O. nervosum Vogel	30	32
O. normale Don	40	47
O. normale Don var. incanum Schmidt et Schlaginweit	40	47
O. nutans Wildenow ex Bentham	40	42
O. oblongatum Link	40	47
O. odorum Salisbury	28	28
O. onites L.	29	29
O. orega Vogel	29	29
O. orientale Miller	40	42
O. pallidum Defontaines	29	29
O. pampaninii Ietswaart	12	12
O. paniculata Spenner	H3	53
O. paniculatum Koch.	H3	53
O. parviflorum Dumond d' Urville	40	47
O. paui Martinez	H11	61
O. petraeum Danin	34	36
O. pruinosum Koch	40	47
O. pseudodictamnus Sieber	4	4
O. pseudo-onites Lindberg	30	31
O. puberulum Klokov	40	42
O. pulchellum Boissier	H7	57
O. pulchrum Boissier et Heldreich	13	13
O. punonense Danin	35	37
O. purpurascens Gilibert	40	42
O. ramonense Danin	36	38
O. rotundifolium Boissier	22	22
O. saccatum Davis	5	5

O. *sardoum* Nyman	40	47
O. *saxatile* Salisbury	4	4
O. *scabrum* Boissier et Heldreich	13	13
O. *semiglaucum* Boissier et Reuter ex Briquet	40	47
O. *siculum* Nyman	40	47
O. *silvestre* Ortega ex Sampaio	40	46
O. *sipyleum* L.	14	14
O. *smyrnaeum* L.	29	29
O. *smyrnaeum* sensu Sibthorp et Smith	40	45
O. *solymicum* Davis	6	6
O. *stoloniferum* Besser ex Reichnbach	40	42
O. *strobilaceum* Mobayen et Gahraman	40	47
O. *suffruticosum* hort. ex Steudel	28	28
O. *symes* Carlström	7	7
O. *syriacum* L.	30	
O. *syriacum* L.	30	30
O. *syriacum* L. var. *aegyptiacum* Täckholm	30	32
O. *syriacum* L. var. *bevanii* Ietswaart	30	31
O. *syriacum* L. var. *sinaicum* Ietswaart	30	32
O. *syriacum* L. var. *syriacum*	30	30
O. *thymiflorum* Reichenbach	40	42
O. *tournefortii* Aiton	2	2
O. *tragoriganum* Zuccagni ex Steudel	29	29
O. *tytthanthum* Gomscharov	40	44
O. *tytthanthum* Gomscharov var. *seravschianum* Borissova	40	44
O. *venosum* Wildenow ex Bentham	40	42
O. *vestitum* Clarke	30	30
O. *vetteri* Briquet et Barbey	15	15
O. *virens* Hoffmannseg et Link	40	46
O. *virens* Hoffmannseg et Link var. *genuinum* Coutinho	40	46
O. *virens* Hoffmannseg et Link var. *macrostachyum* Coutinho	40	46
O. *virens* Hoffmannseg et Link var. *siculum* Bentham	40	47
O. *virens* Hoffmannseg et Link var. *spicatum* Rouy	40	46
O. *virescens* Poiret	40	46
O. *viride* Halácsy	40	47
O. *viride* Halácsy var. *hyrcanum* Bornmüller	40	47
O. *viridulum* Martin-Donos	40	47
O. *vogelii* Greuter and Burdet	25	25
O. *vulgare* L.	40	
O. *vulgare* L. f. *albiflora* Senchovei ex Formánek	40	42
O. *vulgare* L. f. *elongatum* Formánek	40	42
O. *vulgare* L. f. *glabrescens* Beck	40	42
O. *vulgare* L. f. *grecescui* Soó	40	42
O. *vulgare* L. f. *procumbens* Jakus ex Soó et Borhini	40	42
O. *vulgare* L. ssp. *barcense* Jávorka	40	42
O. *vulgare* L. ssp. *glandulosum* Ietswaart	40	43
O. *vulgare* L. ssp. *gracile* Ietswaart	40	44
O. *vulgare* L. ssp. *heracleoticum* Holmboe	40	45
O. *vulgare* L. ssp. *hirtum* Ietswaart	40	45
O. *vulgare* L. ssp. *prismaticum* Gaudin	40	42
O. *vulgare* L. ssp. *prismaticum* Gaudin var. *australe* Gaudin	40	42
O. *vulgare* L. ssp. *prismaticum* Gaudin var. *parviflorum* Gaudin	40	42

Table 3.3 (Continued)

Plant name	No. of species	No. of taxon
O. vulgare L. ssp. *virens* Ietswaart	40	46
O. vulgare L. ssp. *viride* Hayek	40	47
O. vulgare L. ssp. *viridulum* Nyman	40	47
O. vulgare L. ssp. *vulgare*	40	42
O. vulgare L. var. *album* Fraas	40	47
O. vulgare L. var. *americanum* Rafinesque	40	42
O. vulgare L. var. *barcense* Hayek	40	42
O. vulgare L. var. *bracteosum* Petermann ex Soó	40	42
O. vulgare L. var. *creticum* Briquet	40	42
O. vulgare L. var. *exile* Lamotte	40	42
O. vulgare L. var. *formosanum* Hayata	40	42
O. vulgare L. var. *glaucum* Hedge	40	44
O. vulgare L. var. *hirtum* Visiani	40	45
O. vulgare L. var. *humile* Bentham	40	47
O. vulgare L. var. *latebracteum* Beck	40	42
O. vulgare L. var. *laxiflorum* Post	40	47
O. vulgare L. var. *longespicatum* Post	40	47
O. vulgare L. var. *macrostachyum* Brotero	40	46
O. vulgare L. var. *magnilimbis* Boissier	40	47
O. vulgare L. var. *megastachya* Koch	40	45
O. vulgare L. var. *normale* Briquet	40	47
O. vulgare L. var. *pallescens* Martin-Donos	40	42
O. vulgare L. var. *prismaticum* Bentham	40	42
O. vulgare L. var. *puberulum* Beck	40	42
O. vulgare L. var. *purpurascens* Briquet	40	42
O. vulgare L. var. *purpureum* Stokes	40	42
O. vulgare L. var. *rotundifolium* Rafinesque	40	42
O. vulgare L. var. *rufuscens* Stokes	40	42
O. vulgare L. var. *semiglaucum* Boissier ex Briquet	40	47
O. vulgare L. var. *smyrnaeum* Bentham	40	47
O. vulgare L. var. *spicatum* Koch	40	42
O. vulgare L. var. *spiculigerum* Briquet	40	42
O. vulgare L. var. *subglabrum* Schmidt et Schlagintweit	40	42
O. vulgare L. var. *tauricum* Borissova	40	42
O. vulgare L. var. *violacea* Sennen	40	42
O. vulgare L. var. *virens* Bentham	40	46
O. vulgare L. var. *virens* Koch	40	47
O. vulgare L. var. *virescens* Cariot et St. Lager	40	47
O. vulgare L. var. *viride* Boissier	40	47
O. vulgare L. var. *viridulum* Briquet	40	47
O. wallicianum Bentham	40	47
O. wastoni Schmidt et Schlagintweit	40	42
O. × *adanense* Baser et Duman	H1	51
O. × *adonidis* Mouterde	H2	52
O. × *aplii* Boros	H3	53
O. × *barbarae* Bornmüller	H4	54
O. × *dolichosiphon* Davis	H5	55
O. × *font-queri* Pau	39	41
O. × *font-queri* Pau	H6	56
O. × *hybridinum* Miller	H7	57
O. × *hybridum* Heldeich	H10	60
O. × *indercedens* Rechinger	H8	58

O. × *intermedium* Davis	H9	59
O. × *lirium* Heldreich ex Halacsy	H10	60
O. × *majoricum* Cambessedes	H11	61
O. × *minoanum* Davis	H12	62
O. × *pabotii* Mouterde	H13	63
O. × *symeonis* Mouterde	H14	64
Origanomajorana applii Domin	H3	53
Satureja camphorata Bornmüller	31	33
Schizocalyx smyrnaeus Scheele	29	29
Schizocalyx syriacus Schelle	30	
Thymus majorana Kuntze	28	28

At the time of Ietswaart's revision, chemical data were too sparse to be useful in the taxonomic classification. Today however, there are a number of publications referring to the chemisty of *Origanum*, as will now be described.

VOLATILE OILS

Origanum is known widely in the world of herbs and spices for its volatile oils. Oregano is the commercial name of those species that are rich in the phenolic monoterpenoids, mainly carvacrol, occasionally thymol, while marjoram is the commercial name of those that are rich in bicyclic monoterpenoids *cis*- and *trans*-sabinene hydrate (Figures 3.9 and 3.10). It is quite easy to distinguish the difference between the pungent smell of oregano and the sweet smell of marjoram. It is not surprising that when carvacrol and/or thymol is the main compound, a number of chemically related compounds i.e. γ-terpinene, *p*-cymene, thymol and carvacrol methyl ethers, thymol and carvacrol acetates; also compounds such as *p*-cymenene, *p*-cymen-8-ol, *p*-cymen-7-ol, thymoquinone, and thymohydroquinone are also present (Figure 3.9). Furthermore, when *cis*- and/or *trans*-sabinene hydrate is the main compound, α-thujene, sabinene, *cis*- and *trans*-sabinene hydrate acetates, *cis*- and *trans*-sabinol and sabina ketone can also be found (Figure 3.10). Two more biochemical groups, usually of less significance quantitatively, are present in *Origanum*: that of the acyclic monoterpenoids such as, geraniol, geranyl acetate, linalool, linalyl acetate and β-myrcene (Figure 3.11) and that of bornane type compounds such as camphene, camphor, borneol, and bornyl and isobornyl acetate (Figure 3.12). To the above groups of compounds some sesquiterpenoids, such as β-caryophyllene, β-bisabolene, β-bourbonene, germacrene-D, bicyclogermacrene, α-humulene, α-muurolene, γ-muurolene, γ-cadinene, *allo*-aromadendrene, α-cubebene, α-copaene, α-cadinol, β-caryophyllene oxide and germacrene-D-4-ol should also be added. (Figure 3.13). In the following paragraphs a chemotaxonomic account of *Origanum* taxa is presented, by group and by section.

Volatile compounds in group A

In section *Amaracus*, *O. boissieri* Ietswaart (Baser and Duman, 1998), *O. saccatum* Davis (Tumen *et al.*, 1995) and *O. solymicum* P.H. Davis (Tumen *et al.*, 1994), are *p*-cymene rich, *O. cordifolium* Vogel (Valentini *et al.*, 1991) is rich in α-terpineol, γ-terpinene, and *p*-cymene while *O. calcaratum* Juss. is characterized by *p*-cymene, thymol, and γ-terpinene and *O. dictamnus* by *p*-cymene, thymoquinone and carvacrol (Skoula *et al.*, 1999). In section *Anatolicon*, the essential oil of *O. sipyleum* L., and

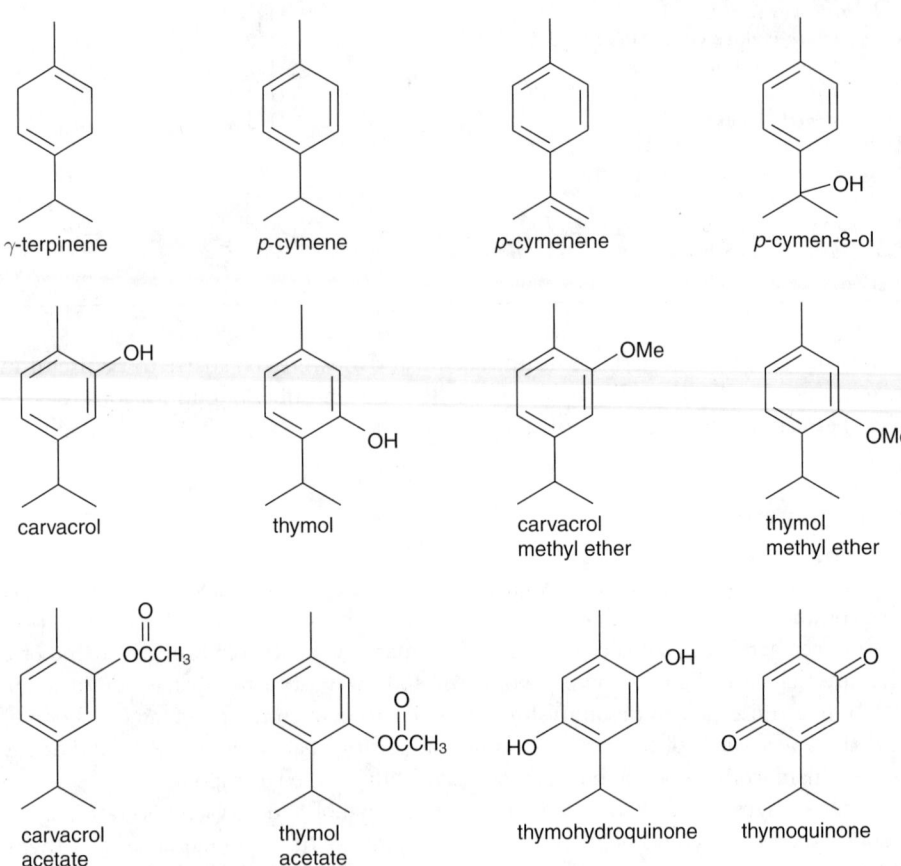

Figure 3.9 Most common cymyl-compounds found in *Origanum* spp.

O. hypercifolium O. Schwarz et P.H. Davis were found to be rich in *p*-cymyl compounds (Baser *et al.*, 1992, Baser *et al.*,1994a) while *O. libanoticum* Boiss (Arnold *et al.*, 2000) is rich in *p*-cymyl and sesquiterpenoid compounds. Thus, it seems that in both these sections taxa are characterized by the *p*-cymyl monoterpenoids. In section *Brevifilamentum*, *O. acutidens* Ietswaart was rich in carvacrol while *O. bargyli* Mouterde, *O. haussknechtii* Boiss. and *O. leptocladum* Boiss. were found to be rich in *p*-cymyl compounds and borneol (Baser *et al.*, 1997a, Baser and Duman, 1998; Baser *et al.*, 1998a; Baser *et al.*, 1996a). *O. rotundifolium* Boiss. contains mainly *cis*-sabinene hydrate, linalyl acetate, α-terpineol, β-caryophyllene, and *trans*-sabinene hydrate (Baser *et al.*, 1995) and the newly found species *O. husnucan-baserii* H. Duman, Z. Aytaç et A. Duran contains borneol, *cis*- and *trans*-sabinene hydrate, β-caryophyllene and germacrene-D (Baser *et al.*, 1998b). It is only in the section *Brevifilamentum* that there is a large accumulation of borneol in the volatile compounds, while sesquiterpenoids are present too. In the whole group A, only *O. rotundifolium* and *O. husnucan-baserii* tend to accumulate sabinyl compounds instead of cymyl as in the other 13 analyzed taxa. In group A, taxa are characterized by a poor essential oil yield (<0.5 per cent) (Table 3.4).

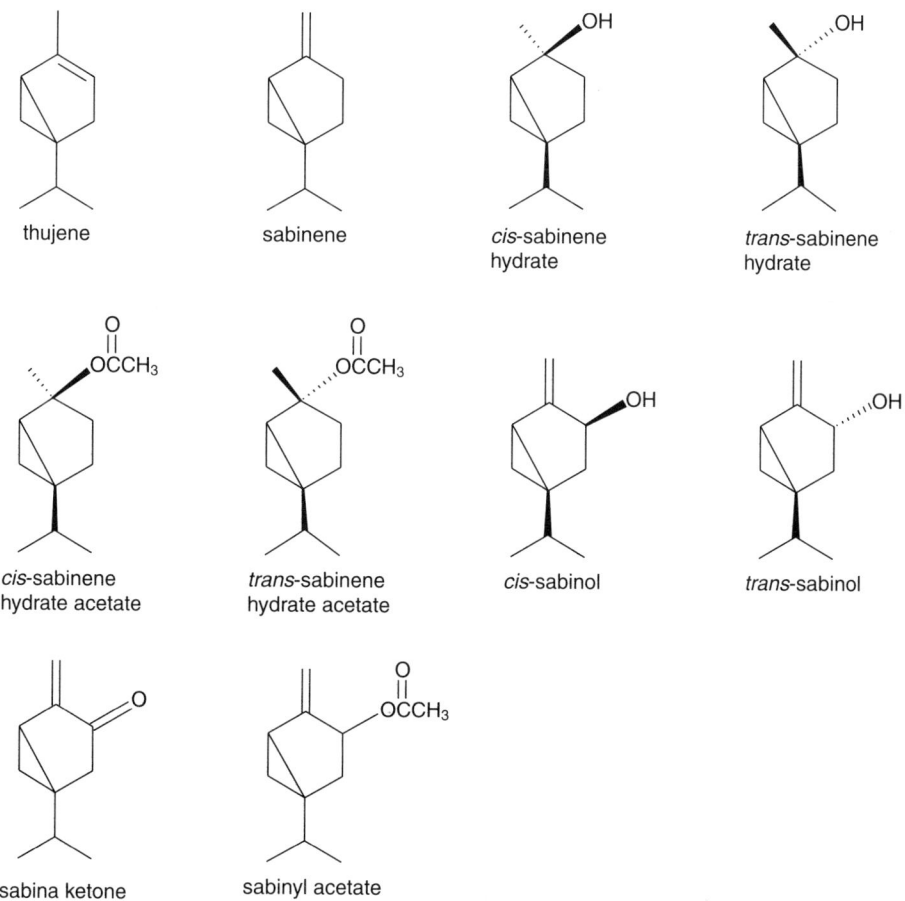

Figure 3.10 Most common sabinyl compounds found in *Origanum* spp.

Volatile compounds in group B

In Section *Chilocalyx*, *O. microphyllum* Vogel (Skoula *et al.*, 1999) and *O. micranthum* Vogel (Baser *et al.*, 1996b) are characterized by *cis*- and *trans*-sabinene hydrate, sabinene, and linalool. On the other hand *O. minutiflorum* Schwarz et Davis (Baser *et al.*, 1991) and *O. bilgeri* Davis (Baser *et al.*, 1996c) are rich in carvacrol and poor in sabinyl compounds. No significant accumulation of bornane compounds and/or sesquiterpenoids has been reported for this section. Taxa in this section usually yield less than 0.5 per cent essential oil. Within the *Majorana* section, *O. syriacum* L. is either carvacrol or thymol rich (Ravid and Putievski, 1983; Beker *et al.*, 1989; Fleisher and Fleisher, 1991; Halim *et al.*, 1991; Tumen and Baser, 1993; Baser *et al.*, 1993a). *O. majorana* is characteristic for its *cis*- and *trans*-sabinene hydrate content (Fischer *et al.*, 1987; Franz, 1990; Arnold *et al.*, 1993; Baser *et al.*, 1993b); however samples collected from Turkey and Cyprus, have been found poor in the sabinene derivatives and to contain mainly carvacrol instead (Arnold *et al.*, 1993; Baser *et al.*, 1993b). *O. onites* L. is characterized by carvacrol, β-bisabolene, and geraniol (Vokou *et al.*, 1988; Skoula *et al.*, 1999);

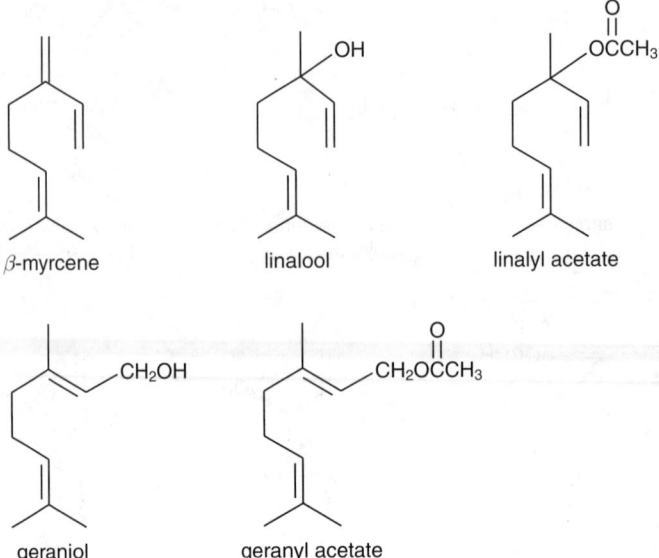

Figure 3.11 Most common acyclic monoterpenoids found in the volatiles of *Origanum* spp.

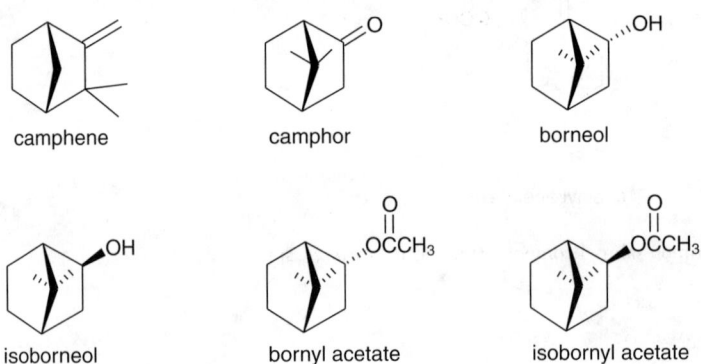

Figure 3.12 Most common bornyl-compounds present in the volatiles of *Origanum* spp.

nevertheless, an almost pure linalool-type has been recorded in Turkey (Baser et al., 1993a). It seems that members in *Majorana* section vary not only at the sectional level, but also within the species; all three biochemical groups, *p*-cymyl, sabinyl and the acyclic monoterpenoids can be distinguished within the *Majorana* section but (at significant concentrations) the sabinyl group and the *p*-cymyl group are mutually exclusive. Throughout both sections of Group B there is compelling evidence of this clear division into sabinyl rich and carvacrol rich plants with no intermediates observed; additionally no taxa rich in sesquiterpenoids or bornane compounds have been recorded in this section (Table 3.4). Taxa in this section can be very rich in essential oils, yield often exceeding 5 per cent.

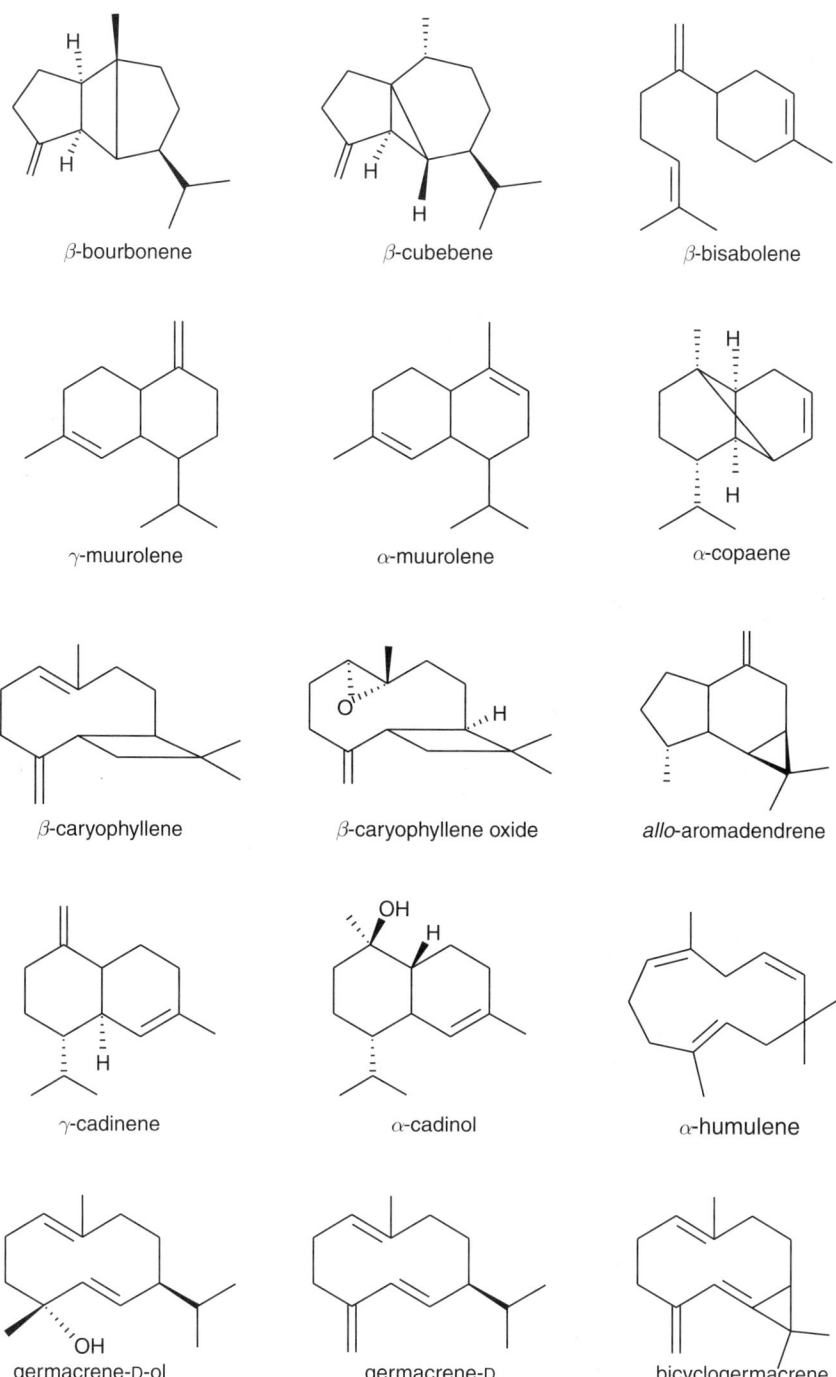

Figure 3.13 Most common sesquiterpenoids found in *Origanum*.

Table 3.4 Distribution of cymyl-, sabinyl-, acyclic-, bornane-, and sesquiterpene-compounds within the infrageneric categories of Ietswaart's classification, with a record on essential oil yield

Group	Section	Taxon	cymyl-	Sabinyl-	Acyclic-	Bornane-	Sesquiterpene-	Essential oil yield[a]
A	Amaracus	O. boissieri	+++[b]	+	−	−	+	Poor
		O. calcaratum	+++	−	−	−	−	Poor
		O. cordifolium	+++	−	−	−	+	Poor
		O. dictamnus	+++	+	−	−	+	Intermediate
		O. saccatum	+++	−	−	−	−	Intermediate
		O. solymicum	+++	−	−	−	−	Poor
	Anatolicon	O. hypericifolium	+++	−	−	−	−	Poor
		O. libanoticum	+++	−	−	−	++	Poor
		O. sipyleum	+++	−	−	−	+	Poor
	Brevifilamentum	O. acutidens	+++	−	−	−	−	Poor
		O. bargyli	+++	−	+	++	+	Intermediate
		O. haussknechtii	++	+	−	+	++	Poor
		O. husnucanbaserii	−	++	−	++	++	Poor
		O. leptocladum	+++	−	−	++	−	Poor
		O. rotundifolium	+	+++	++	+	++	Poor
B	Chilocalyx	O. bilgeri	+++	−	−	−	−	Intermediate
		O. micranthum	+	+++	++	−	+	Intermediate
		O. microphyllum	+	+++	+	−	−	Intermediate
		O. minutiflorum	+++	−	−	−	−	Intermediate
	Majorana	O. majorana I	+	+++	+	−	−	Rich
		O. majorana II	+++	−	−	−	−	Rich
		O. onites I	+++	−	+	+	+	Rich
		O. onites II	−	−	+++	+	+	Rich
		O. syriacum ssp. syriacum	+++	−	−	−	−	Rich
		O. syriacum ssp. bevani	+++	−	−	−	−	Rich
		O. syriacum ssp. sinaicum	+++	−	−	−	−	Rich
C	Campanulaticalyx	O. isthmicum	+++	−	−	−	−	?
		O. ramonense	++	++	+	+	−	Poor
	Elongataspica	O. elongatum	+++	−	−	−	−	Rich
		O. floribundum	+++	−	−	−	−	Rich
	Origanum	O. vulgare ssp. hirtum	+++	−	−	−	−	Rich
		O. vulgare ssp. glandulosum	+++	−	−	−	−	Rich
		O. vulgare ssp. gracile I	+++	−	−	−	−	Intermediate
		O. vulgare ssp. gracile II	−	−	++	−	+++	Poor
		O. vulgare ssp. vulgare	+	−	+++	−	+++	Poor

	O. vulgare ssp. *virens*	–	–	+++	–	+++	Poor
	O. vulgare ssp. *viride* I	+++	–	–	–	–	Intermediate
	O. vulgare ssp. *viride* II	+	++	+++	–	+++	Poor
Prolaticorolla	*O. compactum*	+++	–	–	–	–	Rich
	O. laevigatum	–	–	–	–	+++	Poor
	O. × adanense	+++	–	–	–	+++	Poor
	O. × intercedens	+++	–	–	+	++	Rich
	O. × minoanum	+++	++	–	–	–	Intermediate

Notes
a Essential oil yield <0.5%: Poor; 0.5–2.0% intermediate; 2.0%>: rich.
b 5%< + <10%; 10% <++<30%; 30%< +++ in the essential oils.

Volatile compounds in group C

In the section *Campanulaticalyx* Ietswaart, *O. isthmicum* Danin has been described by smell as carvacrol-rich (Danin and Künne, 1996), whilst *O. ramonense* Danin was shown to contain terpinen-4-ol, α-terpineol and, rather surprisingly, either sabinene hydrates or 1,8-cineole and α- and γ-terpinene, depending on the method of analysis (Danin *et al.*, 1997). In the section *Elongataspica*, *O. elongatum* Emberger et Maire, and *O. floribundum* Munby seem to be clearly rich only in cymyl compounds (Benjilali *et al.*, 1986; Velasco-Negueruela *et al.*, 1991; Houmani *et al.*, 2001). The volatiles reported in the section *Origanum*, although variable, seem to lack sabinyl compounds: subspecies *hirtum* is most commonly carvacrol-rich and less commonly thymol-rich (Kokkini and Vokou, 1989; Baser *et al.*, 1994b; Skoula *et al.*, 1999); subsp. *glandulosum* Ietswaart is rich in cymyl compounds, mainly thymol and carvacrol and their methylethers (Melegari *et al.*, 1995; Houmani *et al.*, 2001); subsp. *gracile* Ietswaart (syn. *O. tyttanthum* Gontsch.) and subsp. *viride* Hayek have been found either rich in acyclic compounds and sesquiterpenoids or carvacrol/-thymol rich (Sezik *et al.*, 1993; Baser *et al.*, 1997b; Arnold *et al.*, 2000); subsp. *vulgare* and subsp. *virens* Ietswaart are rich in acyclic compounds and sesquiterpenoids (Alves-Pereira and Fernandes-Ferriera, 1998; Sezik *et al.*, 1993). From the section *Prolaticorolla*, *O. compactum* Bentham was found to be carvacrol/thymol rich (Benjilali *et al.*, 1986; van Den Broucke and Lemli, 1980) while *O. laevigatum* Boiss. was rich only in sesquiterpenoids (Tucker *et al.*, 1992; Baser *et al.*, 1996d). Thus, with the exception of subsp. *viride*, sabinene compounds are either absent or their presence is uncertain in any unambiguously identified Group C taxa (Table 3.4). However, there are also reports of sabinene group rich plants of *Origanum vulgare* in which the subspecies is unidentified (Lawrence, 1980; Chalchat and Pasquier, 1998).

Group C (12 species examined) is distinguished by being the only group rich in acyclic compounds and/or sesquiterpenoids. Most of the species lack sabinene group compounds, but these compounds are not altogether absent from the section. It does seem that Group C is unique in containing plants with only minor amounts of both cymene- and sabinene groups. The taxonomic difficulties with subspecies of *O. vulgare* seem to extend to the volatiles too. The essential oil yield in this group varies from poor to intermediate and rich, usually the cymyl rich are rich in essential oil, while those that mainly contain acyclic compounds and sesquiterpenoids are poor or intermediate.

Volatile compounds in the hybrids

So far there are reports on the volatile compounds of three hybrids. Their composition has been found to be intermediate to those of their parental species: A sample of *O.* × *adanense* Baser et Duman, the hybrid between *O. bargyli* (A, *Brevifilamentum*) and *O. laevigatum* (C, *Prolatcorolla*), has an essential oil profile rich in sesquiterpenoids (>36 per cent) and in cymyl compounds (>30 per cent), the sesquiterpenoids coming from the parent *O. laevigatum* and the cymyl derivarives from the parent *O. bargyli*. While the bornane and acyclic groups, that are significantly present in *O. bargyli* do not reach more that 4 per cent each, of the total oil in the hybrid (Baser *et al.*, 2000), there is a balanced composition derived from the two parental species. The composition of *Origanum* × *minoanum* Davis, the hybrid between *O. microphyllum* (B, *Chilocalyx*) and *O. vulgare* ssp. *hirtum* (C, *Origanum*), follows a similar pattern (Skoula *et al.*, 1999). The volatiles are composed mainly of cymyl compounds (65 per cent), less than in the 'typical' parent *O. vulgare* ssp. *hirtum*, and by sabinyl compounds (reaching 15 per cent) derived from the other parent *O. microphyllum*, which though has usually more than 60 per cent sabinyl compounds (Skoula *et al.*, 1999). Here there is a skew in the composition towards *vulgare* ssp. *hirtum*. In the third case of *O.* × *intercedens* Rech., a hybrid between *O. onites* (B, *Majorana*) and *O. vulgare* ssp. *hirtum* (C, *Origanum*) the two parental species do not differ in their principal components; in the hybrid, cymyl compounds reach 65 per cent, less than in 'typical' *O. vulgare* ssp. *hirtum*, but more than in 'typical' *O. onites* while the sesquiterpenoids and the acyclic compounds reached 11 per cent and 7 per cent, obviously inherited from *O. onites* (Bosabalidis and Skoula, 1998; Skoula *et al.*, 1999) (Table 3.4).

Table 3.4 shows that taxa of *Origanum* appear to be either rich in sabinene derivatives or rich in *p*-cymene oils, but never intermediate. This is a generalisation which appears to be true with regard to all the *Origanum* species whose volatiles have so far been examined. The presence of low levels of sabinene group compounds in most of the species examined suggests that this pathway is suppressed rather than absent in *Origanum*. The suppression of the sabinene pathways by one or more components of the cymene pathway offers a simple explanation in the case of *O.* × *minoanum*. With regard to the essential oil yield, Table 3.4 shows that in group A taxa in all sections *Amaracus*, *Anatolicon* and *Brevifilamentum* are mostly poor in essential oils, 4/5 of the taxa are poor (<0.5 per cent v/w) while 1/5 of the taxa have intermediate essential oil yield, ranging from 0.5 to 2.0 per cent v/w. In group B, taxa of the section *Chilocalyx*, all have intermediate essential oil yield whilst in section *Majorana* all are rich in essential oils. In group C, taxa are more variable; however, there is a remarkable correlation between carvacrol/thymol rich plants and high essential oil yield, as well as between high sesquiterpenoid content and low essential oil yield.

Infraspecific variation of the essential oil yield and composition

Besides the qualitative variation of the volatile compounds at the infrageneric level, there is considerable quantitative variation at the infraspecific level. Various researchers have examined the geographical variation in the essential oil yield and composition particularly in the high yielding commercially important species, part of which has been discussed previously. The seasonal variation of essential oil yield and composition has received less attention. However it seems that *O. vulgare* ssp. *hirtum* plants produce less essential oils during the cool and wet vegetative period and more during the warm and dry flowering period; after flowering essential oil yield decreases as leaves get older and drier. Whilst thymol and/or carvacrol content drops a little during autumn, the two other hydrocarbons, *p*-cymene and γ-terpinene fluctuate enormously, which is not unexpected as they belong to the same pathway with thymol and carvacrol as their end-products (Poulose and Croteau, 1978; Skoula *et al.*, unpublished data).

Usually seasonal variation is attributed to the changes in the environmental conditions such water deficiency, nutrient availability, photoperiod, quality of light, etc. (Trivino and Johnson, 2000). Indeed, the decline in total essential oil and of thymol or carvacrol which occurs in the autumn naturally, can be mimicked by growing *O. syriacum* in short days (Putievsky *et al.*, 1997). Similarly, *O. majorana* grown in controlled conditions under short days yields less essential oils (Circella *et al.*, 1995). Kokkini *et al.* (1997) reported high content of *p*-cymene in the essential oils of wild *O. vulgare* ssp. *hirtum* collected in autumn.

Seasonal effect cannot be separated from the process of development; in this respect it was found that young leaves of *O. vulgare* had more than twice the concentration of essential oils than older leaves in spring (March) but this difference had disappeared by the end of May. In March, the proportion of *p*-cymene was greater compared with carvacrol and this was especially noticeable in older leaves (Kazantzis, 1999).

GLYCOSIDICALLY BOUND VOLATILE COMPOUNDS

Some volatile compounds which are present in plants in free form are also present as glycosidically bound components. Volatile compounds can be released from the non-volatile relatives by enzymatic or chemical reactions during plant maturation, industrial pretreatment or processing, including fermentation by endogenous β-glycosidase. The increase of essential oil yield in some herbs during storage may be explained by the enzymatic hydrolysis of glycosides present in these plants (Crouzet and Chassagne, 1999). The sugar moiety of these glycosides can be monosaccharidic or disaccharidic. Table 3.5 shows the volatile compounds that have been found up to now in glycoside form in *Origanum* taxa (Stahl-Biskup *et al.*, 1993; Mastelic *et al.*, 2000) analyzed in Germany and Croatia. There is no quantitative relation between the volatiles present in the essential oils and those released from the glycoside fraction, even though there are several in common. The most predominant volatile aglycone was thymoquinone (40 per cent). In the glycone fraction, D-(+)-glucose was identified as the main monosacchride, while the disaccharide found was probably (+)-lactose (D-(+)-glucose and D-(+)-galactose were identified after hydrolysis).

The study of the seasonal variation of the glycosidically bound volatiles in *Origanum* plants growing in Germany, showed that plants before the flowering stage contained

Table 3.5 Volatile aglycones present in *Origanum* bound to sugars

	O. majorana	O. vulgare	O. vulgare ssp. hirtum
Aliphatic alcohols			
Hexan-1-ol		+	
cis-Hex-3-en-1-ol		+	+
trans-Hex-3-en-1-ol		+	+
Octan-3-ol		+	
Oct-1-en-3-ol		+	+
Phenylpropanoids			
Benzaldehyde			+
Benzyl alcohol		+	+
Eugenol		+	+
Hydroquinone	+		
Methyl salicylate			+
2-(p-Methoxyphenyl) ethanol			+
2-Phenyl ethanol		+	+
Thymoquinone			+
o-Vanillin		+	
p-Vanillin		+	
Monoterpenoids			
Carvacrol		+	+
p-Cymen-8-ol			+
Geraniol		+	
Linalool	+	+	
Nerol		+	
cis-Sabinene-hydrate	+		
trans-Sabinene-hydrate		+	
α-Terpineol	+	+	
Terpinen-4-ol	+	+	+
Thymol		+	+
Sesquiterpenoids			
α-Cadinol		+	
cis-Nerolodol		+	
Miscellaneous			
Butyl phthalate monoester			+
1-H-Indole			+

more compounds (270 mg/kg fresh plant material) than during flowering and seed development (80 mg/kg) but that does not necessarily indicate metabolic changes since during flowering there is a considerable increase in biomass (Stahl-Biskup *et al.*, 1993). The content in the sample from Croatia was much lower, 20 mg/kg dry plant material (Mastelic *et al.*, 2000); nevertheless considerable antioxidant activity was found to be present in these compounds; thus, they are receiving increasing attention recently (Milos *et al.*, 2000).

Diterpenoids and triterpenoids

Diterpenoids have been reported only from two *Origanum* species, *O. akhdarense* and *O. pampaninii*, whilst they have not been found in *O. dictamnus* and *O. vulgare* nor have been reported from other *Origanum* species. These diterpenes are isoprimara-7,15-dien-19-ol (akhdarenol), isoprimara-15-en-8β,19-diol (akhdardiol), isoprimara-15-en-8β,11α,19-triol (akhdartriol) and isoprimara-15-en-3β,8β,19-triol (isoakhdartriol), all found in *O. akhdarense* while only akhdardiol was found in *O. pampaninii* (Passannanti *et al.*, 1984; Piozzi *et al.*, 1985). Figure 3.14 shows the diterpenoids found in the genus.

The triterpenoids that have been found in *Origanum* are shown in Figure 3.15. The major triterpenoids that have been reported are ursolic and oleanolic acids from *O. dictamnus*, *O. akhdarense*, *O. pampaninii*, *O. majorana*, *O. vulgare*, and *O. compactum*; possibly these are present in other *Origanum* taxa as well, since they are common to most Labiatae (Hegnauer, 1966, 1989; Passannanti *et al.*, 1984; Piozzi *et al.*, 1985; Bellakhdar *et al.*, 1988; Chung *et al.*, 2001). Ursolic acid constituted 1.8 per cent w/w and oleanolic acid constituted 0.4 per cent w/w of the dried leaves of *O. dictamnus* (Revinthi-Moraiti *et al.*, 1985). In *O. majorana* ursolic and oleanolic acid were 0.6 per cent and 0.3 per cent w/w, respectively. In *O. vulgare* ursolic acid was 0.3 per cent w/w. In addition, some rare compounds, methyl 3β-21α-dihydroxyursolic and methyl 3β-21α-dihydroxy-oleanolic acids have been isolated from leaves of *O. dictamnus*, together with methylursolic and methyloleanolic acids. In flowers of *O. dictamnus* methylursolic and methyloleanolic acids were the major triterpenoids, while the hydroxy-derivatives were present too (Piozzi *et al.*, 1986). In *O. compactum*, 21α-dihydroxyursolic and 21α-dihydroxyoleanolic acids, β-amyrin, betulin and betulinic acid have also been found (Bellakhdar *et al.*, 1988). The triterpenic alcohol uvaol was isolated from *O. pampaninii* and was also found present in the flowers of *O. dictamnus* but absent from the leaves (Passannanti *et al.*, 1984; Piozzi *et al.*, 1986).

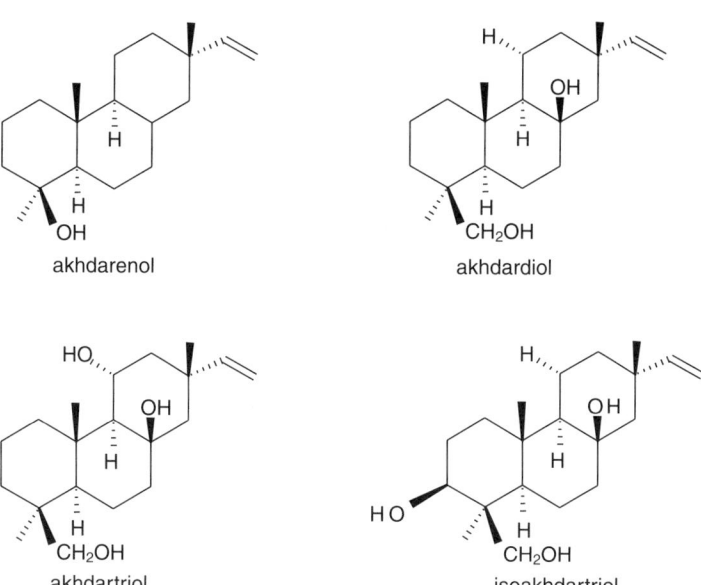

Figure 3.14 Diterpenoids found in *Origanum* spp.

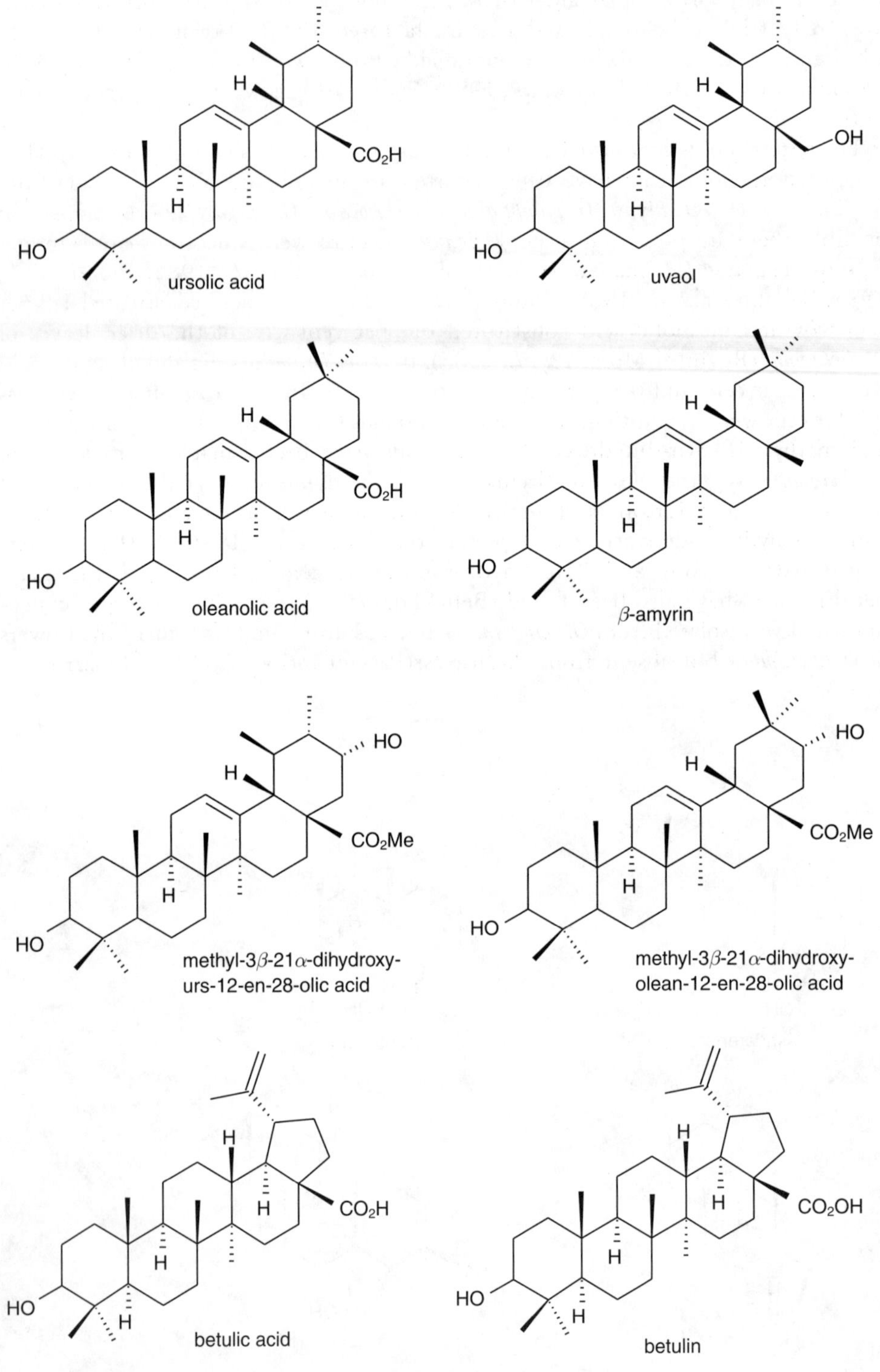

Figure 3.15 Triterpenoids found in *Origanum* spp.

Lipids and fatty acids

Fatty acids occur mainly in plants in bound form, esterified to glycerol, as fats or lipids. Lipids, generally, comprise up to 7 per cent of the dry weight in leaves of all higher plants and are important as membrane constituents in the chloroplasts and mitochondria. There are two categories of lipids: the neutral lipids (triglycerides) and the polar lipids (phospholipids and glycolipids). In *Origanum dictamnus*, the total lipid fraction constituted 9.7 per cent w/w of dried leaves; non-polar lipids were 7.8 per cent and included sterols, steryl esters, fatty alcohols, free fatty acids, waxes, traces of triglycerides and triterpenic acids; the triterpenic acid was ursolic acid that constituted 1.8 per cent w/w of the dried leaves. 1.9 per cent were polar lipids, the majority being glycolipids: monogalactosyl-diglycerides, digalactosyl-diglycerides, sulpholipids, cerebrocides and polygalactosyl-diglycerides. Phospholipids includes: phosphatidic acid, phosphatidyl-ethanolamine, -glycerol, -inositol, -serine, and -choline. (Revinthi-Moraiti *et al.*, 1985). Glycolipids and essential fatty acids have been found in the aerial part of *O. vulgare* ssp. *gracile* (*O. tytthanthum*); 15-methyl-hexadecanoic acid was found among the free fatty acids (Asilbekova *et al.*, 2000). Glycolipids, phospholipids, and sterols were isolated from the aerial parts of *O. onites*; linolenic and palmitic acids dominate in the acids of the lipids, while the main components of the hydrocarbons are nonacosane and of the sterols, sitosterol (Azcan *et al.*, 2000).

PHENOLS, PHENOLIC ACIDS, HYDROXYCINNAMIC ACIDS, AND QUINONES

Few simple phenols are common in plants; hydroquinone (the reduced form of benzoquinone) is the most common and has been reported to be present in *Origanum majorana* (Subramanian *et al.*, 1972). As well as hydroquinone the presence of hydroquinone monomethyl ether has been detected in the aqueous alcoholic extract of dry *O. majorana*, together with their respective glucosides arbutin (hydroquinone-O-β-glucoside) and methylarbutin (Assaf *et al.*, 1987).

Phenolic acids such as *p*-hydroxybenzoic, protocatechuic, vanillic, syringic, gallic, and hydroxycinnamic acids such as *p*-coumaric, caffeic, their methylated derivatives such as ferulic and sinapic acid are present in all angiosperms. Mirovich *et al.* (1989) identified cinnamic, caffeic, *p*-hydroxybenzoic, vanillic, syringic and protocatechuic acids in *O. vulgare*. Kikuzaki and Nakatani (1989) isolated a new compound, 2-caffeoyloxy-3-[2-(4-hydroxybenzyl)-4,5-dihydroxy] phenylpropionic acid, from leaves of *O. vulgare*.

Esterified forms of hydroxycinnamic acids such as chlorogenic and rosmarinic acid are also common in many angiosperms. Rosmarinic acid has been isolated from *Rosmarinus officinalis* and *Salvia officinalis*. Rosmarinic and chlorogenic acids tend to accumulate in all plants of the subfamily *Nepetoideae* (Hegnauer, 1989). Reschke (1983) found that *O. vulgare* had a higher content of rosmarinic acid (1.6 per cent w/w) than *Rosmarinus* and *Salvia*. The concentration of rosmarinic acid in *Origanum* is highly variable, probably reflecting genetic variability. However Eguchi *et al.* (1996) proposed a method for selecting high-phenolic-producing clones using *Pseudomonas* spp. because *Pseudomonas* stimulated phenolics and rosmarinic acid and their concentration in various clonal lines was directly correlated to tolerance to the *Pseudomonas*.

The absence of references for other *Origanum* species on the above phenolic compounds does not mean that the compounds are not present in them. Much more remains to be done on these compounds and the increasing interest in natural antioxidants is likely to lead to more reports in the near future.

FLAVONOIDS

Flavonoids, although phenolic compounds are discussed separately. Compared with the essential oils, the flavonoids of *Origanum* have received significantly less attention. It is only in recent years, that antioxidant compounds have become highly appreciated, hence more light is expected to be thrown on flavonoids that are always present, together with other phenolic compounds, in herbal infusion and other plant preparations (Vekiari et al., 1993; Moller et al., 1999). For example, Kanazawa et al. (1995) reported that galangin and quercetin extracted from *Origanum* have shown specific antimutagenic activity against dietary carcinogens.

Two classes of flavonoids are recognized in plants of the *Labiatae*: free flavonoids and the flavonoid glycosides. Both classes are present in *Origanum*. Free flavonoids are often highly methylated. Methyl ethers of flavones, flavonols, flavanones and dihydroflavones are insoluble in water and are found on the plant surfaces, in the waxy layer or in the glands. The other class of flavonoids includes the O-glycosides and the C-glycosides, that are water soluble and are accumulated in the vacuoles of plants.

Free flavonoids

Free flavonoids have been identified in many members of the genus. Table 3.6 and Figures 3.16 show the free flavones, flavanones, dihydroflavonols and flavonols that have been found in *Origanum* species. As is characteristic of the Labiatae (Tomas-Barberan et al., 1988a), significantly more flavones (15) than flavonols (7) have been found. The amount of research work that has been carried out on *Origanum* does not yet offer sufficient information to make generalisations and draw conclusions on flavonoid distribution with regard to infrageneric taxonomic groups; however it is quite clear that there is an abundance of 6-substituted flavonoids; almost half of the flavones (7 out of 15) and half of the flavonols (3 out of 7) bear either a hydroxy- or a methoxy group in the sixth position. It also worth mentioning here the presence of 8-substituted compounds in the free flavone group (3 out of 15) of *Origanum*, which are uncommon elsewhere.

Tomas-Barberan et al. (1988b) report that within the genus *Origanum* species of the section *Origanum* do not accumulate external flavonoids. This, disagrees with other workers that report several free flavonoids of *O. vulgare*, from Russia (Antonesku et al., 1983; Peshkova and Mirovich, 1984; Mirovich, 1987), from Japan (Kanazawa et al., 1995), from China (Zheng et al., 1997) and from Greece (Vekiari et al., 1993); unfortunately in all these studies no subspecies are named. On the other hand Tomas-Barberan et al. (1988b) also state that species in the sections *Amaracus*, *Anatolicon* and *Majorana* do accumulate external flavonoids, and this is confirmed by other researchers (Passannanti et al., 1984; Voirin et al., 1984; Harvala and Skaltsa, 1986; Souleles, 1990). Inter-sectional hybrids seem to accumulate free flavones too (Palomino et al., 1997; Bosabalidis et al., 1998).

Not much information is yet available on the quantity of the various classes of free flavonoids accumulated in the plants, or on the quantitative and qualitative variation

Table 3.6 Free flavonoids found in *Origanum* spp.

Trivial name	Compound	Species
Flavones		
Chrysin	5,7-di-OH-flavone	O. vulgare
Negletein	5,6-di-OH-7-OMe-flavone	O. vulgare
Mosloflavone	5-OH-6,7-di-OMe-flavone	O. vulgare
Apigenin	5,7,4'-di-OH-flavone	O. dictamnus
		O. vulgare
		O. × majoricum
Genkwanin	5,4'-OH-7-OMe-flavone	O. × intercedens
Acacetin	5,7-di-OH-4'-OMe-flavone	O. majorana
Apigenin-4,7-dimethylether	5-OH-7,4'-di-OMe-flavone	O. pampanini
Cirsimaritin	5,4'-di-OH-6,7-di-OMe-flavone	O. akhadarensis
Luteolin	5,7,3',4'-tetra-OH-flavone	O. dictamnus
		O. vulgare
		O. × majoricum
Chrysoeriol	5,7,4' tri-OH-3'-OMe-flavone	O. × majoricum
Thymusin	5,6,4'-tri-OH-7,8-tri-OMe-flavone	O. onites
		O. × intercedens
6-Hydroxy-luteolin-7,3'-dimethylether	5,6,4'-tri-OH-7,3'-di-OMe-flavone	O. boissieri
		O. vetteri
		O. acutidens
		O. leptocladum
		O. scabrum
		O. onites
		O. syriacum
		O. × intercedens
6-Hydroxy-luteolin-7,3',4'-dimethylether	5,6-di-OH-7,3',4'-tri-OMe-flavone	O. onites
Thymonin	5,6,4'-tri-OH-7,8,3'-tri-OMe-flavone	O. majorana
		O. boissieri
		O. × intercedens
		O. cordifolium
		O. saccatum
		O. vetteri
		O. acutidens
		O. leptocladum
		O. scabrum
		O. onites
5-Dismethylnobeletin	5-OH-6,7,8,3',4'-OMe-flavone	O. pampanini
Flavonols		
Galangin	3,5,7-tri-OH-flavone	O. vulgare
Kaempferol	3,5,7,4'-tetra-OH-flavone	O. vulgare
Penduletin	5,4'-di-OH-3,6,7-tri-OMe-flavone	O. majorana
Quercetin	3,5,7,3',4'-penta-OH-flavone	O. dictamnus
Axillarin	5,7,3',4'-tetra-OH-3,6-di-OMe-flavone	O. majorana
Chrysosplenol-D	5,3',4'-tri-OH-3,6,7-tri-OMe-flavone	O. majorana
Retusin	5-OH-3,7,3',4'-OMe-flavone	O. vulgare
Flavanones		
Naringenin	5,7,4'-OH-flavanone	O. vulgare
		O. × majoricum
Eryodictyol	5,7,3',4'-OH-flavanone	O. dictamnus
		O. vulgare

Table 3.6 (Continued)

Trivial name	Compound	Species
Dihydroflavonols		
Aromadendrin/ dihydrokampferol	3,5,7,4'-OH-dihydroflavonol	O. compactum
Taxifolin/dihydroquercetin	3,5,7,3',4'-OH-dihydroflavonol	O. vulgare

within and between populations of *Origanum* species. Mirovich (1987) reported on the differences in the flavonoid content of *O. vulgare* with regard to the growing site, to the developmental stage and to the plant organ. Kanazawa *et al.* (1995) gave an indication of the flavonol content in *O. vulgare*, reporting that the galangin and the quercetin content were 3.4 mg and 10.9 mg respectively, in 100 g plant material. Palomino *et al.* (1997) found that the maximum content of flavonoid in *O.* × *majoricum*, from Balearic islands, occurs in May.

Flavonoid glycosides

A number of O-glycosides and C-glycosides have been found in *Origanum* (Table 3.7). Up to the present, 21 compounds have been reported, 14 of them being flavone

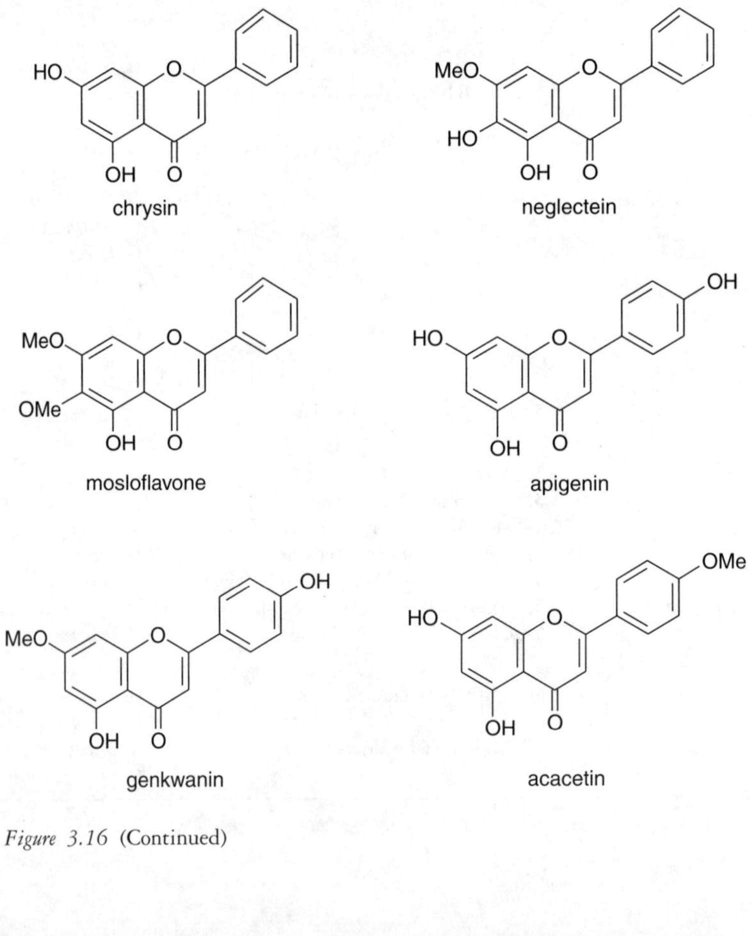

Figure 3.16 (Continued)

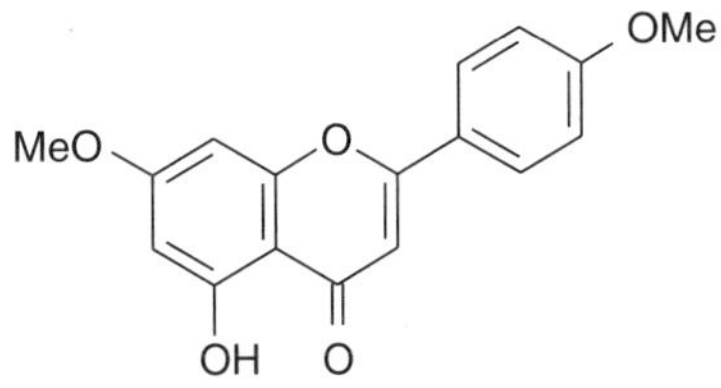

Figure 3.16 (Continued)

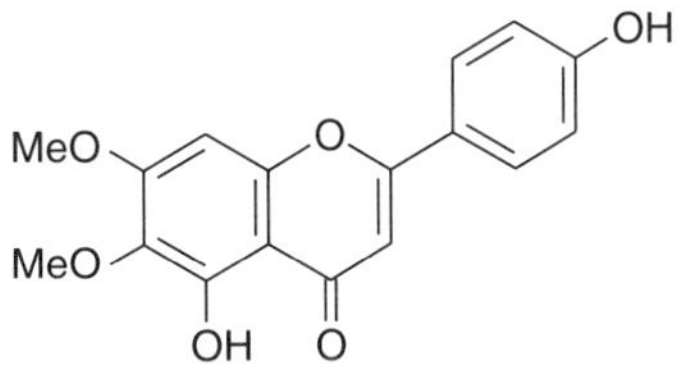

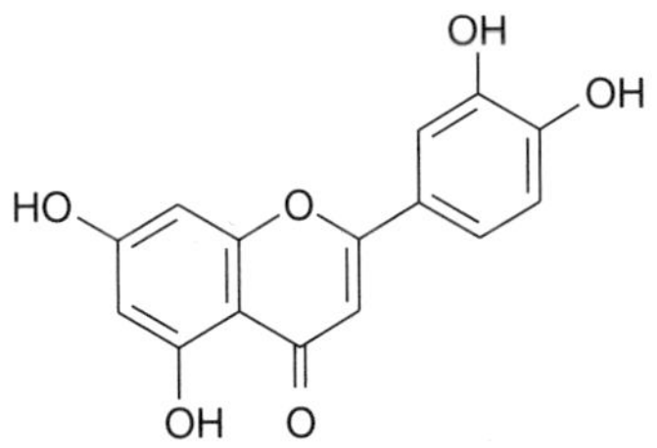

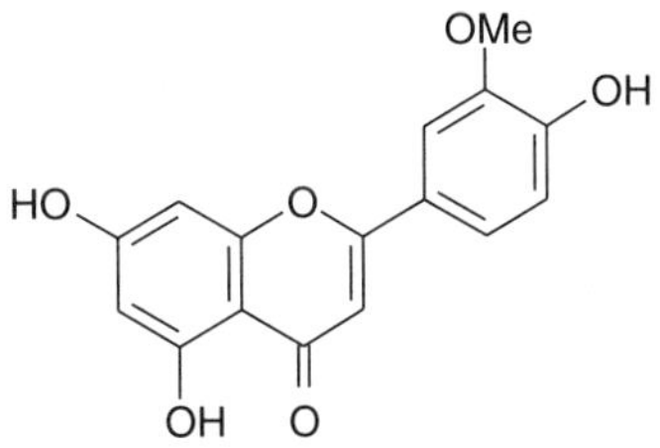

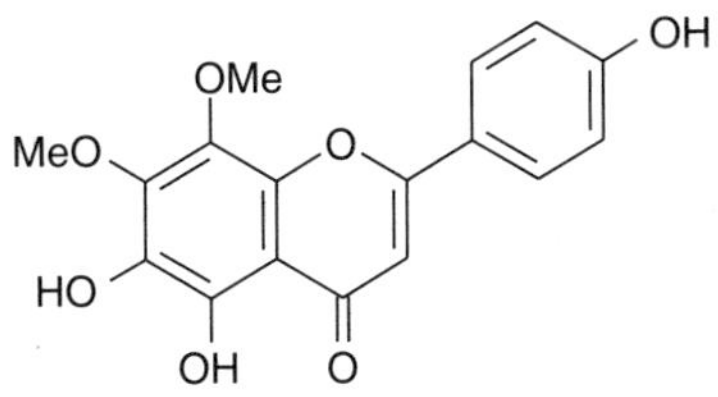

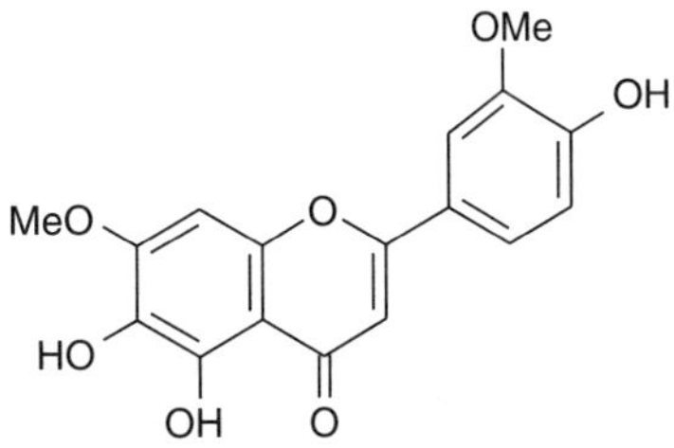

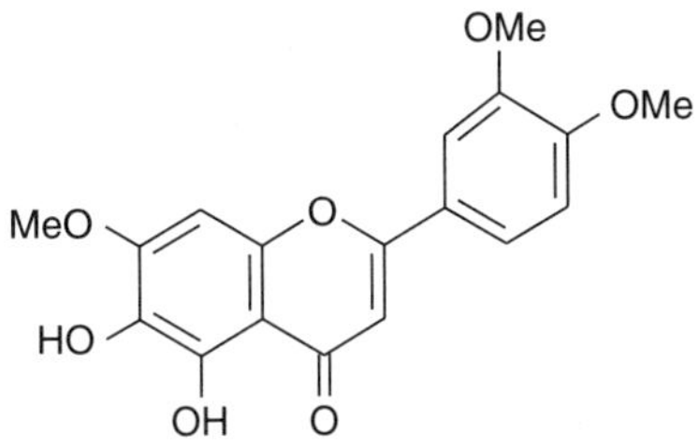

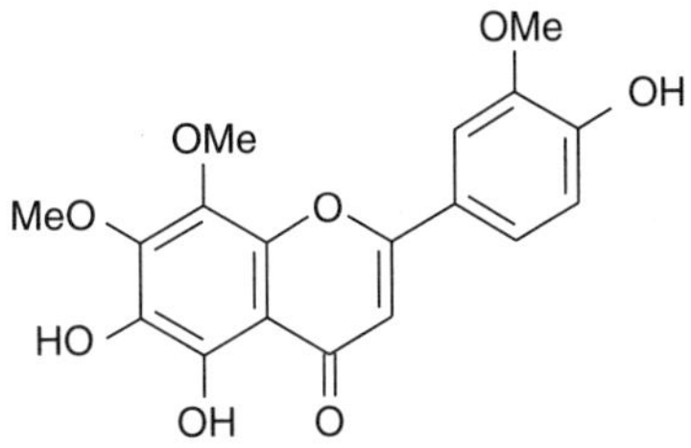

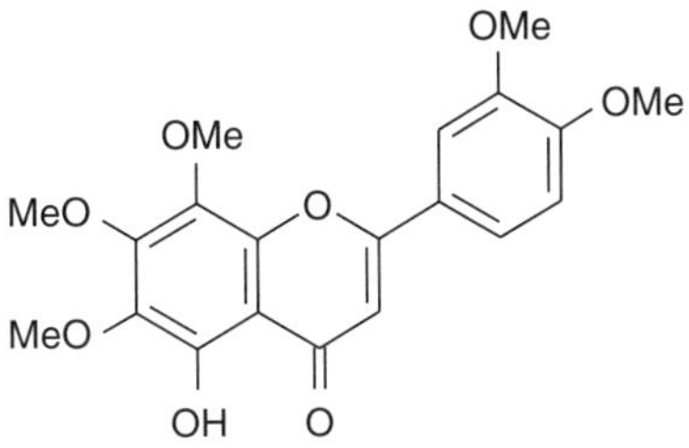

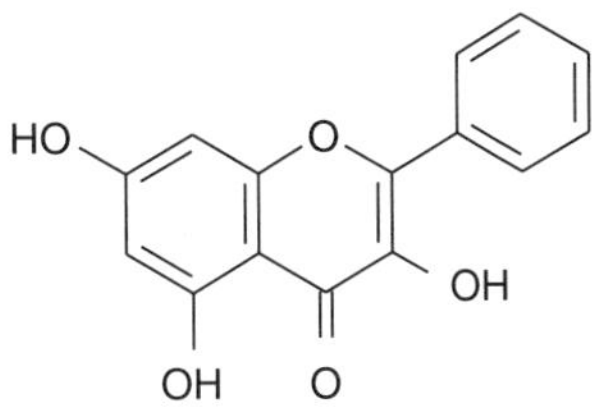

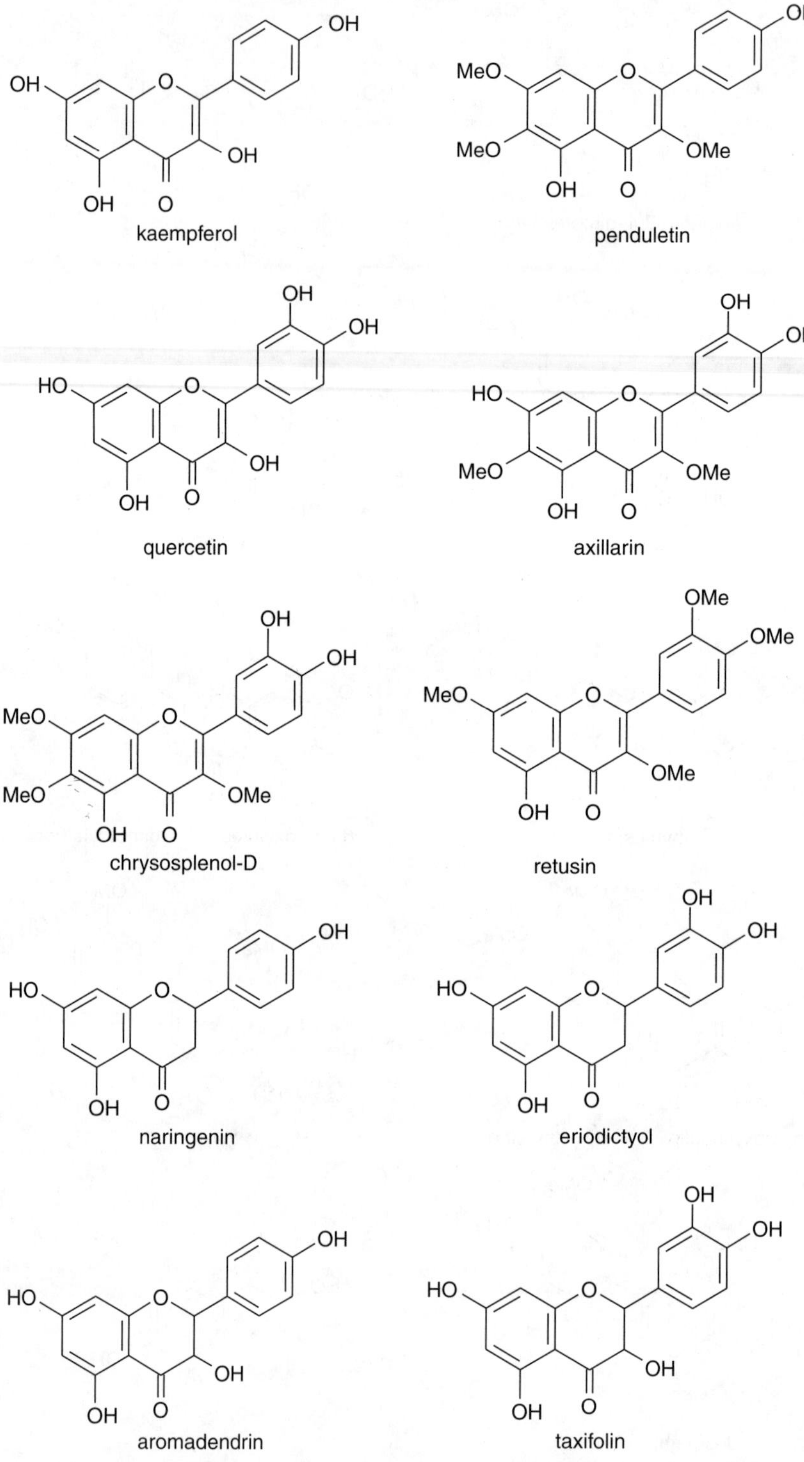

Figure 3.16 Free flavone, flavonol, flavanone and dihydroflavonol structures found in *Origanum* spp.

Table 3.7 Flavonoid glycosides found in *Origanum* spp.

Trivial name (aglycon name)	Compound	Species
Flavone-O-glycosides		
Cosmosiin (apigenin)	5,7,4'-tri-OH-flavone, 7-O-glucoside	O. boissierii
		O. dictamnus
		O. hypercifolium
		O. isthmicum
		O. ramonense
		O. sipyleum
		O. vulgare
		O. dayi
(apigenin)	5,7,4'-tri-OH-flavone, 7-O-glucuronide	O. boissierii
		O. dayi
		O. isthmicum
		O. majorana
		O. microphyllum
		O. onites
		O. ramonense
		O. syriacum
		O. vetteri
(negletein)	5,6-di-OH-7-OMe-flavone, 6-O-rhamnosyl (1 → 2) fucoside	O. vulgare
Cynaroside (luteolin)	5,7,3'4'-tetra-OH-flavone, 7-O-glucoside	O. majorana
		O. akhdarense
		O. boissierii
		O. calcaratum
		O. cordifolium
		O. dayi
		O. dictamnus
		O. hypercifolium
		O. isthmicum
		O. laevigatum
		O. leptocladum
		O. libanoticum
		O. microphyllum
		O. ramonense
		O. rotundifolium
		O. saccatum
		O. scabrum
		O. sipyleum
		O. syriacum
		O. vetteri
		O. vulgare ssp. hirtum
		O. vulgare ssp. virens
		O. vulgare ssp. viride
		O. vulgare ssp. vulgare
(luteolin)	5,7,3',4'-tetra-OH-flavone, 7-O-glucuronide	O. vulgare ssp. vulgare
(luteolin)	5,7,3',4'-tetra-OH-flavone, 7-O-diglucuronide	O. akhadarense
		O. boissierii
		O. calcaratum
		O. cordifolium
		O. dayi
		O. dictamnus

Table 3.7 (Continued)

Trivial name (aglycon name)	Compound	Species
		O. hypercifolium
		O. isthmicum
		O. laevigatum
		O. libanoticum
		O. majorana
		O. microphyllum
		O. onites
		O. ramonense
		O. rotundifolium
		O. saccatum
		O. scabrum
		O. syriacum
		O. vulgare ssp. hirtum
		O. vulgare ssp. virens
		O. vulgare ssp. viride
		O. vulgare ssp. vulgare
(luteolin)	5,7,3'4'-tetra-OH-flavone, 7-O-rutinoside	O. majorana
		O. compactum
		O. ehrenbergii
		O. laevigatum
(luteolin)	5,7,3'4'-tetra-OH-flavone, 7,4'-O-diglucoside	O. akhadarense
		O. amanum
		O. boissierii
		O. calcaratum
		O. compactum
		O. cordifolium
		O. dictamnus
		O. hypercifolium
		O. laevigatum
		O. libanoticum
		O. microphyllum
		O. rotundifolium
		O. saccatum
		O. scabrum
		O. vulgare ssp. hirtum
		O. vulgare ssp. hirtum
		O. vulgare ssp. virens
		O. vulgare ssp. viride
(diosmetin)	5,7,3'-tri-OH-4'-OMe-flavone, 7-O-glucuronide	O. majorana
(chrysoeriol)	5,7,4'-tri-OH-3'-OMe-flavone, 7-O-glucuronide	O. micranthum
Scuttelarin (scuttelarein)	5,6,7,4'-tetra-OH-flavone, 7-O-glucuronide	O. boissierii
		O. dayi
		O. saccatum
(hispidulin)	5,7,4'-di-OH-6-OMe-flavone, 7-O-glucuronide	O. majorana
(scuttelarein-4'-methylether)	5,6,7-tri-OH-4'-OMe-flavone, 7-O-glucoside	O. akhadarense
		O. calcaratum
		O. majorana
		O. onites
		O. rotundifolium
		O. saccatum

		O. scabrum
		O. sipyeum
		O. syriacum
		O. vetteri
(6-hydroxyluteolin)	5,6,7,3',4'-tetra-OH-flavone, 7-O-glucoside	O. dayi
		O. isthmicum
		O. majorana
		O. microphyllum
		O. onites
		O. ramonense
		O. syriacum
		O. vetteri
Flavanone-O-glycosides		
(eryodictyol)	5,7,3',4'-tri-OH-flavanone, 7-O-glucoside	O. dictamnus
(isosakuranetin)	5,7,-di-OH-4'-OMe-flavanone, 7-O-neohesperidoside	O. vulgare
Flavone-C-glycosides		
Isovitexin (apigenin)	5,7,4'-tri-OH-flavone, 6-C-glucoside	O. dictamnus
Vitexin (apigenin)	5,7,4'-tri-OH-flavone, 8-C-glucoside	O. dictamnus
		O. majorana
Vicenin-2 (apigenin)	5,7,4'-tri-OH-flavone, 6,8-C-diglucoside	O. dictamnus
		O. onites
		O. majorana
		O. microphyllum
		O. syriacum
Orientin (luteolin)	5,7,3',4'-tetra-OH-flavone, 8-C-glucoside	O. dictamnus
Isoorientin (luteolin)	5,7,3',4'-tetra-OH-flavone, 6-C-glucoside	O. dictamnus

7-O-glycosides, five flavone-C-glycosides and two flavonone-7-O-glycosides, whereas no flavonol glycoside has been found.

Luteolin is the most common aglycone of the 7-O-glycosides followed by apigenin; most sugar moieties are glucosides and glucuronides. The 7-O-glycosides of luteolin occur in almost every taxon of *Origanum* analyzed, more than half of the aglycones (5 out of 9) bear either a hydroxy- or a methoxy-group in the sixth position (Husain et al., 1982; Peshkova and Mirovich, 1984; Barberan, 1986; Mirovich, 1987). The 6-substituted aglycons occur in members of the A and B infrageneric groups as well as in the section Campanulaticalyx of the C infrageneric group (Subramanian et al., 1972; Husain et al., 1982; Gil-Munoz, 1993). With one exception (Zheng et al., 1997) no 6-substituted aglycone has been found in any member of the section *Origanum*.

The less common C-glycosides are either apigenin or luteolin based. C-glycosides have been reported from *O. dictamnus* (section *Amaracus*, group A) and from members of the infrageneric group B; none has been reported from any taxon of the group C (Husain and Markham, 1981; Skaltsa and Harvala, 1987; Tomas-Barberan et al., 1988b; Gil-Munoz, 1993). In all identified compounds the sugar moiety is glucose and the sugars are linked to the C-6 or C-8 position.

Anthocyanins

Anthocyanins, a specific class of flavonoids responsible for pink and purple pigmentation are present in petals, calyces, and bracts of most *Origanum*, sometimes they appear in leaves too. Although the presence of colour in petals, bracts or calyces is used for infraspecific taxonomic identification, no information on the anthocyanins from *Origanum* was found in the literature. Anthocyanin substituted with both aromatic and aliphatic acids are likely to be present, since such pigments have been shown to occur widely in the Labiatae (Saito and Harborne, 1992).

REFERENCES

Alves-Pereira, I.M.S. and Fernandes-Ferriera, M. (1998) Essential oils and hydrocarbons from leaves and calli of *Origanum vulgare* ssp. *virens*. *Phytochemistry* 48(5), 795–799.

Antonescu, V., Sommer, L., Prodescu, I. and Barza, P. (1983) *C.A.* 99. 19689X.

Arnold, N., Bellomaria, B. and Valentini, G. (2000) Composition of the essential oil of three different species of *Origanum* in the Eastern Mediterranean. *J. Essent. Oil Res.* 12, 192–196.

Arnold, N., Bellomaria, B., Valentini, G. and Arnold, H.J. (1993) Comparative study of essential oil from three species of *Origanum* growing wild in the Eastern Mediterranean Region. *J. Essent. Oil Res.* 5, 71–77.

Asilbekova, D.T., Glushenkova, A.I., Azcan, N., Ozek, T. and Baser, K.H.C. (2000) Lipids of *Origanum tytthanthum*. *Chem. Nat. Compounds* 36(2), 124–127.

Assaf, M.H., Ali, A.A., Makboul, M.A., Beck, J.P. and Anton, R. (1987) Preliminary study of phenolic glycosides from *Origanum majorana*; quantitative estimation of arbutin; cytotoxic activity of hydroquinone. *Planta Medica*. 53(4), 343–345.

Azcan, N., Kara, M., Asilbekova, D.T., Ozek, T. and Baser, K.H.C. (2000) Lipids and essential oil of *Origanum onites*. *Chem. Nat. Compounds* 36(2), 132–136.

Barberan, F.A.T. (1986) The flavonoids from *Labiatae*. *Fitoterapia* 57, 67–95.

Baser, K.H.C. and Duman, H. (1998) Composition of the essential oils of *Origanum boissieri* Ietswaart and *O. bargyli* Mouterde. *J. Essent. Oil Res.* 10(1), 71–72.

Baser, K.H.C., Tumen, G. and Sezik, E. (1991) The essential oil of *Origanum minutiflorum* O. Schwarz and P.H. Davis. *J. Essent. Oil Res.* 3(6), 445–446.

Baser, K.H.C., Ozek, T., Kurkcuoglu, M. and Tumen, G. (1992) Composition of the essential oil of *Origanum sipyleum* of Turkish origin. *J. Essent. Oil Res.* 4(2), 139–142.

Baser, K.H.C., Ozek, T., Tumen, G. and Sezik, E. (1993a) Composition of the essential oils of Turkish *Origanum* species with commercial importance. *J. Essent. Oil Res.* 5(6), 619–623.

Baser, K.H.C., Kirimer, N. and Tumen, G. (1993b) Composition of the essential oil of *Origanum majorana* L. from Turkey. *J. Essent. Oil Res.* 5(5), 577–579.

Baser, K.H.C., Ermin, N., Kurkcuoglu, M. and Tumen, G. (1994a). Essential oil of *Origanum hypericifolium* O. Schwarz et P.H. Davis. *J. Essent. Oil Res.* 6(6), 631–633.

Baser, K.H.C., Ozek, T., Kurkcuoglu, M. and Tumen, G. (1994b) The essential oil of *Origanum vulgare* subsp. *hirtum* of Turkish origin. *J. Essent. Oil Res.* 6(1), 31–36.

Baser, K.H.C., Ozek, T. and Tumen, G. (1995) Essential oil of *Origanum rotundifolium* Boiss. *J. Essent. Oil Res.* 7(1), 95–96.

Baser, K.H.C., Ermin, N., Ozek, T., Demircakmak, B., Tumen, G. and Duman, H. (1996a) Essential oils of *Thymbra sintenisii* Bornm. et Aznav. subsp. *isaurica* P.H. Davis and *Origanum leptocladum* Boiss. *J. Essent. Oil Res.* 8(6), 675–676.

Baser, K.H.C., Ozek, T., Kurkcuoglu, M. and Tumen, G. (1996b) Essential oil of *Origanum micranthum* Vogel. *J. Essent. Oil Res.* 8(2), 203–204.

Baser, K.H.C., Tumen, G. and Duman, H. (1996c) Essential oil of *Origanum bilgeri* P.H. Davis. *J. Essent. Oil Res.* 8(2), 217–218.

Baser, K.H.C., Ozek, T., Kurkcuoglu, M. and Tumen, G. (1996d) Essential oil of *Origanum laevigatum* Boiss. *J. Essent. Oil Res.* 8(2), 185–186.

Baser, K.H.C., Tumen, G. and Duman, H. (1997a) Essential oil of *Origanum acutidens* (Hand.-Mazz.) Ietswaart. *J. Essent. Oil Res.* 9(1), 91–92.

Baser, K.H.C., Demircakmak, B., Nuriddinov, K.R., Nigmatullaev, A.M. and Aripov, Kh.N. (1997b) Composition of the essential oil of *Origanum tyttanthum* Gontsch. from Uzbekistan. *J. Essent. Oil Res.* 9(5), 611–612.

Baser, K.H.C., Kurkcuoglu, M. and Tuman, G. (1998a) Composition of the essential oil of *Origanum haussknechtii* Boiss. *J. Essent. Oil Res.* 10(2), 227–228.

Baser, K.H.C., Kurkcuoglu, M., Duman, H. and Aytac, Z. (1998b) Composition of the essential oil of *Origanum husnucan-baserii* H. Duman, Z. Aytaç et A. Duran, a new species from Turkey. *J. Essent. Oil Res.* 10(4), 419–421.

Baser, K.H.C., Duman, H. and Aytac, Z. (2000) Composition of the essential oil of *Origanum × adanense* Baser et Duman. *J. Essent. Oil Res.* 12(4), 475–477.

Beker, R., Dafni, A., Eisikowitch, D. and Ravid, U. (1989) Volatiles of two chemotypes of *Majorana syriaca* L. (Labiatae) as olfactory cues for the honeybee. *Oecologia* 79(4), 446–451.

Bellakhdar, J., Passannanti, S., Paternostro, M.P. and Piozzi, F. (1988) Constituents of *Origanum compactum*. *Planta Medica* 1, 94.

Benjilali, B., Richard, H.M.J. and Baritaux, O. (1986) Etude des huiles essentielles de deux espèces d' Origan du Maroc: *Origanum compactum* Benth. et *Origanum elongatum* Emb. et Maire. *Lebensmittel Wissenschaft and Technologie* 19, 22–26.

Bentham, G. (1834) *Labiatarum genera et species*. Ridgway and Sons, London.

Bentham, G. (1848) Labiatae. In A. de Cantolle (ed.), *Prodromus Systematis Naturalis Regni Vegetalis*. Treuttel and Wurtz, Paris, Vol. 12, pp. 191–197.

Bosabalidis, A., Gabrieli, C. and Niopas, I. (1988) Flavone aglycones in glandular hairs of *Origanum × intercedens*. *Phytochemistry* 49(6), 1549–1553.

Bosabalidis, A. and Skoula, M. (1998) A comparative study of the glandular trichomes of the upper and lower leaf surfaces od *Origanum × intercedens* Rech. *J. Essent. Oil Res.* 10, 277–286.

Briquet, J. (1895) Labiatae. In A. Engler and K. Prantl (eds), *Die naturrlichen pflanzenfamilien*. W. Engelmann, Leipzig, pp. 183–375.

van Den Broucke, C.O. and Lemli, J.A. (1980) Chemical investigation of the essential oil of *Origanum compactum*. *Planta Medica* 38(3), 264–266.

Cantino, P.D. (1992) Towards a phylogenetic classification of the *Labiatae*. In R.M. Harley and T. Reynolds (eds), *Advances in Labiate Science*. Royal Botanic Gardens Kew, Richmond, Surrey, UK, pp. 27–37.

Cantino, P.D., Harley, R.M. and Wagstaff, S.J. (1992) Genera of *Labiatae*: Status and classification. In R.M. Harley and T. Reynolds (eds), *Adv. Labiate Sci*. Royal Botanic Gardens Kew, Richmond, Surrey, UK, pp. 511–522.

Carlstrom, A. (1984) New species of *Alyssum*, *Consolida*, *Origanum* and *Umbilicus* from SE Aegean sea. *Willdenowia* 14, 15–26.

Chalchat, J.C. and Pasquier, B. (1998) Morphological and chemical studies of origanum clones: *Origanum vulgare* L. ssp. *vulgare*. *J. Essent. Oil Res.* 10(2), 119–125.

Chung, Y.K., Heo, H.J., Kim, E.K., Kim, H.K., Huh, T.L., Lim, Y., Kim, S.K. and Shin, D.H. (2001) Inhibitory effect of ursolic acid purified from *Origanum majoran*a L. on the acetylcholinesterase. *Mol. Cells* 11(2), 137–143.

Circella, G., Franz, C., Novak, J. and Resch, H. (1995) Influence of day length and leaf insertion on the composition of marjoram essential oil. *Flavour Fragrance J.* 10, 371–374.

Crouzet, J. and Chassagne, D. (1999) Glycosidically bound volatiles in plants. In R. Ikan (ed.), *Naturally Occurring Glycosides*. John Wiley & Sons, Chichester, UK, pp. 225–274.

Danin, A. (1990) Two new species of *Origanum* (Labiatae) from Jordan. *Willldenowia* 19, 401–405.

Danin, A. and Künne, I. (1996) *Origanum jordanicum* (Labiatae), a new species from Jordan, and notes on the other species of sect. Campanulaticalyx. *Willdenowia* 25, 601–611.

Danin, A., Ravid, U., Umano, K. and Shibamoto, T. (1997) Essential oil composition of *Origanum ramonense* Danin leaves from Israel. *J. Essent. Oil Res.* 9, 411–417.

Duman, H., Baser, K.H.C. and Aytec, Z. (1998) Two new species and a new hybrid from Anatolia. *Tr. J. Bot.* 22, 51–55.

Duman, H., Aytec, Z., Ekici, M., Karaveliogullari, E.A., Donmez, A. and Duran, A. (1995) Three new species (Labiatae) from Turkey. *Flora Mediterr.* 5, 221–228.

Eguchi, Y., Curtis, O.F. and Shetty, K. (1996) Interaction of hyperhydricity-preventing *Pseudomonas* sp. with oregano (*Origanum vulgare*) and selection of high phenolics and rosmarinic acid-producing clonal lines. *Food Biotechnol.* 10, 191–202.

Fischer, N., Nitz, S. and Drawert, F. (1987) Original flavour compounds and the essential oil composition of marjoram (*Majorana hortensis* Moench). *Flavour Fragrance J.* 2(2), 55–61.

Fleisher, A. and Fleisher, Z. (1991) Chemical composition of *Origanum syriacum* L. essential oil. Aromatic plant of the Holy Land and the Sinai, Part V. *J. Essent. Oil Res.* 3, 121–123.

Franz, C. (1990) Sensorial versus analytical quality of marjoram. *Herba-Hungarica* 29(3), 79–86.

Gil-Munoz, M.I. (1993) Contribution al estudio fitochimico y quimiosistematico de flavonoides en la familia *Labiatae*. Ph.D. Thesis, University of Murcia, 293pp.

Greuter, W., Burdet, H.M. and Long, G. (1986). *Med-Checklist, Vol. 3*. Editions de Conservatoire de Jardin Botaniques de la Ville de Geneve, 1986.

Halim, A.F., Mashaly, M.M., Zaghloul, A.M., Abd-El-Fattah, H. and Pooter, de, H.L. (1991) Chemical constituents of the essential oils of *Origanum syriacum* and *Stachys aegyptiaca*. *Int. J. Pharmacognosy* 29(3), 183–187.

Harvala, C. and Skaltsa, H. (1986) Contribution a l'etude chimique *d'Origanum dictamnus* – 1 recommunication. *Plantes Medicinales et Phytotherapie* 20(4), 300–304.

Hedge, I.C. (1992) A global survey of the biogeography of the Labiatae. In R.M. Harley and T. Reynolds (eds), *Advances in Labiate Science*. Royal Botanic Gardens Kew, Richmond, Surrey, UK, pp. 7–17.

Hegnauer, R. (1966) Chemotaxonomie der Pflanzen. Eine Übersicht über die Verbreitung und die systematiche Bedeutung der Pflanzenstoffe, Band 4, Birkhåusen Verlag Basel und Stuttgart.

Hegnauer, R. (1989) Chemotaxonomie der Pflanzen. Eine Übersicht über die Verbreitung und die systematiche Bedeutung der Pflanzenstoffe, Band 8, Birkhåusen Verlag Basel, Boston, Berlin.

Houmani, Z., Azzoudj, S., Naxakis, G. and Skoula, M. (2002) The essential oil composition of Algerian zaâtar: *Origanum* spp. and *Thymus* spp. *J. Herbs Spices Med. Plants* (in press).

Husain, S.Z. and Markham, K.R. (1981) The glycoflavone vicenin-2 and its distribution in related genera within the *Labiatae*. *Phytochem.* 20(5), 1171–1173.

Husain, S.Z., Heywood, V.H. and Markham, K.R. (1982) Distribution of flavonoids as chemotaxonomic markers in the genus *Origanum* L. and related genera in *Labiatae*. In N. Margaris, A. Koedam, and D. Vokou (eds), *Aromatic Plants: Basic and Applied Aspects*. Martinus Nijhoff Publishers, The Hague, pp. 141–152.

Ietswaart, J.H. (1980) A taxonomic revision of the genus *Origanum* (Labiatae). Leiden Botanical Series 4. Leiden University Press, The Hague.

Kanazawa, K., Kawasaki, H., Samejima, K., Ashida, H. and Danno, G. (1995) Specific desmutagens (antimutagens) in oregano against dietary carcinogen, Trp-P-2, are galangin and quercetin. *J. Agric. Food Chem.* 43, 404–409.

Kaufmann, M. and Wink, M. (1994) Molecular systematics of the Nepetoideae (Family Labiatae): phylogenetic implications from *rbc*L gene sequences. *Z. Naturforsh* 49, 635–635.

Kazantzis, A. (1999) Seasonal ontogenetical and UV-B effect of essential oil composition in *Origanum vulgare* ssp. *hirtum*, M.Sc. Thesis, Chania, Greece.

Kikuzaki, H. and Nakatani, N. (1989) Structure of a new antioxidative phenolic acid from oregano (*Origanum vulgare* L.). *Agric. Biol. Chem.* 53(2), 519–524.

Kokkini, S., Karousou, R., Dardioti, A., Krigas, N. and Lanaras, T. (1997) Autumn essential oils of Greek oregano. *Phytochemistry* 44(5), 883–886.

Kokkini, S. and Vokou, D. (1989) Carvacrol-rich plants in Greece. *Flavour Fragrance J.* 4(1), 1–7.

Lawrence, B.M. (1980) Essential Oils 1979–1980, Allured Publishing Corporation. Carol Stream, Illinois, USA, pp. 79–80.

Linnaeus, C. (1754) Genera Plantarum, facsmile edition 1960: 256 Engelmamn (Cramer), Weinheim, Wheldon & Wesley, Codicote.

Mastelic, J., Milos, M. and Jerkovic, I. (2000) Essential oil and glycosidically bound volatiles of *Origanum vulgare* ssp. *hirtum* (Link) Ietswaart. *Flavour Fragrance J.* 15, 190–194.

Melegari, M., Severi, F., Bertoldi, M., Benvenuti, S., Circetta, G., Morone Fortunato, I., Bianchi, A., Leto, C. and Carrubba, A. (1995) Chemical characterization of essential oils of some *Origanum vulgare* L. subspecies of various origin. *Riv. Ital. Eppos* 16, 21–28.

Milos, M., Mastelic, J. and Jerkovic, I. (2000) Chemical composition and antioxidant effect of glycosidically bound volatile compounds from oregano (*Origanum vulgare* ssp. *hirtum*). *Food Chem.* 71, 79–83.

Mirovich, M.V. (1987) Studies on the phenolic compounds of common *Origanum*. *Nauchnye Trudy, Vsesoyuznyi Nauchno-Issledovatel'skii Institut Farmatsii* 25, 105–109.

Mirovich, V.M., Peshkova, V.A., Shatokhina, R.K. and Fedoseev, A.P. (1989) Phenolcarboxylic acids of *Origanum vulgare*. *Chem. Nat. Compounds* 25(6), 722–723.

Moller, J.K.S., Lindberg Madsen, H., Aaltonen, T. and Skibsted, L.H. (1999) Dittany (*Origanum dictamnus*) as source of water-extractable antioxidants. *Food Chem.* 64, 215–219.

Palomino, O.M., Gomez-Serranillos, P., Carretero, E. and Cases, A. (1997) Variation in the flavonoid content of *Origanum* × *majoricum* in different plant stages by HPLC. *Planta Medica* 63(6), 584.

Passannanti, S., Paternostro, M., Piozzi, F. and Barbagallo, C. (1984) Diterpenes from the genus *Amaracus*. *J. Nat. Prod.* 47(5), 885–889.

Peshkova, V.A. and Mirovich, V.M. (1984) Flavonoids of *Origanum vulgare*. *Khimiya Prirodnykh Soedinenii* 4, 522.

Piozzi, F., Paternostro, M. and Passannanti, S. (1985) A minor diterpene from *Amaracus akhdarensis*. *Phytochemistry* 24(5), 1113–1114.

Piozzi, F., Paternostro, M., Passannanti, S. and Gacsbaitz, E. (1986) Triterpenes from *Amaracus dictamnus*. *Phytochemistry* 25(2), 539–541.

Poulose, A.J. and Croteau, R. (1978) Biosynthesis of aromatic monoterpenes: conversion of γ-terpinene to *p*-cymene and thymol in *Thymus vulgaris* L. *Arch. Biochem. Biophys.* 187(2), 307–314.

Putievsky, E., Dudai, N. and Ravid, U. (1997) Cultivation, selection Putievsky, E., Dudai, N. and Ravid, U. Cultivation, selection and conservation of oregano species in Israel. In S. Padulosi, (ed.), *Oregano. Proceedings of the IPGRI International Workshop on Oregano, 8–12 May 1996, CIHEAM, Valenzano, Bari, Italy*. IPGRI, Rome, Italy. pp. 103–110,

Ravid, U. and Putievsky, E. (1993) Constituents of essential oils from *Majorana syriaca, Coridothymus capitatus* and *Satureja thymbra*. *Planta Medica* 49(4), 248–249.

Reschke, A. (1983) Capillary gas chromatographic determination of rosmarinic acid in herbs. *Zeitschrift fur Lebensmittel-Untersuchung und -Forschung* 176(2), 116–119.

Revinthi-Moraiti, K., Komaitis, M.E., Evangelatos, G. and Kapoulas, V.V. (1985) Identification and quantitatve determination of the lipids of dried *Origanum dictamnus* leaves. *Food Chem.* 16, 15–24.

Saito, N. and Harborne, J.B. (1992) Correlations between anthocyanin type, pollinator and flower colour in the *Labiatae*. *Phytochemistry* 31, 3009–3015.

Sezik, E., Tumen, G., Kirimer, N., Ozek, T. and Baser, K.H.C. (1993) Essential oil composition of four *Origanum vulgare* subspecies of Anatolian origin. *J. Essent. Oil Res.* 5(4), 425–431.

Skaltsa, H. and Harvala, C. (1987) Contribution a l'etude chimique d' *Origanum dictamnus* 2eme communication (glucosides des feulles). *Plantes Medicinales et Phytotherapie* 21(1), 56–62.

Skoula, M., Gotsiou, P., Naxakis, G. and Johnson, C.B. (1999) A chemosystematic investigation on the mono- and sesquiterpenoids in the genus *Origanum* (Labiatae). *Phytochemistry* 52(4), 649–657.

Souleles, C. (1990) Sur les flavonoides d'*Origanum dubium*. *Plantes Medicinales et Phytotherapie* 24(3), 175–178.

Stahl-Biskup, E., Inert, F., Holthuijzen, J., Stengele, M. and Schulz, G. (1993) Glycosidically bound volatiles – A review 1986–1991. *Flavour Fragrance J.* 8, 61–81.

Subramanian, S.S., Nair, A.G.R., Rodriguez, E. and Mabry, T.J. (1972) Polyphenols of the leaves of *Majorana hortensis*. *Curr. Sci.* 41(6), 202–204.

Tomas-Barberan, F.A., Greyer-Barkmeijer, R., Gil, M. and Harborne, J.B. (1988a) Distribution of 6-hydroxy, 6-methoxy- and 8-hydroxyflavone glycosides in the *Labiatae*, the *Scrophulariaceae* and related families. *Phytochemistry* 27(8), 2631–2645.

Tomas-Barberan, F.A., Husain, S.Z. and Gil, M.I. (1988b) The distribution of methylated flavones in the *Lamiaceae*. *Biochem. Syst. Ecol.* 16(1), 43–46.

Trivino, M.G. and Johnson, C.B. (2000) Season has a major effect on the essential oil yield response to nutrient supply in *Origanum majorana*. *J. Horticultural Sci. Biotechechnol.* 75(5), 520–527.

Tucker, A.O. and Marciarello, M.J. (1992) The essential oil of *Origanum laevigatum* Boiss. (Labiatae). *J. Essent. Oil Res.* 4(4), 419–420.

Tumen, G. and Baser, K.H.C. (1993) The essential oil of *Origanum syriacum* L. var. *bevanii* (Holmes) Ietswaart. *J. Essent. Oil Res.* 5(3), 315–316.

Tumen, G., Ermin, N., Ozek, T. and Baser, K.H.C. (1994) Essential oil of *Origanum solymicum* P.H. Davis. *J. Essent. Oil Res.* 6(5), 503–504.

Tumen, G., Baser, K.H.C., Kirimer, N. and Ozek, T. (1995) Essential oil of *Origanum saccatum* P.H. Davis. *J. Essent. Oil Res.* 7(2), 175–176.

Valentini, G., Arnold, N., Bellomaria, B. and Arnold, H.J. (1991) Study of the anatomy and of the essential oil of *Origanum cordifolium*, an endemic of Cyprus. *J. Ethnopharmacol.* 35(2), 115–122.

Vekiari, S.A., Oreopoulou, V., Tzia, C. and Thomopoulos, C.D. (1993) Oregano flavonoids as lipid antioxidants. *J. Am. Chem. Soc.* 70, 483–487.

Velasco-Negueruela, A., Perez-Alonso, M. and Burzaco, A. (1991) Materias aromatizantes de origen vegetal: aceites esenciales de *Thymus riatarum* y *Origanum elongatum*. *Anales de Bromatologia* 43(4), 395–400.

Voirin, B., Favre-Bonvin, J., Indra, V. and Nair, A.G.R. (1984) Structural revision of the flavone majoranin from *Majorana hortensis*. *Phytochemistry* 23, 2973–2975.

Vokou, D., Kokkini, S. and Bessiere, J.M. (1988) *Origanum onites* (Lamiaceae) in Greece: distribution, volatile oil yield, and composition. *Econ. Bot.* 42(3), 407–412.

Wagstaff, S.J., Olmstead, R.G. and Cantino, P.D. (1995). Parsimony analysis of cpDNA restriction site variation in subfamily *Nepetoideae* (Labiatae). *Am. J. Bot.* 82(7), 886–892.

Zheng, S., Wang, X., Gao, L., Shen, X. and Liu, Z. (1997) Studies on the flavonoid compounds of *Origanum vulgare* L. *Indian J. Chem.* 36, 104–106.

4 The Turkish *Origanum* species

K. Hüsnü Can Baser

INTRODUCTION

The genus *Origanum* L. (Tribe Mentheae, Family Labiatae) is represented in Turkey by 22 species or 32 taxa, 21 being endemic to Turkey (Table 4.1). The rate of endemism among the Turkish *Origanum* species is 63 per cent. Out of 52 known taxa of *Origanum*, 32 are distributed in Anatolia, meaning 60 per cent of all *Origanum* taxa are recorded to grow in Turkey. This high rate is suggestive that the gene centre of *Origanum* is Turkey (Ietswaart, 1980; Davis, 1982; Lawrence, 1984; Kokkini, 1997).

Origanum species are known as "kekik" in Turkey. Herbal parts of *Origanum* species are aromatic and are used as condiment or herbal tea. Dried *Origanum* species are also used for the production of essential oil (Origanum oil, kekik yagı) and an aromatic water or hydrosol (Origanum water, kekik suyu) (Aydin *et al.*, 1993; Baser, 1995; Baytop and Baser, 1995; Tumen *et al.*, 1995; Baser, 1998; Baytop, 1999).

Trade of *Origanum* in Turkey

Turkey has, in recent years, become a major supplier of oregano herb of which *Origanum* consists of over 90 per cent of oregano exports. *Origanum onites* (Izmir kekigi, bilyalı kekik, Turkish oregano) tops the list of commercial *Origanum* species of Turkey. It is obtained both from wild and cultivated plants. The other *Origanum* species collected from wild for commercial use are: *O. majorana* (Beyaz kekik, white oregano), *Origanum vulgare* subsp. *hirtum* (Istanbul kekigi, kara kekik, Greek oregano), *Origanum minutiflorum* (Yayla kekigi, toka kekigi, Sütçüler kekigi), *Origanum syriacum* var. *bevanii* (Israeli oregano). Recently, *Origanum acutidens*, another endemic species has entered the Turkish market.

The export of oregano from Turkey has been steadily increasing in recent years (Figure 4.1). In 1991, Turkey had exported 4633 tons of oregano for a return of just over USD8 million, the value per kg of oregano being only USD1.74. In 1994, dried oregano exports reached USD16.1 million for 6.5 million kilograms with a unit value of USD2.5 per kg. In 1999, oregano exports of Turkey exceeded the 7500 ton mark for a return of USD16.6 million. The unit value that year was USD2.2.

It is estimated that, annually, over 10 000 tons of dried oregano are harvested from wild sources in Turkey. Thousand tons are believed to be used domestically as condiment or herbal tea, and the rest is either used for essential oil production or exported. *O. onites* is cultivated in the Aegean part of Turkey in fields totalling over

Table 4.1 Origanum species of Turkey

Sectio **Amaracus** (Gleditsch) Bentham
 1. *O. boissieri* Ietswaart [E]
 2. *O. saccatum* Davis [E]
 3. *O. solymicum* Davis [E]

Sectio **Anatolicon** Bentham
 4. *O. hypericifolium* Schwartz et Davis [E]
 5. *O. sipyleum* L. [E]

Section **Brevifilamentum** Ietswaart
 6. *O. acutidens* (Hand.-Mazz.) Ietswaart [E]
 7. *O. bargyli* Mouterde
 8. *O. brevidens* (Bornm.) Dinsmore [E]
 9. *O. haussknechtii* Boiss. [E]
 10. *O. leptocladum* Boiss. [E]
 11. *O. rotundifolium* Boiss.
 12. *O. munzurense* Kit Tan et Sorger [E]
 13. *O. husnucan-baseri* H. Duman, Z. Aytaç et A. Duran [E]

Sectio **Longitubus** Ieatswaart
 14. *O. amanum* Post [E]

Sectio **Chilocalyx** (Briq.) Ietswaart
 15. *O. bilgeri* Davis [E]
 16. *O. micranthum* Vogel [E]
 17. *O. minutiflorum* Schwartz et Davis [E]

Sectio **Majorana** (Miller) Benth.
 18. *O. majorana* L. [Syn.: *O. dubium* Boiss.]
 19. *O. onites* L. [Syn.: *O. smyrnaeum* L.]
 20. *O. syriacum* var. *bevanii* (Holmes) Ietswaart [Syn.: *O. bevani* Holmes]

Sectio **Origanum** L.
 21. *O. vulgare* L. subsp. *vulgare* [Syn.: *O. creticum* L.]
 22. *O. vulgare* L. subsp. *gracile* (Koch) Ietswaart [Syn.: *O. tyttanthum* Gontsch.]
 23. *O. vulgare* L. subsp. *hirtum* (Link) Ietswaart [Syn.: *O. heracleoticum* L.]
 24. *O. vulgare* L. subsp. *viride* (Boiss.) Hayek [Syn.: *O. heracleoticum* L.]

Sectio **Prolaticorolla** Ietswaart
 25. *O. laevigatum* Boiss. [E]

Hybrids
 26. *O.* × *dolichosiphon* P.H. Davis [*O. amanum* Post × *O. laevigatum* Boiss.] [E]
 27. *O.* × *intermedium* P.H. Davis [*O. sipyleum* L. × *O. onites* L.] [E]
 28. *O.* × *symeonis* Mouterde [*O. syriacum* L. × *O. laevigatum* Boiss.] [E]
 29. *O.* × *intercedens* Rech. fil. [*O. vulgare* L. subsp. *hirtum* (Link) Ietswaart × *O. onites* L.]
 30. *O.* × *vulgare* L. subsp. *hirtum* (Link) Ietswaart × *O. micranthum* Vogel [E]
 31. *O.* × *adanense* Baser et Duman [*O. laevigatum* Boiss. × *O. bargyli* Mouterde] [E]
 32. *O.* × *majoricum* Cambess [*O. vulgare* L. subsp. *virens* (Hoffm. et Link) Ietswaart × *O. majorana* L.]

Note
[E] = Endemic.

6300 ha in the provinces of Denizli (4000 ha), Izmir (2000 ha) and Isparta (300 ha). Organic farming of *O. onites* is also practiced in Western provinces of Turkey.

Although reliable figures are hard to come by but at present Turkey exports >30 tons of Oregano oil.

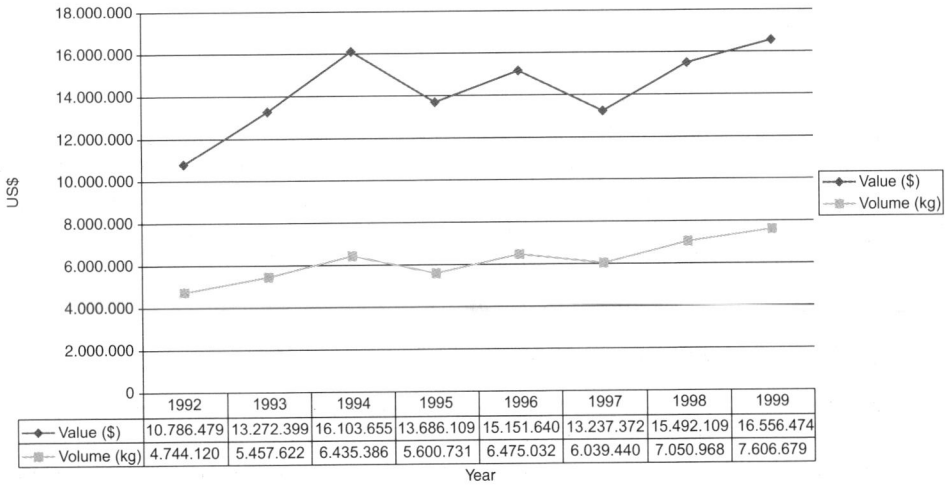

Figure 4.1 Oregano exports of Turkey

World trade of Oregano

Turkey has a dominant position in the worldwide trade of oregano. Mexico has lost its leading position over the last 15 years. Mexican oregano which is obtained not from *Origanum*, but *Lippia* species is used mainly in the United States. However, oregano imports to United States from Turkey have also increased in recent years. Latest estimates as shown in Table 4.2 put worldwide production of oregano at about 9000 tons. In 1999, over 2/3 of this amount was produced in Turkey alone. The reason that the market favours Turkish oregano is due to the expertize and skill developed in Turkey to process wild crafted oregano. The final product is clean and conforms to accepted international standards. It does not contain less than 2.5 per cent essential oil and is free from microbial contamination and foreign matter including insect fragments. Due to the good reputation of Turkish oregano, annually *c.* 650 tons of oregano is imported into Turkey for re-export after processing (Olivier, 1997; Kayhan, 2000).

Table 4.2 Recent estimates on the world oregano production (Kayhan, 2000)

	1985	1990	1995	1999
Turkey	1925	2261	4125	6655
Mexico	2735	2418	1646	1689
Others*	1170**	392	502	531
World total	5830	5158	6273	8875

Notes
* Israel and Greece.
** Albania, Morocco, Egypt.

Domestic uses of Oregano in Turkey

Oregano (kekik) is used as a condiment to flavour soups, salads, olives, chicken and red meat dishes, especially grilled or fried beef steak and lamb chops. It is also an ingredient of several sauces used as seasoning.

Oregano is also a popular herbal tea in Turkey. Oregano water rich in carvacrol is taken orally for gastrointestinal disorders, to reduce blood cholesterol and glucose levels and also for cancer. Oregano oil rich in carvacrol is used as a painkiller in rheumatism by rubbing externally on the painful limbs (Aydin et al., 1993; Baser, 1995; Baytop and Baser, 1995; Baytop, 1999).

Biological activities of Turkish Oregano

Pharmacological activities of the essential oils of *O.onites* and *O.minutiflorum* were tested. Both the oils reduced the tonus of rat stomach fundus. They also reduced the carbachol-induced contractions on isolated rat ileum and inhibited the spontaneous activity of sheep ureter. They exhibited analgesic activity in hot-plate test on mice as well. Acute toxicity of the oils were determined on albino mice. Acute toxicity tests resulted in sedation and anaesthesia followed by respiratory failure and death. LD_{50} values for the oils were found as 1.6 mL/kg for *O. onites* and as 2.4 mL/kg for *O. minutiflorum* (Cingi et al., 1992).

Origanum onites oil was shown to have hypotensive activity at 0.5 mL i.v. on rats, while origanum water showed hypertensive activity. The following positive pharmacological effects were observed with origanum water on rats: bile flow, barbiturate sleeping time, isolated ileum and aorta experiments. origanum water inhibited gastrointestinal contractions and showed choleretic effect. Acute and chronic toxicity of origanum water were tested on mice and rats. LD_{50} was found to be more than 21.9 mL/kg i.p. on mice indicating the safe use of origanum water (Aydin et al., 1997, 1997a,b,c).

Pure carvacrol ex origanum oil (99.3 per cent purity) was also tested on rat ileum and was found to possess strong antispasmodic activity. The oil of *O. onites* and carvacrol (1 mg/kg i.p.) were shown to have analgesic activity in tail-flick test on mice (Aydin et al., 1996).

Carvacrol was tested for lung tumours induced by DMBA on rats *in vivo* and was found to have strong antitumour activity at $0.1 \text{ mg} \cdot \text{kg}^{-1}$ i.p. Evidences for anti-angiogenic effects of carvacrol were observed suggesting a possible link with calcium metabolism (Zeytinoglu et al., 1998).

Additional evidence for its antitumour activity was obtained when carvacrol was tested on CO25 myoblast cells bearing a mutant human N-ras oncogene during the transformation and differentiation processes. It was found to inhibit DNA synthesis while there was no change in protein synthesis. These findings suggested that the antitumour effect of carvacrol was not due to cytotoxicity but possibly due to prevention of the prenylation of several proteins including ras (Zeytinoglu et al., 2000).

Carvacrol containing oils of *O. onites*, *Satureja hortensis* and *Thymus serpyllum* completely inhibited the growth of the following yeast causing food spoilage: *Candida tropicalis*, *Hansenula anomala*, *Kloeckera apiculata*, *Pichia membranea*, *Rhodotorula glutinis* and *Saccharomyces cerevisiae* (Kivanc and Akgul, 1989).

Table 4.3 Some biological activities of carvacrol tested at TBAM (Baser, 1998)

• Insecticidal activity	Against common grain insects (100% mortality at 10% concentration)
• Plant growth inhibition	*Lemna* test at 50 ppm. 100% inhibition and bleaching effect
• Antileishmanial activity	*Leishmania major* promastigotes *in vitro* IC$_{50}$ 3.125 µg/ml
• Bactericidal activity	Agar well diffusion protocol, 100 µg/ml. (*Bacillus cereus* 100% inh., *Pseudomonas aeruginosa* 90% inh., *Shigella boydii* 80% inh.)
• Fungicidal activity	Agar tube dilution method, 200 µg/ml against human, animal and plant pathogens (*Candida albicans* 100%, *Aspergillus niger* 100%, *Microsporum canis, Fusarium solani*).
• Mast cell degranulation activity	*Origanum onites* oil was used in this pharmacological test. The test revealed antihistaminic activity of the oil

Several oregano spices including *O. onites* and *O. vulgare* subsp. *hirtum* inhibited the following food-borne bacteria: *Bacillus cereus, Escherichia coli, Enterobacter aerogenes, Proteus vulgaris, Pseudomonas aeruginosa, Salmonella aureus, S. thypi* (Akgul and Kivanc, 1998).

O. onites exhibited complete inhibition of the following food-borne fungi: *Aspergillus flavus, Aspergillus niger, Geotrichum candidum, Mucor* sp., *Penicillium roquefortii* and some other *Penicillium* sp. The combined use of oregano and table salt (Sodium chloride) showed a synergistic antifungal effect (Akgul and Kivanc, 1998a).

The essential oils of *O. onites* and *O. vulgare* subsp. *hirtum* were found to be effective against all the molds and bacteria tested except for *P. aeruginosa*. The other microorganisms tested were *Bacillus subtilis, Staphylococcus aureus, E. coli, A. niger, Aspergillus candidus, P. chrysogenum, Fusarium* sp. (Dortunc and Cevikbas, 1992).

The oil of *O. minutiflorum* as well as carvacrol and thymol resulted in complete inhibition of the following fungi: *Fusarium moniliforme, Rhizoctonia solani, Sclerotinia sclerotiorum, Phytophthora capsici* (Muller-Riebau *et al.*, 1995).

Carvacrol and thymol at 100–400 ppm completely inhibited the growth of most food-borne bacteria. Carvacrol showed the highest antibacterial activity. The combination of carvacrol, cuminaldehyde, sodium chloride and ascorbic acid showed the greater inhibitory effect against all the bacteria tested including *P. aeruginosa* (Kivanc and Akgul, 1998).

Carvacrol (500 mg) at 28 °C completely inhibited the growth and toxin production of *A. flavus* and *Aspergillus parasiticus*. Aflatoxin production was inhibited to a greater degree than that for the growth. *A. flavus* was more sensitive to carvacrol than *A. parasiticus* (Akgul *et al.*, 1991).

Some other biological activities of carvacrol and *O. onites* oil tested at TBAM are summarized in Table 4.3 (Baser, 1998).

Essential oils of Turkish *Origanum* taxa

Twenty-five *Origanum* taxa growing in Turkey have been chemically studied by our group at TBAM.

Origanum onites

O. onites (Syn. *O. smyrnaeum* L.) is the most widely traded *Origanum* species in Turkey. It consists of over 80 per cent of all the oregano exports of Turkey. This species is mainly wildcrafted, but also cultivated in areas exceeding 6000 ha in Denizli, Izmir and Isparta provinces in Western Turkey. During our research into the essential oils of *O. onites*, carvacrol (67–82 per cent) was found as main constituent in 13 samples. However, four samples from Antalya and Mugla provinces in Southern Turkey yielded oils rich in linalool (80–92 per cent). Mixed types were also encountered in three samples containing carvacrol (36–66 per cent) and linalool (15–52 per cent) as main constituents. In one sample, carvacrol (57 per cent) and thymol (12 per cent) were found as main constituents (Table 4.4) (Ogutveren *et al.*, 1992; Baser *et al.*, 1993; Tumen *et al.*, 1995; Kirimer *et al.*, 1995; Boydag *et al.*, 2000) Composition of the oil of *O. onites* of Turkish origin has also been previously reported (Gurgen, 1948; Tanker, 1965; Marquard *et al.*, 1966; Ceylan, 1976; Buil *et al.*, 1977; Taysi *et al.*, 1978; Vomel and Ceylan, 1979; Scheffer *et al.*, 1986; Akgul and Bayrak, 1987).

Lipids of *O. onites* have also been studied by our group. The so-called "cyclone powder" which is a waste material of industrial origanum herb processing in origanum mills has at present no commercial value. This fine powder was worked up for the recovery of essential oil and lipids. Cyclone powder yielded by hydrodistillation 1.7 per cent essential oil which contained carvacrol (63 per cent) and thymol (13 per cent) as main constituents. The yield of total lipids varied between 4.2 and 8.4 per cent according to the solvent system used in the extraction. This extract was fractionated into neutral, glyco- and phospholipids and their components were characterized. Main fatty acid components in origanum herb and the cyclone powder were linolenic acid (45 per cent, 19 per cent), palmitic acid (23 per cent, 39 per cent), linoleic acid (14 per cent, 13 per cent), oleic acid (8 per cent, 14 per cent). Possible use of origanum lipids in shampoos and other cosmetic preparations was suggested (Azcan, 1998; Azcan *et al.*, 2000; Azcan *et al.*, 2000a).

Kekik water

Kekik water has in recent years become a commercial commodity in Turkey. It is the aromatic water obtained after removing essential oil from the distillate of oregano herbs. Although kekik water has been known and produced mainly in western and southern parts of Turkey in villages for use as a household remedy, rising demand for it especially in urban areas has forced commercial origanum oil distillers to market it in

Table 4.4 Essential oil composition of Turkish *O. onites*

	Oil Yield (%)	Carvacrol (%)	Thymol (%)	γ-terpinene (%)	p-cymene (%)	Linalool (%)
Carvacrol-type* (13 samples)	1.1–4.7	66.5–81.9	0.2–1.9	1.6–8.7	3.0–10.9	0.04–1.9
Linalool-type (4 samples)	2.6–4.0	0.3–13.4	0.0–0.05	0.03–0.1	0.01–0.3	79.9–91.9**
Mixed-type (3 samples)	2.5–2.8	36.3–66.0	0.04–3.6	1.3–4.1	1.5–4.9	14.8–51.8
*Exception (1 sample)	2.4	57.4	11.6	5.2	8.1	0.1

Note
** Antalya, Mugla.

cities. Reliable statistics are hard to come by on the local consumption of kekik water in Turkey, but its monthly consumption can be estimated as 50 tons (Aydin et al., 1997).

For the preparation of kekik water in villages, a cauldron is placed over a fireplace. Plant material is placed in the cauldron and water is added to cover the plant line. At the inner centre of the cauldron an elevated pot is placed over the plant material. The lid of the cauldron is turned upside down on top and the shallow part is filled with cold water. During distillation, aromatic vapour condenses on the cold surface of the lid and collects in the pot. Distillation is terminated when the pot is filled with the distillate. Essential oil floating on the distillate is scooped out with a spoon and transferred to a bottle. The aromatic water so prepared is cooled and consumed (Baser, 1995; Aydin et al., 1997).

A more sophisticated distillation still which is also used for the production of essential oil, aromatic water and alcoholic drinks especially in mountain villages consists of two copper or tin vessels fitting on top of the other. The bottom vessel (c. 9 L) contains plant material and water. The smaller capacity (c. 3 L) top vessel has a conical shape inside. Outside the cone cooling water is run through the inlet and outlet pipes. Bottom circle of the cone is lined with a trough in which the distillate accumulates and run out through a pipe (Baytop and Baser, 1995; Aydin et al., 1997).

Several kekiks including *Origanum*, *Thymbra* and *Satureja* species are used in the preparation of Kekik water, the commercial kekik water is produced mainly from *O. onites*.

GC/MS analysis of the SPME-trapped volatiles of kekik water *ex O. onites* gave the following composition: Carvacrol (63.8 per cent), thymol (9.6 per cent), cis-p-menth-4-ene-1,2-diol (8.7 per cent) and linalool (5.6 per cent) as main constituents. *n*-Hexane and chloroform extracts of kekik water contained besides carvacrol as main constituent also 3,7-dimethyl-1-octen-3,7-diol, cis-p-menth-3-ene-1,2-diol and cis-p-menthan-1,8-diol (Aydin et al., 1997). Recently, we have repeated the experiment with kekik water *ex O. onites* of a different origin. *n*-Hexane extraction, headspace and immersion SPME studies were carried out. Headspace SPME experiments were conducted with or without the application of heat. Main constituent was determined as carvacrol in all the experiments. Minimum and maximum carvacrol contents (82.5 per cent and 89.8 per cent) were found in not-heated and heated samples, resp. Immersion SPME technique gave a carvacrol content of 88.1 per cent (Boydag et al., 2000).

n-Hexane extract of the distillate gave a carvacrol content of 89.8 per cent similar to that of the heated sample by headspace-SPME. Chloroform extract of the distillate already extracted with *n*-hexane gave a totally different composition with cis-p-menth-4-ene-1,2-diol (87.7 per cent) as the main constituent. Carvacrol content in this extract was 12.3 per cent (Boydag et al., 2000).

Origanum vulgare subsp. *hirtum*

This species (syn.: *O. heracleoticum* L.) is distributed in the Western and rarely Southern coastal provinces of Turkey. It is used mainly as a condiment and herbal tea and not commonly used for essential oil production. Thirty-eight samples of this species have been studied by our group. 1.1–6.5 per cent of essential oil were obtained with carvacrol, a monoterpenic phenol isomeric with thymol, being the main constituent (23–79 per cent) of the oils. The results are summarized in Table 4.2. During the work, a simple exercise was carried out to see whether there was a correlation between the oil yield and the carvacrol content. It is known that thymol and carvacrol are biosynthesized from γ-terpinene through p-cymene. Therefore, for each oil, carvacrol + thymol (**A**) and p-cymene + γ-terpinene (**B**) percentage values were totalled and the ratio of A/B was calculated. When these ratios were extrapolated against the yield a quite meaningful correlation was observed. The conclusion was *the larger the ratio, the higher the carvacrol (+ thymol) content, the better the oil yield* (Tumen et al., 1995; Baser et al., 1993, 1994). Essential oil composition of *O. vulgare* subsp. *hirtum* of Turkish origin has also been reported by other groups (Tanker, 1965; Scheffer et al., 1986; Akgul and Bayrak, 1987).

Oil yield (%)	A/B ratio	Exception
1.1–5.1	1.1–6.4	7.4 (3.6%)
5.2–6.5	7.3–9.1	–

Origanum minutiflorum

O. minutiflorum is an endemic species with limited distribution on Western Taurus mountains stretching in a narrow band between Isparta and Antalya provinces. It is mainly collected in Sütçüler and Çandir towns of Isparta. Wild-crafting of this endemic herb has caused some concern among the environmentalists, however, an ingenious unwritten protocol between the local Forestry Department and five village cooperatives has averted the treat of extinction. The protocol requires the collectors not to start collecting the herb before a certain date which is decided and announced every year. It is always the end of August and the beginning of September. This is the period the plant reaches maturity and sets seed. After the permission is granted the peasants move to the mountain plains with their families and start harvesting and drying. They are given clear instructions to cut the stem with a sharp knife or chisel and not to use bare hands to pull the plant to avoid uprooting. This practice has been successful and is shown as an example for sustainable wild-harvesting. The crop is then purchased by the village cooperatives and sold to merchants. The plant is rich in essential oil as well as carvacrol and therefore is a commodity of trade (Table 4.5) (Baser et al., 1991, 1993; Tumen et al., 1995; Muller-Riebau, 1995).

Origanum syriacum var. *bevanii*

This species which grows in south-eastern Mediterranean parts of Turkey is also wild crafted for export. Essential oil from a sample collected in Kahraman Maras contained 3.7 per cent oil rich in carvacrol (43 per cent), thymol (25 per cent) and γ-terpinene (13 per cent) (Baser et al., 1993; Tumen and Baser, 1993; Tumen et al., 1995; Kirimer et al., 1995). In previous studies, carvacrol was also reported as main constituent from other Turkish materials (Marquard et al., 1966; Scheffer et al., 1986; Akgul and Bayrak, 1987). Oils of Israeli origin were reported to contain either thymol (60 per cent) or

Table 4.5 Main components of the essential oil of *O. minutiflorum*

No. samples studied	Oil yield (%)	Carvacrol (%)	Thymol (%)	γ-terpinene (%)	p-cymene (%)	Linalool (%)
6	1.1–3.8	42–84	0.02–3	3–11	4–17	0.06–0.2

carvacrol (80 per cent) as main components, while carvacrol was the main constituent (77 per cent) in the oil of Egyptian origin (Fleisher and Fleisher, 1991).

Origanum majorana

Origanum majorana has two chemotypes in Turkey. The type commonly known as Sweet marjoram or Mercanköşk in Turkish, which contains terpinen-4-ol, *cis*-sabinene hydrate and linalool in its oil is cultivated in gardens as a culinary herb in the Aegean region of Turkey. The type growing wild in Antalya and İçel provinces in Southern Turkey, however, is known as Beyaz kekik (White oregano) and yields 5–8 per cent oil rich in carvacrol. This is a steady character and no chemotypes of this species has been encountered (Table 4.6) (Baser *et al.*, 1993, 1993a; Tumen *et al.*, 1995; Kirimer *et al.*, 1995). Previous reports also confirm this finding (Marquard *et al.*, 1966; Sarer *et al.*, 1982, 1985).

This taxon was named *O. dubium* Boiss. before Ietswaart included it under *O. majorana* in 1980 in his treatise and while writing the chapter *Origanum* in the Flora of Turkey and the East Aegean Islands two years later (Ietswaart, 1980; Davis, 1982). However, these two species were kept separate in the Flora of Cyprus of Meikle 3 years later emphasizing the oil-rich and carvacrol-rich nature of *O. dubium* giving reference to Ietswaart's revision (Meikle, 1985).

Chemical results have led us to consider that the status of *O. dubium* as a distinct species may be reinstated after careful examination of its morphological, anatomical, caryological, palinological, cytological and chemical features. Such a revision is hoped to alleviate a lot of confusion since *O. majorana* is widely known as Sweet marjoram. Beyaz kekik, in Turkey, is the preferred material for origanum oil production due to high oil yield and high carvacrol content.

A hybrid of *O. majorana* and *O. vulgare* subsp. *virens*, named *O. majoricum* has been cultivated in home gardens of Western Turkey as a culinary herb like sweet marjoram. This is an introduced hybrid to Turkey since *O. vulgare* subsp. *virens* is not a native plant. Its essential oil composition is similar to that of *O. majorana* of sweet marjoram type (Table 4.6).

Origanum acutidens and *O. bilgeri*

Other *Origanum* species rich in carvacrol are *O. acutidens* and *O. bilgeri*. Both species are endemic and the latter has a very narrow distribution near Geyik dagı in Antalya province. *O. acutidens*, however, has a wider distribution in Eastern Anatolia. This

Table 4.6 Main components of the essential oil of *O. majorana* and its hybrid

	No. of samples studied	Oil yield (%)	Carvacrol (%)	Thymol (%)	γ-terpinene (%)	p-cymene (%)	Linalool (%)
Carvacrol type	5	5.2–7.8	32–84	0.5–2	0.7–12	4–12	0.2–12
Sweet marjoram type			Cis-Sabinene hydrate (%)	Terpinen-4-ol (%)	Linalool (%)	Carvacrol (%)	
	4	1.1–2.3	30–44	8–14	2–5	0.6–2	
O. × majoricum	6	0.7–1.9	24–37	6–13	2–6	7–18	

Table 4.7 Main components of the essential oils of O. acutidens and O. bilgeri

Species	Oil yield (%)	Carvacrol (%)	p-cymene (%)	γ-terpinene (%)	Thymol (%)	No. of samples studied
O. acutidens	1.4	66	14–17	2–4	4–12	2
O. bilgeri	1.3	66	6	5	4	1

species has recently found its way into local markets in Turkey. Both species contain 1.3 per cent essential oil with carvacrol (66 per cent) as the main constituent (Table 4.7) (Baser et al., 1996, 1997).

Origanum hypericifolium

Origanum hypericifolium is a rare endemic species confined to south-west Anatolia. Air dried herbal parts collected before flowering are locally used in Burdur: Gölhisar, Evciler village as a condiment at breakfast and for flavouring meat. They also use it as herbal tea to cure the common cold, for stomach ache and for debility. The plants are never collected while flowering. This was quite obvious from the analysis of the oils. The best oil yield and highest carvacrol content was obtained from the plant collected before flowering whereas for plants collected during early flowering and flowering stages, the oil yield was poor and the amount of *p*-cymene was found to increase while carvacrol content was reduced (Table 4.8) (Baser et al., 1994a).

Origanum sipyleum

Origanum sipyleum is a polymorphic endemic species growing in Western half of Turkey. The analysis of essential oils obtained from 12 samples collected in different provenance showed that no single chemical could characterize the species. Although, thymol and carvacrol were present in all the samples examined main constituents varied from *p*-cymene, γ-terpinene, β-caryophyllene, myrcene to thymol methylether. Oil yield varied between 0.1–1.7 per cent (Table 4.9) (Baser et al., 1992). Its essential oil composition was previously reported (Sezik and Basaran, 1989).

Table 4.8 Main components of the essential oils of O. hypericifolium

	Oil yield (%)	Carvacrol (%)	p-cymene (%)	γ-terpinene (%)	Thymol (%)
Before flowering	2.5	64.3	11.7	11.1	0.4
Early flowering					
I	1.4	47.3	22.9	8.4	8.3
II	0.9	33.9	36.1	10.1	0.6
Flowering					
I	0.8	10.0	47.8	17.3	0.9
II	0.6	8.2	62.1	1.4	7.6
III	3.3	1.0	36.0	40.0	5.0

Table 4.9 Main components of the essential oil of O. sipyleum

Main components (%)	Samples studied
p-Cymene (28–40)	4
γ-terpinene (23–34)	4
β-caryophyllene (17–22)	2
Myrcene (37)	1
Thymol methylether (20)	1
Thymol (tr–16)	12
Carvacrol (1–12)	12

Origanum × adanense and its parents

Parents of this recently described endemic hybrid were identified as O. bargyli and O. laevigatum (Duman et al., 1998). The analysis of their essential oils confirmed the identity of the parents (Baser et al., 2000). One of the parents, O. laevigatum is highly poor in oil which contains sesquiterpene hydrocarbons, bicyclogermacrene (38 per cent) and germacrene-D (22 per cent) as main constituents. Carvacrol content of the oil is 0.6 per cent (Tucker and Maciarello, 1992; Baser et al., 1996a). The other parent, O. bargyli, on the other hand, yields over 1 per cent oil containing a much higher percentage of carvacrol (15 per cent) and lower percentages of bicyclogermacrene (2.4 per cent) and germacrene-D (0.6 per cent) (Baser and Duman, 1998). The hybrid O. × adanense yields relatively poor oil (0.2 per cent) which contains carvacrol (17.3 per cent), bicyclogermacrene (9.3 per cent) and germacrene-D (3 per cent) as major constituents (Table 4.10) (Baser et al., 2000).

Origanum × dolichosiphon and it parents

Origanum × dolichosiphon is an endemic hybrid of O. amanum and O. laevigatum in Turkey. The material collected from Hatay province was distilled to yield 0.04 per cent oil which contained bicyclogermacrene (20 per cent), β-caryophyllene (13 per cent) and germacrene-D (11 per cent) as main constituents. O. laevigatum is the only Origanum species containing bicyclogermacrene as main constituent in its oil. As seen in O. × adanense, this hybrid also seems to have inherited the bicyclogermacrene-rich oil charactered from O. laevigatum. Since we have not yet analyzed the oil of O. amanum, I am not in a position to comment on its chemical constituents.

Table 4.10 Main constituents of the essential oil of O. × adanense and its parents

	Oil yield (%)	Carvacrol (%)	Bicyclogermacrene (%)	Germacrene-D (%)
O. bargyli	1.1	14.7	2.4	0.6
O. × adanense	0.2	17.3	9.3	3.0
O. laevigatum	0.03	0.6	37.9	21.7

Origanum × intercedens and its parents

Origanum × intercedens is a hybrid from *O. onites* and *O. vulgare* subsp. *hirtum*, first described from the island of Evoia in Greece. It is an Eastern Mediterranean element also growing in Western Turkey. The hybrid is rich in oil which contains carvacrol (46 per cent) as main constituent (Table 4.11). The Greek material was reported to contain carvacrol (85 per cent) also as the main constituent (Kokkini and Vokou, 1993).

Table 4.11 Main components of the oil of *O. × intercedens* and its parents

	Oil yield (%)	Carvacrol (%)	Thymol (%)	γ-terpinene (%)	p-cymene (%)
O. vulgare ssp. *hirtum*	3.2	52.1	13.2	12.1	7.4
O. × intercedens	4.3	45.9	2.4	13.8	9.3
O. onites	2.6	66.5	1.2	5.5	7.3

Notes
1. All the materials were collected in July 1995 from Mugla: Kiran village.
2. In the only previous study, carvacrol (85%) was reported as main constituent in the oil of *O. × intercedens* from Greece.

Table 4.12 *Origanum* species of Turkey with oils rich in *p*-cymene

Species	Oil yield (%)	p-cymene (%)	Samples studied	References
boissieri [E]	0.3	43	1	Baser and Duman, 1998
haussknechtii [E]	0.2	16	1	Baser *et al.*, 1998b
leptocladum [E]	0.2	48	1	Baser *et al.*, 1996b
saccatum [E]	1.3–1.4	61–84	2	Tumen *et al.*, 1995a
solymicum [E]	0.2	53	1	Tumen *et al.*, 1994

Note
[E] = Endemic.

Table 4.13 Main components of the monoterpene-rich Origanum oils of Turkey

Species	Oil yield (%)	Main components (%)	Sample studied	References
husnucan-baseri [E]	0.13	Borneol (20)	1	Baser *et al.*, 1998c, Kurkcuoglu *et al.*, 1997
micranthum [E]	0.5	Linalyl acetate (12)	2	Baser *et al.*, 1996c
	0.4	Linalyl acetate (10*)		
rotundifolium	0.3	*cis*-Sabinene hydrate (22)	1	Baser *et al.*, 1995
vulgare ssp. *viride***	0.08	Linalool (21) Germacrene-D (11)	1	Sezik *et al.*, 1993

Notes
* Other major components are: linalool (9–11%), α-terpineol (10–11%), carvacrol (0.1–11%).
** In one sample germacrene-D (16%) was the main constituent.
[E] = Endemic.

Table 4.14 Origanum species of Turkey with oils rich in sesquiterpenes (Sezik et al., 1993)

Species	Oil yield (%)	Main components (%)	Sample studied
vulgare ssp. viride*	0.13	Germacrene-D (16)	1
vulgare ssp. vulgare	0.08	Germacrene-D (18–22) β-caryophyllene (18–20)	4
vulgare ssp. gracile	0.04	β-caryophyllene (18–20)	2

Note
* In one sample linalool (21%) was the main constituent together with Germacrene-D (11%).

Table 4.15 Cumulative results of previous reports on Origanum oils of Turkey

Species	Oil yield (%)	Main components (%)	References
onites	1.8–3.4	Carvacrol (47–74) Carvacrol + thymol (51–74)	Buil et al., 1977 Akgul and Bayrak, 1987 Ceylan, 1976 Scheffer et al., 1986 Tanker, 1965 Taysi et al., 1978 Vomel and Ceylan, 1979 Gurgen, 1948 Marquard et al., 1966
vulgare ssp. hirtum	2.3–5.4	Carvacrol (52–61)	Akgul and Bayrak, 1987 Scheffer et al., 1986 Tanker, 1965
majorana	7.5	Carvacrol (65–74)	Marquard et al., 1966 Sarer et al., 1982 Sarer et al., 1985
syriacum var. bevanii	2.4–3.4	Carvacrol (63–64)	Akgul and Bayrak, 1987 Scheffer et al., 1986 Marquard et al., 1966
leptocladum	1.0	Carvacrol (43)	Tanker et al., 1986*
saccatum	1.4	p-cymene (24)	Sezik and Basaran, 1986**
laevigatum	0.005	Bicyclogermacrene (25) Germacrene-D (21)	Tucker and Maciarello, 1992***

Notes
* Oil of this species is rich in p-cymene (48%).
** Our studies indicated much higher percentage of p-cymene.
*** This plant was raised in the United States from seeds collected in Turkey.

p-Cymene-rich Origanum oils of Turkey

Several endemic *Origanum* species of Turkey have been shown to contain p-cymene as main constituent in their oils (Table 4.12).

Origanum oils of Turkey rich in other monoterpenes

Origanum species of Turkey with oils rich in other monoterpenes are shown in Table 4.13. Among the species included in this table, the recently described *Origanum*

husnucan-baseri is characterized by a unique composition of its oil with borneol (20 per cent), α-terpineol (12 per cent) and *trans*-sabinene hydrate (11 per cent) as major constituents. Carvacrol (1 per cent) is a minor component of its oil (Duman *et al.*, 1995; Kurkcuoglu *et al.*, 1997; Baser *et al.*, 1998c).

The oil of *O. micranthum*, an endemic species, is characterized by the presence of linalyl acetate, linalool and α-terpineol (Baser *et al.*, 1996c).

Origanum oils of Turkey rich in sesquiterpenes

Oil-poor subspecies of *O. vulgare* contained sesquiterpene hydrocarbons such as β-caryophyllene and germacrene-D as main constituents (Table 4.14). *Origanum laevigatum*, also an oil-poor species mentioned earlier, contains bicyclogermacrene and germacrene-D as main constituents (Tucker and Maciarello, 1992; Baser *et al.*, 1996a).

Previous reports on the composition of Origanum oils of Turkey

Although the most comprehensive study on the oils of Turkish *Origanum* species has been carried out by our group, the oil compositions of eight *Origanum* taxa were previously reported. Cumulative results of these reports are given in Table 4.15.

CONCLUSIONS

The genus *Origanum* is a main contributor to the oregano trade. Turkey has, in recent years, become a major supplier of *Origanum* herb and its oil to the world markets. Due to its high quality, the Turkish *Origanum* products are highly esteemed and preferred.

These facts have prompted us to study the chemistry and biological activities of Turkish *Origanum* species as reported in this chapter. The results suggest that diverse biological activities and the trade potential of *Origanum* herbs and oils as well as their main constituents with simple chemistry should not be underestimated. Especially, the monoterpenic phenol carvacrol, the main constituent of commercial oreganos should be given special emphasis since its low toxicity and surprisingly high and diverse biological activities render this simple molecule a promising lead for the development of novel medicines not only for humans but also for animals and plants. Carvacrol-rich oils are also used for their high potency especially as antibacterial and antifungal agents. The use of carvacrol-rich oils as ingredients in animal feed and for the preservation of food against bacterial or fungal spoilage and as antioxidants appears to increase (Nguyen *et al.*, 1991; Deryabin, 1991; Michl, 1993; DeLuca *et al.*, 1999; Ninkov, 1999; Baraka, 2000; Nitsas, 2000).

In a recent paper, an attempt was made towards the chemosystematic investigation of *Origanum* taxa based on essential oil constituents in the published literature. The authors have concluded that most *Origanum* species are rich either in the so-called sabinyl compounds or cymyl compounds but never both (Skoula *et al.*, 1999). Although the occurrence of chemotypes is not common in the genus *Origanum*, several hybrids have been discovered. However, the present data on oil compositions of *Origanum* taxa are not sufficient enough to draw conclusions for the use of essential oil constituents in the chemotaxonomy of the genus *Origanum*.

ACKNOWLEDGEMENT

I am grateful to all my colleagues who have enthusiastically contributed to the *Origanum* story reported in this chapter.

REFERENCES

Akgul, A. and Bayrak, A. (1987) Constituents of essential oils from *Origanum* species growing wild in Turkey. *Planta Med.* 51, 114.

Akgul, A. and Kivanc, M. (1988) Inhibitory effects of six Turkish thyme-like spices on some common food-borne bacteria. *Die Nahrung* 32, 201–203.

Akgul, A. and Kivanc, M. (1988a) Inhibitory effects of selected Turkish spices and oregano components on some food-borne fungi. *Intern. J. Food Microbiol.* 6, 263–268.

Akgul, A., Kivanc, M. and Sert, S. (1991) Effect of carvacrol on growth and toxin production by *Aspergillus flavus* and *Aspergillus parasiticus*. *Sciences des Aliments* 11, 361–370.

Aydin, S., Ozturk, Y. and Baser, K.H.C. (1993) Ethnopharmacological investigations on *Origanum onites* L. growing in the Aegean region, 10th Symposium on Plant Drugs, 20–22 May 1993, İzmir, Turkey.

Aydin, S., Baser, K.H.C. and Ozturk, Y. (1997a) The chemistry and pharmacology of *Origanum* (Kekik) Water. In Ch. Franz, A. Mathe and G. Buchbauer (eds), *Proceedings of the 27th International Symposium on Essential Oils*, Essential Oils: Basic and Applied Research, 8–11 September 1996, Vienna, Austria, pp. 52–60.

Aydin, S., Ozturk, Y. and Baser, K.H.C. (1997) Effect of the essential oil of *Origanum onites* L. on cardiovascular system. In M. Coskun (ed.), *Proceedings of the 11th Symposium on Plant Drugs*, Ankara, pp. 332–338.

Aydin, S., Ozturk, Y. and Baser, K.H.C. (1997b) Cardiovascular actions of kekik (*Origanum onites* L.) aqueous distillate. In M. Coskun (ed.), *Proceedings of the 11th Symposium on Plant Drugs*, Ankara, pp. 339–344.

Aydin, S., Ozturk, Y. and Baser, K.H.C. (1997c) Choleretic effect of *Origanum* water ex *Origanum onites* L. In M. Coskun (ed.), *Proceedings of the 11th Symposium on Plant Drugs*, Ankara, pp. 345–351.

Aydin, S., Ozturk, Y., Beis, R. and Baser, K.H.C. (1996) Investigation of *Origanum onites*, *Sideritis congesta* and *Satureja cuneifolia* oils for analgesic activity. *Phytother. Res.* 10, 342–344.

Azcan, N. (1998) *Origanum onites* L. ve kekik siklon tozunun lipitleri ve kekik siklon tozunun degerlendirilmesi, Ph.D. Dissertation, Anadolu University, Eskisehir, Turkey.

Azcan, N., Kara, M., Ozek, T. and Baser, K.H.C. (2000) The evaluation of *Origanum* cyclone powder, an industrial waste, 4th National Congress of Chemical Engineering, 4–7 September 2000, Istanbul, Turkey.

Azcan, N., Kara, M., Asilbekova, D.T., Ozek, T. and Baser, K.H.C. (2000a) Lipids and essential oil of *Origanum onites* L. *Khim. Prir. Soedin.* 106.

Baraka, M.W. (2000) Herbal compositions for diabetes and method of treatment. *United States Patent* 6, 042, 834, 28 March 2000.

Baser, K.H.C. (1995) Essential oils from aromatic plants which are used as herbal tea in Turkey. In K.H.C. Baser (ed.), *Proceedings of the 13th International Congress of Flavours, Fragrances and Essential Oils*, Istanbul, Turkey, 15–19 October 1995, AREP Publ., Istanbul, 2, 67–79.

Baser, K.H.C. (1998) The Turkish Oregano, 3rd MEDUSA Workshop, 27–28 April 1998, Coimbra, Portugal.

Baser, K.H.C., Tumen, G. and Duman, H. (1996) The essential oil of *Origanum bilgeri* P.H. Davis. *J. Essent. Oil Res.* 8, 217–218.

Baser, K.H.C., Tumen, G. and Duman, H. (1997) The essential oil of *Origanum acutidens* (Hand.-Mazz.) Iestwaart. *J. Essent. Oil Res.* 9, 91–92.

Baser, K.H.C., Tumen, G., Ozek, T. and Kürkcuoglu, M. (1992) Composition of the essential oil of *Origanum sipyleum* of Turkish origin. *J. Essent. Oil Res.* 4, 139–142.

Baser, K.H.C., Ermin, N., Kurkcuoglu and Tumen, G. (1994) The essential oil of *Origanum hypericifolium* O. Schwarz et P.H. Davis. *J. Essent. Oil Res.* 6, 631–633.

Baser, K.H.C., Duman, H. and Aytac, Z. (2000) Composition of the Essential Oil of *Origanum* × *adanense* Baser et Duman. *J. Essent. Oil Res.* 12, 475–477.

Baser, K.H.C., Ozek, T., Kurkcuoglu, M. and Tumen, G. (1996a) The essential oil of *Origanum laevigatum* Boiss. *J. Essent. Oil Res.* 8, 185–186.

Baser, K.H.C. and Duman, H. (1998) Composition of the essential oils of *Origanum boissieri* Ietswaart and *Origanum bargyli* Mauterde. *J. Essent. Oil Res.* 10, 71–72.

Baser, K.H.C., Kurkcuoglu, M. and Tumen, G. (1998b) Composition of the essential oil of *Origanum haussknechtii* Boiss. *J. Essent. Oil Res.* 10, 227–228.

Baser, K.H.C., Kurkcuoglu, M., Duman, H. and Aytac, Z. (1998c) Composition of the essential oil of *Origanum husnucan-baseri* H. Duman, Z. Aytaç et A. Duran, a new species from Turkey. *J. Essent. Oil Res.* 10, 419–421.

Baser, K.H.C., Ozek, T., Kurkcuoglu, M. and Tumen, G. (1996c) The essential oil of *Origanum micranthum* Vogel. *J. Essent. Oil Res.* 8, 203–204.

Baser, K.H.C., Ozek, T. and Tumen, G. (1995) The essential oil of *Origanum rotundifolium* Boiss. *J. Essent. Oil Res.* 7, 95–96.

Baser, K.H.C., Ozek, T., Tumen, G. and Sezik, E. (1993) Composition of the essential oils of Turkish *Origanum* species with commercial importance. *J. Essent. Oil Res.* 5, 619–623.

Baser, K.H.C., Kirimer, N. and Tumen, G. (1993a) Composition of the essential oil of *Origanum majorana* L. from Turkey. *J. Essent. Oil Res.* 5, 577–579.

Baytop, T. and Baser, K.H.C. (1995) On essential oils and aromatic waters used as medicine in Istanbul between 17th and 19th centuries. In K.H.C. Baser (ed.), *Proceedings of the 13th International Congress of Flavours, Fragrances and Essential Oils*, Istanbul, Turkey, 15–19 October 1995, AREP Publ., Istanbul, 2, 99–107.

Baytop, T. (1999) Türkiye'de Bitkiler ile Tedavi, Geçmiste ve Bugün (Therapy with Medicinal Plants in Turkey, Past and Present), Nobel Tip Kitabevleri, Istanbul.

Boydag, I., Kurkcuoglu, M. and Baser, K.H.C. (2000) The headspace and immersion type SPME trapping of volatiles in the aromatic water of *Origanum onites*, 31st International Symposium on Essential Oils, 10–13 September 2000, Hamburg, Germany.

Buil, P., Garnero, J., Guichard, G. and Konur, Z. (1977) Sur quelques huiles essentielles enprovenance de Turquie. *Revista Ital.* 59, 379–384.

Ceylan, A. (1976) *Origanum smyrnaeum* L. 'da verim ve ontogenetik varyabilite. *Ege Üni. Ziraat Fak. Dergisi* 13(2), 139–143.

Cingi, M.I., Kirimer, N., Sarikardasoglu, I., Cingi, C. and Baser, K.H.C. (1992) Pharmacological activities of the essential oils of *Origanum onites* and *Origanum minutiflorum*. In K.H.C. Baser (ed.), *Proceedings of the 9th Symposium on Plant Drugs*, Eskisehir, pp. 10–15.

Davis, P.H. (1982) Flora of Turkey and the East Aegean Islands, Vol. 7, Edinburgh University Press, Edinburgh.

DeLuca, D.L., Sparks, W.S., Ronzio, R.A. and DeLuca, D.R. (1999) Oregano for the treatment of internal parasites and protozoa. *United States Patent* 5, 955, 086, 21 September 1999.

Deryabin, A.M. (1991) Medicinal agent and method for treatment of mastitis in animals and humans. *United States Patent* 5, 061, 491.

Dortunc, T. and Cevikbas, A. (1992) Investigations on the antibacterial and antifungal effects of some volatile oils. *J. Pharm. Univ. Marmara* 8, 117–128.

Duman, H., Baser, K.H.C. and Aytac, Z. (1998) Two new species and a hybrid from Anatolia. *Tr. J. Botany* 22, 51–55.

Duman, H., Aytac, Z., Ekici, M., Karaveliogullari, F.A., Donmez, A. and Duran, A. (1995) Three new species (Labiatae) from Turkey. *Flora Mediterranea* 5, 221–228.

Fleisher, A. and Fleisher, Z. (1991) Chemical composition of *Origanum syriacum* L. essential oil. Aromatic plants of the Holy Land and the Sinai, Part V. *J. Essent. Oil Res.* 3, 121–123.

Gurgen, A. (1948) Türkiyenin önemli eteri yaglari üzerinde arastirmalar II. *Ankara Yüksek Ziraat Enst. Dergisi* 9(2), 332–360.

Ietswaart, J.C. (1980) A Taxonomic Revision of the Genus *Origanum* (Labiatae), Leiden Botanical Series, Leiden University Press, The Hague.

Kayhan, C. (2000) KÜTAS, Izmir, Personal communication.

Kirimer, N., Baser, K.H.C. and Tumen, G. (1995) Carvacrol rich plants in Turkey. *Khim. Prir. Soedin.* pp. 49–54, *Chem. Nat. Comp.* 31, 37–41.

Kivanc, M. and Akgul, A. (1988) Effects of some essential oil components on the growth of food-borne bacteria and synergism with some food ingredients. *Flav. Fragr. J.* 3, 95–98.

Kivanc, M. and Akgul, A. (1989) Inhibitory effects of spice essential oils or yeasts, Doga. *Tu. J. Agri. Forest* 13, 68–71.

Kokkini, S. and Vokou, D. (1993) The hybrid *Origanum* × *intercedens* from the island of Nisyros (SE Greece) and its parental taxa; Comparative study of essential oils and distribution. *Biochem. Syst. Ecol.* 21, 397–403.

Kokkini, S. (1997) Taxonomy, Diversity and Distribution of *Origanum* species. In S. Padulosi (ed.), *Oregano*, IPGRI, Rome, pp. 2–12.

Kurkcuoglu, M., Baser, K.H.C. and Duman, H. (1997) The essential oils of three new Labiatae taxa from Turkey: *Origanum husnucan-baseri*, *Sideritis gulendamii* and *Salvia aytachii*. In Ch. Franz, A. Mathe and G. Buchbauer (eds), *Proceedings of the 27th International Symposium on Essential Oils*, Essential Oils: Basic and Applied Research, 8–11 September 1996, Vienna, Austria, pp. 229–232.

Marquard, R., Muller, T., Ceylan, A., Bayram, E. and Otan, H. (1966) *Origanum*-wildsammlungen aus der Türkei: Gehalte und Zusammensetzung des aetherischen öls. *Arznei- und Gewürzepflanzen* 1, 134–137.

Lawrence, B.M. (1984) The botanical and chemical aspects of Oregano. *Perfumer and Flavorist*, 9, 41–51.

Meikle, R.D. (1985) Flora of Cyprus, 2, The Bentham-Moxon Trust, Royal Botanic Gardens, Kew.

Michl, R.J. (1993) Dental material and method for the control of caries and paradentitis. *United States Patent* 5, 213, 615.

Muller-Riebau, F., Berger, B. and Yegen, O. (1995) Chemical composition and fungitoxic properties to phytopathogenic fungi of essential oils of selected aromatic plants growing wild in Turkey. *J. Agric. Food Chem.* 43, 2262–2266.

Nguyen, U., Frakman, G. and Evans, D.A. (1991) Process for extracting antioxidants from Labiatae herbs. *United States Patent* 5, 017,397.

Ninkov, D. (1999) Pharmaceutical compositions suitable for use against histomoniasis. *United States Patent* 5, 990, 178, 23 November 1999.

Nitsas, F.A. (2000) Pharmaceutical compositions containing herbal-based active ingredients; methods for preparing same and uses of same for medical and veterinary purposes. *United States Patent* 6, 106, 838, 22 August 2000.

Ogutveren, M., Erdemgil, F.Z., Kurkcuoglu, M., Ozek, M. and Baser, K.H.C. (1992) Composition of the essential oil of *Origanum onites*. In *Proceedings of the 8th Turkish National Symposium on Chemistry and Chemical Engineering*, Marmara University Publication No. 518, Istanbul, 2, 119–124.

Olivier, G.W. (1997) The world market of oregano. In S. Padulosi, (ed.), *Oregano*, IPGRI, Rome, 142–146.

Sarer, E., Scheffer, J.J.C. and Baerheim Svendsen, A. (1982) Monoterpenes in the essential oil of *Origanum majorana*. *Planta Med.* 46, 236–239.

Sarer, E., Scheffer, J.J.C., Janssen, A.M. and Baerheim Svendsen, A. (1985) Composition of the essential oil of *Origanum majorana* grown in different localities in Turkey. In A. Baerheim

Svendsen and J.J.C. Scheffer (eds), Essential Oils and Aromatic Plants, Martinus Nijhoff/Dr. W. Junk Publishers, Dortrecht, the Netherlands, pp. 209–212.

Scheffer, J.J.C., Looman, A., Baerheim Svendsen A. and Sarer, E. (1986) The essential oils of three *Origanum* species grown in Turkey. In E.-J. Brunke (ed.), Progress in Essential Oil Research, Walter de Gruyter, Berlin, New York, pp. 151–156.

Sezik, E. and Basaran, A. (1989) The volatile oil of *Origanum sipyleum* L. *Acta Pharmaceutica Turcica* 31(4), 129–133.

Sezik, E., Tümen, G., Kirimer, N. and Baser, K.H.C. (1993) Essential oil composition of four *Origanum vulgare* subspecies of Anatolian origin. *J. Essent. Oil Res.* 5, 425–431.

Sezik, E. and Basaran, A. (1986) Phytochemical investigations on the plants used as folk medicine and herbal tea in Turkey. IV. The volatile oil of *Origanum saccatum* L. *J. Fac. Pharm. Gazi* 3, 177–184.

Skoula, M., Gotsiou, P., Naxakis, G. and Johnson, C.B. (1999) A chemosystematic investigation on the mono- and sesquiterpenoids in the genus *Origanum* (Labiatae). *Phytochemistry* 52, 649–657.

Tanker, M. (1965) Deux succedannes du thym l'*Origanum heracleoticum* L. et la *Majorana onites* (L.) Benth. *Ist. Ecz. Fak. Mec.* 1, 32.

Tanker, N., Ilisulu, F., Koyuncu, M. and Coskun, M. (1986) Phytochemical screening of plants from the Ermenek-Mut-Gülnar (Turkey) area. III Labiatae. *Int. J. Crude Drug Res.* 24(4) 177–182.

Taysi, V., Vomel, A. and Ceylan, A. (1978) Erfahrungen mit arzneipflanzenanbau in der Türkei. *Deutsch. Apoth. Zeitung* 118, 399–403.

Tucker, A.O. and Maciarello, M.J. (1992) The essential oil of *Origanum laevigatum* Boiss. (Labiatae). *J. Essent. Oil Res.* 4, 419–420.

Tumen, G. and Baser, K.H.C. (1993) The essential oil of *Origanum syriacum* L. var. *bevanii* (Holmes) Ietswaart. *J. Essent. Oil Res.* 53, 315–316.

Tumen, G., Baser, K.H.C., Kirimer, N. and Ozek, T. (1995) The essential oil of *Origanum saccatum* P.H. Davis. *J. Essent. Oil Res.* 7, 175–176.

Tumen, G., Ermin, N., Ozek, T. and Baser, K.H.C. (1994) The essential oil of *Origanum solymicum* P.H. Davis. *J. Essent. Oil Res.* 6, 503–504.

Tumen, G., Baser, K.H.C. and Kirimer, N. (1995a) The essential oils of Turkish *Origanum* species: a treatise. In K.H.C. Baser (ed.), *Proceedings of the 13th International Congress of Flavours, Fragrances and Essential Oils*, Istanbul, Turkey, 15–19 October 1995, AREP Publ., Istanbul, 2, 200–210.

Vomel, A. and Ceylan, A. (1979) Ökologische grundlagen zur einfahrung des arzneipflanzenbaues im Ege gebiet der Türkei, Ege Üniv. Ziraat Fak. Dergisi, Vamik Taysi Özel Sayısı, Aralık, pp. 63–105.

Zeytinoglu, M., Aydin, S., Ozturk, Y. and Baser, K.H.C. (1998) Inhibitory effects of carvacrol on DMBA induced pulmonary tumorigenesis in rats. *Acta Pharmaceutica Turcica* 40, 93–98.

Zeytinoglu, H., Incesu, Z. and Baser, K.H.C. (2000) The inhibition of DNA synthesis by carvacrol in mouse myoblast cells bearing a human N-ras oncogene, *Phytomedicine* 7, Supplement II, 123.

5 The chemistry of the genus *Lippia* (Verbenaceae)

Cesar A.N. Catalan and Marina E.P. de Lampasona

INTRODUCTION

The genus *Lippia* Houst. consists of approximately 200 species of which 46 have been chemically examined. Because most of the species are aromatic, the studies on the chemistry of this genus are mostly related with the composition of the essential oils and only a very few ones devoted to the non-volatile constituents. An outstanding feature of *Lippia* is the difference observed in the essential oil composition reported for the same species from different geographic origins. The mono- and sesquiterpenoids found in the essential oils for all but two of the *Lippia* species investigated so far are quite common and widespread in the plant kingdom, the exceptions being *L. integrifolia* (Gris.) Hieron. which produces ketones based on the unique sesquiterpene skeletons named lippifoliane and integrifoliane and *L. dulcis* Trev. which contains (+)-hernandulcin, a sesquiterpenoid 1500 times sweeter than sucrose. Iridoids glucosides, phenylpropanoids, naphthoquinoids and flavonoids are the four types of significant non-volatile secondary metabolites reported in *Lippia*.

TERPENOIDS

A review on the essential oil constituents of the genus *Lippia* covering literature until 1993 has been published (Terblanché and Kornelius, 1996) and therefore in the present review only publications on this subject which appeared since then up to date will be considered. The synonymy employed here follows the classification of Moldenke (Moldenke, 1959).

The 46 *Lippia* species chemically investigated so far, the products described and the corresponding references are listed in Table 5.1.

Three types of *L. adoensis* Hochst. (syn. *L. multiflora* Mold.) collected from different geographic areas of Togo were characterized, one of which is rich in neral and geranial (23–89 per cent) and the other two types, poor in neral and geranial, contained either 1,8-cineole (16–63 per cent) or thymol (15–40 per cent) plus *p*-cymene (15–20 per cent) as the major constituents (Garneau *et al.*, 1996; Koumaglo *et al.*, 1996, 1996a). A collection from Benin (Menut *et al.*, 1995) yielded an essential oil rich in myrtenol (27 per cent), 1,8-cineole (12 per cent) and linalool (12 per cent) while plants growing in the Bangui area of the Central African Republic (Menut *et al.*, 1995a) contained 70 per cent of the uncommon 6,7-epoxymyrcene (1) (Figure 5.1). A paper on the composition and *in vitro* antimalarial activity of the volatile components of *L. multiflora* showed that the essential oil (EO) inhibited

Table 5.1 Species of *Lippia* chemically studied

Species (synonymy)	Compounds	References
L. adoensis Hochst. (syn. *L. multiflora* Mold.; *L. grandifolia* Martius et Schau.; *L. schimperi* Wolp.)	EO	Terblanché et al., 1996 Koumaglo et al., 1996a Koumaglo et al., 1996 Garneau et al., 1996 Menut et al., 1995a Menut et al., 1995 Valentin et al., 1995 Pelissier et al., 1994 Demissew, 1993 Kanko et al., 1996, 1999
	Phenylpropanoid	Taoubi et al., 1997
L. affinis aristata Schau.	EO	Terblanché et al., 1996
L. affinis sidoides Cham.	EO	Terblanché et al., 1996
L. alba (Mill.) N.E. Brown (syn. *L. asperifolia* A. Rich; *L. asperifolia* Poepp.; *L. balsamea* Mart.; *L. capensis* (Thunb.) Spreng.; *L. citrata* Cham. et Schlecht; *L. crenata* Sessé et Moc.; *L. geminata* HBK; *L. globiflora* Kuntze; *L. lantanoides*; *L. panamensis* Turcz.; *L. trifolia* Sessé et Moc.; *L. virgata* Sessé et Moc.)	EO	Terblanché et al., 1996 Leclerq et al., 1999 Viana et al., 1998 Pino et al., 1997 Frighetto et al., 1998 Pino et al., 1996, 1996a Retamar, 1994 de Abreu Matos et al., 1996 Gomes et al., 1993 Zoghbi et al., 1998
L. alnifolia Schau. (syn. *L. brasiliensis* A.S. Muller)	EO	Terblanché et al., 1996
L. americana L. (syn. *L. floribunda* HBK; *L. hemisphaerica*; *L. pauciserrata* Turcz.; *L. pyramidata* Crantz)	EO	Terblanché et al., 1996
L. aristata Schau. (syn. *L. arguta* Mart.; *Lantana aristata* (Schau.) Briq.)	EO	Terblanché et al., 1996
L. canescens Kunth	Flavonoids	Tomas-Barberán et al., 1987
L. carviodora Meikle	EO	Terblanché et al., 1996
L. carviodora Meikle var. *minor* Meikle	EO	Terblanché et al., 1996
L. citriodora Kunth (syn. *L. triphylla* (L'Hér.) Kuntze; *Aloysia triphylla* (L'Hér.) Britt.)	EO	Terblanché et al., 1996 Zrira and Benjillali, 1998 Özek et al., 1996 Djerrari et al., 1993 Zygadlo et al., 1994 Bellakhdar et al., 1993, 1994
	Phenylpropanoids	Nakumura et al., 1997
	Flavonoids	Skaltsa and Shammas, 1988 Tomas-Barberán et al., 1987
L. chevalieri Moldenke	EO	Menut et al., 1993
L. chamaedrifolia Stued. (syn. *L. chamaedryoides* Stued.; *Aloysia chamaedrifolia* Cham.)	EO	Terblanché et al., 1996

L. dauensis (Chiov.) Chiov. (syn. *Lantana dauensis* Chiov.)	EO	Terblanché *et al.*, 1996
L. dulcis Trev. (syn. *L. asperifolia* Benth.; *L. asperifolia* Reichenb.; *L. dulcis* var. *mexicana* Wehmer; *Phyla scaberrima* (A.L. Juss.) Moldenke)	EO	Terblanché *et al.*, 1996 Kaneda *et al.*, 1992 Souto-Bachiller *et al.*, 1997 Souto-Bachiller *et al.*, 1996
L. fissicalyx Tronc.	EO	Terblanché *et al.*, 1996
L. grandis Martius and Schau.	EO	Terblanché *et al.*, 1996
L. gracilis HBK	EO	Lemos *et al.*, 1992
L. grata Schau.	EO	Terblanché *et al.*, 1996
L. graveolans HBK (syn. *L. amentacea* M.E. Jones; *L. berlandieri* Millsp.; *L. berlandieri* Schau.; *L. graveolans* Schau.; *L. tomentosa* Sessé et Moc.)	EO Iridoids Flavonoids	Terblanché *et al.*, 1996 Rastrelli *et al.*, 1998 Dominguez *et al.*, 1989
L. grisebachiana Mold. (syn. *L. lantanaefolia* Griseb.)	EO	Terblanché *et al.*, 1996 Retamar *et al.*, 1981
L. hastulata (Griseb.) Hier. (syn. *Acantholippia hastulata* Griseb.)	EO	Terblanché *et al.*, 1996
L. integrifolia (Griseb.) Hier.	EO	Terblanché *et al.*, 1996 Fricke *et al.*, 1999 de Lampasona *et al.*, 1999 Catalán *et al.*, 1994, 1995 Catalán *et al.*, 1983 Catalán *et al.*, 1993, 1993a Catalán *et al.*, 1991, 1992 Zygadlo *et al.*, 1995a Duschatzky *et al.*, 1998
L. javanica (Burm. f.) Spreng. (syn. *L. asperifolia* L.C. Rich.; *L. asperifolia* var. *anomala* Moldenke; *L. scabra* Hochst.)	EO	Terblanché *et al.*, 1996 Chagonda *et al.*, 2000
L. juneillana (Mold.) Tronc.	EO	Terblanché *et al.*, 1996 Duschatzky *et al.*, 1999 Juliani *et al.*, 1998 Juliani *et al.*, 1994 Zygadlo *et al.*, 1995 Zygadlo and Grosso, 1995 Duschatzky *et al.*, 1998
L. ligustrina (Lag.) Britton (syn. *Aloysia ligustrina* (Lag.) Small; *Junellia ligustrina* (Lag.) Moldenke; *Verbena ligustrina* Lag.)	EO	Terblanché *et al.*, 1996
L. lycioides (Cham.) Steud. (syn. *L. lagustrina* Britton; *Aloysia gratissima* (Gill. et Hook.) Tronc.; *Aloysia lycioides* Cham.)	EO	Terblanché *et al.*, 1996 Zygadlo *et al.*, 1995 Zygadlo and Grosso, 1995
L. microphylla Cham.	EO	Lemos *et al.*, 1992
L. micromera Schauer in DC.	EO	Pino *et al.*, 1998 Tucker *et al.*, 1993
L. nodiflora L. Greene (syn. *L. cuneifolia* Zipp.; *L. nodiflora* (L.) Michx.; *L. repens* Spreng.; *Phyla chinensi* Lour; *Phyla nodiflora* (L.) Greene)	EO Flavonoids	Terblanché *et al.*, 1996 Tomas-Barberán *et al.*, 1987 Nair *et al.*, 1973

Table 5.1 (Continued)

Species (synonymy)	Compounds	References
L. origanoides HBK (syn. *L. berterii* Spreng.)	EO	Terblanché et al., 1996
L. oatesii Rolfe	EO	Chagonda et al., 2000
L. polystachia Gris. (syn. *Aloysia polystachia* (Gris.) Moldenke)	EO	Terblanché et al., 1996; Zygadlo et al., 1995; Zygadlo and Grosso, 1995
L. rugosa A. Chev.	EO	Menut et al., 1993
L. savoryi Meikle	EO	Menut et al., 1993
L. scaberrima Sond.	EO	Terblanché et al., 1998
L. sellowii Briq. (syn. *L. affinis* Briq.; *L. spiraeoides* Mart.; *Aloysia gratissima* (Gill. et Hook.) Tronc. var. *sellowii* (Briq.) Moldenke; *Aloysia sellowii* (Briq.) Moldenke	EO	Terblanché et al., 1996
L. seriphioides A. Gray (syn. *L. foliosa* Phil.; *Acantholippia seriphioides* (A. Gray) Moldenke)	EO	Terblanché et al., 1996
L. sidoides Cham. (syn. *L. multicapitata* Mart.)	EO	Terblanché et al., 1996; De Abreu Matos et al., 1999
	Naphthoquinoids	Macampira et al., 1986
L. somalensis Vatke	EO	Terblanché et al., 1996
L. stoechadifolia HBK (syn. *Phyla stoechadifolia* (L.) Small)	EO	Terblanché et al., 1996
L. thymoides Martius and Schau.	EO	Terblanché et al., 1996
L. trifida C. Gay (syn. *L. gracilis* R.A. Phil.; *L. hispida* Gay; *L. parvifolia* Gardner; *Acantholippia trifida* Clos.; *Acantholippia trifida* (C. Gay) Moldenke)	EO	Terblanché et al., 1996
L. turbinata Gris. (syn. *L. aprica* R.A. Phil.; *L. disepala* R.A. Phil.; *L. poleo* Lillo)	EO	Terblanché et al., 1996; Duschatzky et al., 1998; Zygadlo et al., 1995
L. ukambensis Vatke	EO	Terblanché et al., 1996
	Fatty acids, sterols, triterpenoids	Chogo and Grank, 1982
L. wilmsii H.H.W. Pearson	EO	Terblanché et al., 1996; Mwangi et al., 1989

growth, mostly at the trophozoite–schizont step, of the parasite (Valentin *et al.*, 1995). The hypotensive effect of the compounds extracted from *L. multiflora* with methanol (Chanh *et al.*, 1988) and water (Naomesi *et al.*, 1985) has been verified.

The most outstanding feature of the genus *Lippia* is the perplexing difference observed in the essential oil composition reported for the same species collected at different places and *Lippia alba* is one of the typical examples. On the other hand, it is worth to note that the oils obtained from the same plant stock remain chemically constant in successive crops (Soler *et al.*, 1986; Souto-Bachiller *et al.*, 1996). As the biosynthesis of mono- and sesquiterpenoids is enzymatically controlled, the above facts indicate that the genus possesses a rich genetic diversity. *L. alba* of the Peruvian Amazon (Leclerq *et al.*, 1999) contained carvone (63 per cent), germacrene-D (5.6 per cent) and limonene (5 per cent) as the major constituents; plants from Cuba (Pino *et al.*, 1996; Pino *et al.*, 1997) yielded carvone (33 per cent), limonene (31 per cent) and

Chemistry of the genus Lippia 131

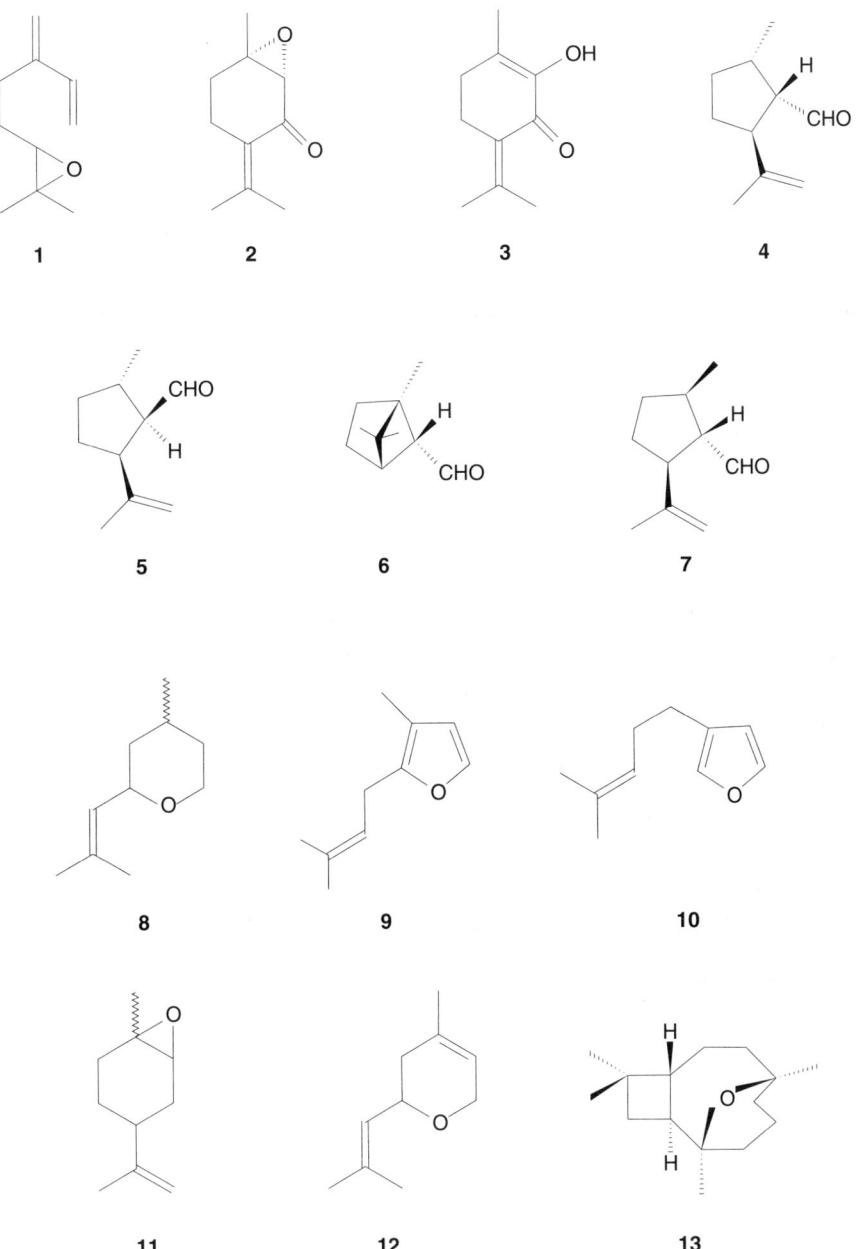

Figure 5.1 Terpenoids of Lippia.

β-guaiene (13 per cent); plants collected in Sao Paulo, Brazil (Frighetto et al., 1998) gave linalool (78 per cent), 1,8-cineole (6.5 per cent) and β-caryophyllene (2.7 per cent) while six collections from north-eastern Brazil (de Abreu Matos et al., 1996) corresponded to two different chemotypes: one rich in carvone (42–55 per cent) and limonene (23–30 per cent), being chemically similar to the Cuban (Pino et al., 1996,

1997) type, and the other rich in geranial (27–30 per cent) and neral (36–41 per cent). Plants collected in Tucumán, Argentina, in spring and in autumn (Catalán et al., 1977) gave piperitone (37 per cent and 24 per cent resp.), limonene (34 per cent and 47 per cent resp.) and 1,8-cineole (10 per cent and 13 per cent resp.); *L. alba* collected on the shores of Parana and Uruguay rivers in Entre Rios province, Argentina (Retamar, 1994) produced high levels of lippione (1,2-epoxypulegone) (2).

The absolute configuration of the natural dextrorotatory lippione (2) was established as 1S,2S by optical rotatory dispersion measurements (Shimizu et al., 1966). Lippiaphenol (3), considered to be an artefact formed by acid treatment of 2, has been detected as a natural product (Ricciardi et al., 1981) in the steam distilled essential oil of *L. turbinata*. Lippione (2) is an insect neurotoxin acting by inhibition of the enzyme acetylcholinesterase (Grundy and Still, 1985; Grundy and Still, 1985a). Both the epoxy and the keto groups appear to be necessary for irreversible inhibition (Grundy and Still, 1985a).

Lippia citriodora Kunth (syn. *Aloysia triphylla* Britton) is well known with the common name of 'lemon grass', 'lemon verbena' or 'vervain'. The plant is cultivated in many countries for preparing teas and alcoholic drinks. More than 65 components have been identified in the essential oil (Terblanché et al., 1996), the main constituents being geranial, neral, 6-methyl-5-hepten-2-one, 1,8-cineole, limonene, β-caryophyllene and caryophyllene oxide. However, as it is typical in the genus, the composition of the oil can change significatively according to the region where the plants grow (Bellakhdar et al., 1993; Djerrari et al., 1993; Bellakhdar et al., 1994; Zygadlo et al., 1994; Özek et al., 1996; Terblanché et al., 1996; Zrira and Benjillali, 1998). Some interesting minor components isolated from the oil are the photocitral isomers 4–7, the oxides 8–12 (von Roman Kaiser and Lamparsky, 1976) and the new sesquiterpenoid caryophyllane-2,6-β-oxide (13) (von Roman Kaiser and Lamparsky, 1976a). Monographs with information on the acute oral and dermal toxicity, irritation, phototoxicity, sensitization and regulatory status of verbena oil (*L. citriodora*) (Ford et al., 1992) and verbena absolute (Ford et al., 1992a) have appeared and it is recommended that the essential oil not be used in fragrances (Ford et al., 1992).

L. dulcis Trev. is under extensive investigation as a source of the intensely sweet bisabolane sesquiterpene (+)-hernandulcin (14) (Compadre et al., 1985, 1986; Sauerwein et al., 1991, 1991a) (Figure 5.2) which is the prototype of a new class of sweetening

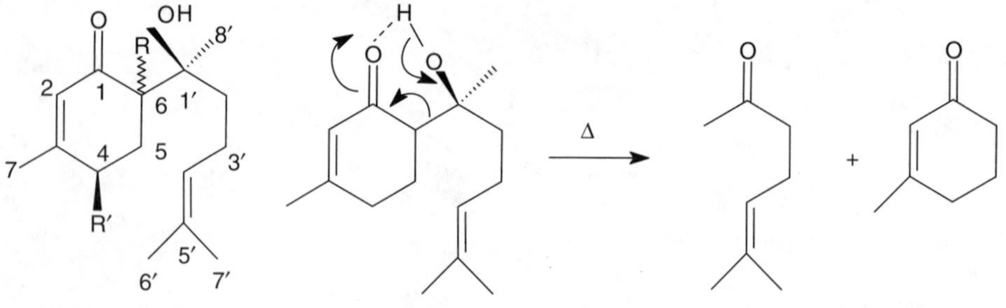

Figure 5.2 Sesquiterpenoids of *Lippia dulcis*.

agents. The herb is endemic to tropical America where it has been used in traditional medicine since centuries ago for the treatment of coughs, bronchitis, asthma and colics (Compadre et al., 1986). By solid–liquid extraction of aerial parts of plants collected in Panama (Kaneda et al., 1992) and Puerto Rico (Souto-Bachiller et al., 1997), sweet (+)-hernandulcin (14) and its non-sweet C-6 epimer (−)-epihernandulcin (15) were the main isolated constituents. Both (+)- and (±)-hernandulcin have been synthesized. (+)-Hernandulcin identical to the natural compound was synthesized from (R)-(+)-limonene (Mori and Kato, 1986; Mori and Kato, 1986a) establishing the absolute configuration of 14 as 6S,1'S. The natural ketone has also been obtained by both hairy root cultures (Sauerwein et al., 1991a) and shoot cultures (Sauerwein et al., 1991) of *L. dulcis*. It is interesting to note that the remaining three stereoisomers (6S,1'R; 6R,1'R and 6R,1'S) of (+)-hernandulcin (14) are devoid of sweetness. The racemic form of 14 has also been described (Bubnov and Gurkii, 1986; Compadre et al., 1987).

Some populations of *L. dulcis* contain only trace amounts (Compadre et al., 1987; Souto-Bachiller et al., 1997) of 14; thus, steam distillation of plant material purchased in Mexico city yielded a slightly sweet essential oil where the major constituent was camphor – a bitter monoterpene ketone – whilst hernandulcin could not be detected chromatographically (Compadre et al., 1986). The isomeric hernandulcins are relatively non-volatile and thermally sensitive compounds (Compadre et al., 1987). Consequently, special care must be taken to analyse them by gas chromatography (Souto-Bachiller et al., 1997) since pyrolysis of 14 in the injector system produces 6-methyl-5-hepten-2-one (17) and 3-methyl-2-cyclohexen-1-one (18) by a McLafferty rearrangement (Compadre et al., 1987; Souto-Bachiller et al., 1997) as shown in Figure 5.2.

A new sweet compound from *L. dulcis*, (+)-4β-hydroxyhernandulcin (16), was isolated in 1992 (Kaneda et al., 1992) in addition to 14, (−)-epihernandulcin (15), 17 and acteoside (verbascoside) (19) (Figure 5.3), a known bitter phenylpropanoid glucoside with analgesic (Nakumura et al., 1997) and protein kinase C inhibitory properties (Herbert et al., 1991). Acteoside (19) and isoacteoside (20) also exhibited moderate to

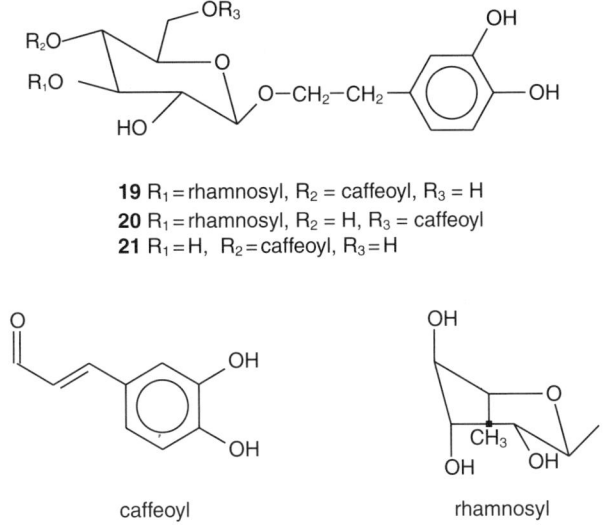

Figure 5.3 Phenylpropanoids isolated from *Lippia*.

weak cytotoxic activity (Pettit et al., 1990) against the murine P-388 lymphocytic leukemia.

The essential oil of *Lippia gracilis* H.B.K. collected in Ceará, Brazil (Lemos et al., 1992) contained myrcene (2.9 per cent), *p*-cymene (10.7 per cent), 1,8-cineole (9.4 per cent), γ-terpinene (2.1 per cent), α-terpinolene (9.1 per cent), α-terpineol (0.9 per cent), 4-terpenylacetate (10.1 per cent), thymol (30.6 per cent), carvacrol (11.8 per cent), thymolacetate (1.1 per cent), α-copaene (1.8 per cent), β-cubebene (1.2 per cent) and germacrene A (0.3 per cent), whilst leaves collected in Piani State, Brazil (de Abreu Matos et al., 1999) yielded α-thujene (0.6 per cent), myrcene (2.0 per cent), *p*-cymene (19.2 per cent), 1,8-cineole (0.8 per cent), *trans*-sabinene hydrate (1.7 per cent), linalool (0.4 per cent), terpinen-4-ol (1.2 per cent), methylthymol (6.2 per cent), methylcarvacrol (0.6 per cent), thymol (4.8 per cent), carvacrol (47.7 per cent), thymylacetate (2.3 per cent), α-copaene (0.7 per cent), β-caryophyllene (1.1 per cent), *trans*-α-bergamotene (0.6 per cent), 4-methoxythymol ? (0.6 per cent), 4-methoxycarvacrol ? (1.8 per cent), β-bisabolene (0.7 per cent) and caryophyllene oxide (2.5 per cent). As expected, owing to the high content of phenolic compounds, the oil showed antimicrobial activity (Lemos et al., 1992).

Lippia graveolans H.B.K. is a strongly aromatic herb widely used in traditional medicine (Caceres, 1996) and the essential oil composition of several different collections has been reported (Terblanché et al., 1996). The flavanones naringenin (22) and pinocembrin (23) as well as the naphtoquinoid derivative lapachenol (24) were isolated (Dominguez et al., 1989) from aerial parts and roots of *L. graveolans* collected in Mexico whilst leaves of plant material from Guatemala afforded ten iridoids and secoiridoid glucosides (Rastrelli et al., 1998). The major constituents were the novel iridoid glucosides caryoptosidic acid (25) and its 6'-O-*p*-coumaroyl (26) and 6'-O-caffeoyl (27) derivatives (Rastrelli et al., 1998). Minor components were the known iridoids caryoptoside (28), loganin (29), loganic acid (30), secologanin (31), secoxyloganin (32), dimethylsecologanoside (33) and epiloganic acid (34) (Figure 5.4).

In the essential oil of *Lippia grisebachiana* obtained by steam distillation of flowers and leaves collected in Tucumán, Argentina, the following constituents were identified (Retamar et al., 1981): α-pinene (1.0 per cent), camphene (3.6 per cent), β-pinene (2.0 per cent), myrcene (8.2 per cent), limonene (1.8 per cent), 1,8-cineole (2.3 per cent), *p*-cymene (1.6 per cent), methylheptenone (4.1 per cent), citronellal (0.4 per cent), piperitone (5.4 per cent), linalool (6.4 per cent), linalylacetate (23.4 per cent), bornylacetate (2.8 per cent), pulegone (4.7 per cent), dihydrocarvone (7.8 per cent), α-terpineol (5.1 per cent), isoborneol (4.3 per cent), geranylacetate (2.2 per cent), geraniol (3.0 per cent) and caryophyllene (5.3 per cent).

Lippia integrifolia (Griseb.) Hieron. is an aromatic herb native to South America; its habitat extends from Central and Northern Argentina to Bolivia and Chile (Cabrera, 1993). *Lippia integrifolia* is known with the vernacular names of 'incayuyo', 'te del inca' and 'poleo' whose aerial parts are sold to prepare a decoction used as emmenagogue, diuretic, stomachic and nervine (Toursarkissian, 1980). The essential oil showed to be a rich source of sesquiterpenoids, some of them with novel skeletons. In 1983 the crystalline bicyclohumulendione (35) was isolated from the essential oil (Catalán et al., 1983) and its stereochemistry settled definitively in 1991 by a single crystal X-ray analysis (Catalán et al., 1991). This diketone, named integrifolian-1,5-dione (35), is based on a new sesquiterpene skeleton: [4R,9S,10R]-4,7,7,10-tetramethylbicyclo[8.1.0]undecane, which was christened integrifoliane (Catalán et al., 1991;

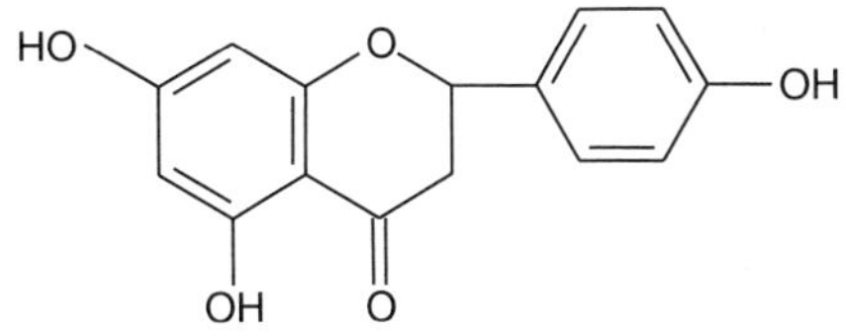

Figure 5.4 Constituents of *Lippia graveolans* H.B.K.

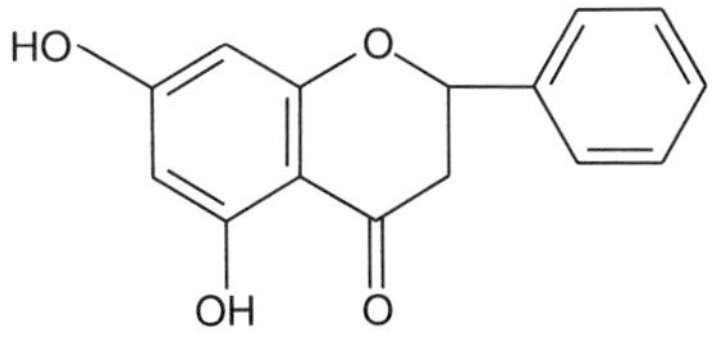

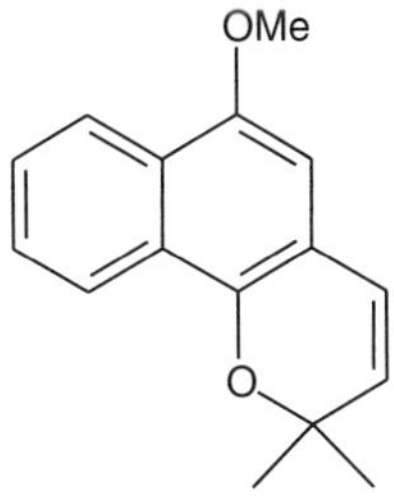

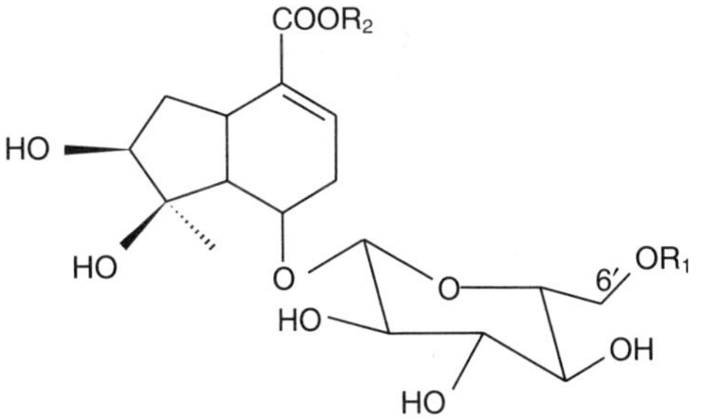

Catalán *et al.*, 1993). The major constituent of the oil was the tricyclosesquiterpene ketone 36 possessing other novel skeleton (Catalán *et al.*, 1991) named lippifoliane. A series of additional new compounds were isolated latter (Catalán *et al.*, 1992; Catalán *et al.*, 1993; Catalán *et al.*, 1994; Catalán *et al.*, 1995; de Lampasona *et al.*, 1999) from the oxygenated sesquiterpene fraction of the oil: 4,5-seco-african-4,5-dione (37) (Catalán *et al.*, 1992), *trans*-humul-9(E)-en-2,6-dione (38) (Catalán *et al.*, 1993a),

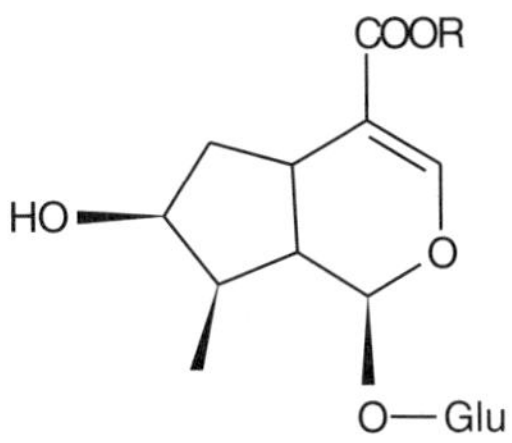

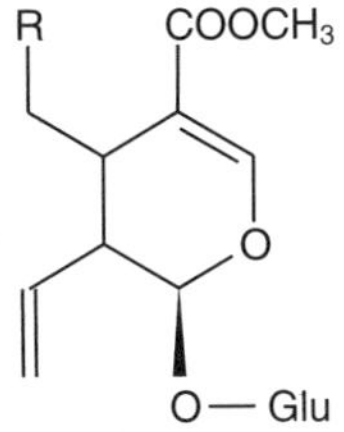

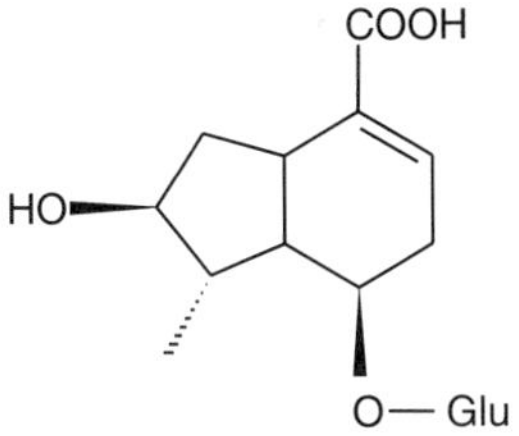

1,6-*cis*- and 1,6-*trans*-lippifolian-1α-ol-5-one (**39**) and (**40**) resp. (Catalán *et al.*, 1994), lippifoli-1(6)-en-4β-ol-5-one (**41**) (Catalán *et al.*, 1994), the africanane derivative 2β,4β,9α-2,6,6,9-tetramethyltricyclo[6.3.0.0^{2,4}]undec-1(8)-en-7,11-dione (**42**), 3α-hydroxy-6-asteriscene (**43**) and 1α,7β,9α-1-hydroxy-3,6,6,9-tetramethylbicyclo[5.4.0] undec-3-en-8-one (**44**) in addition to the known S-(+)-*trans*-nerolidol (**45**) and spathulenol (**46**). (Figure 5.5). Recently (Fricke *et al.*, 1999) three new (**47–49**) and one known (**50**) (Cullmann and Becker, 1998) africanane derivatives, the new asterisca-

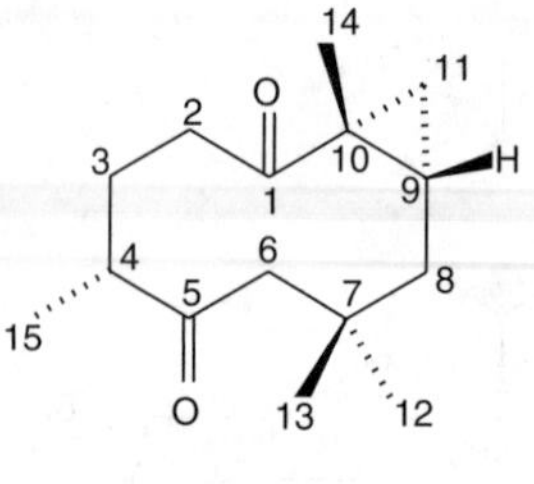

35

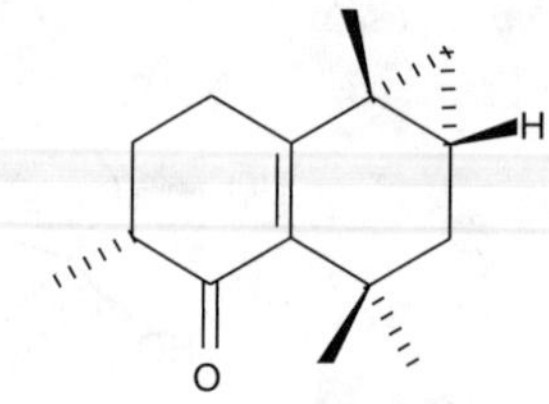

36

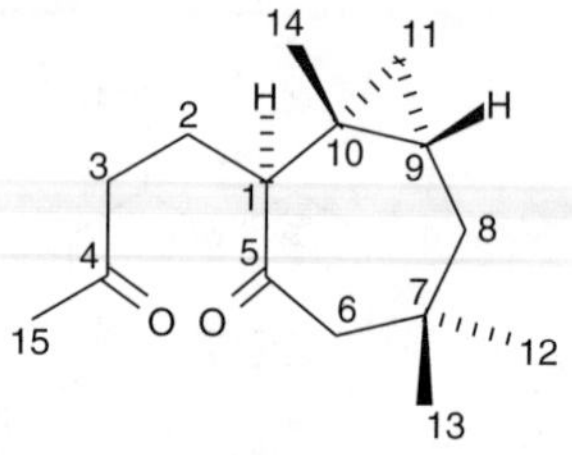

37

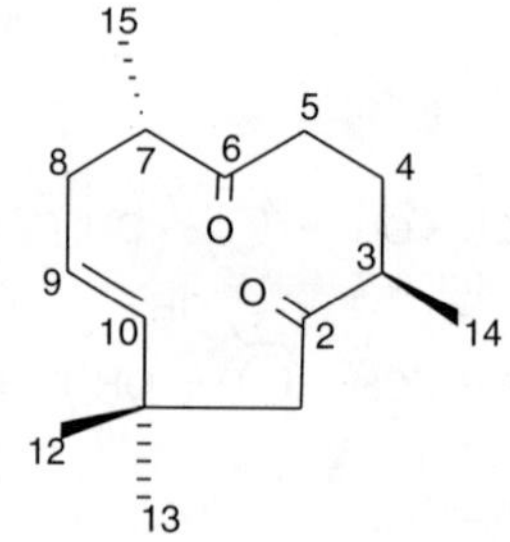

38

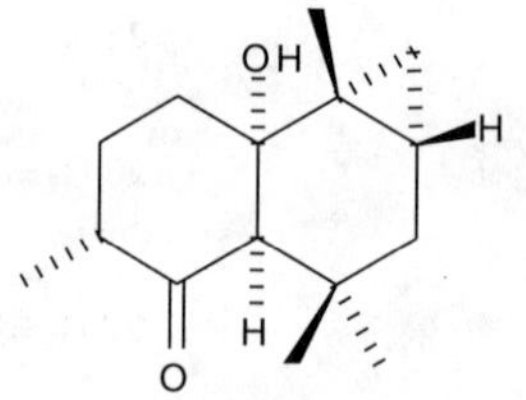

39

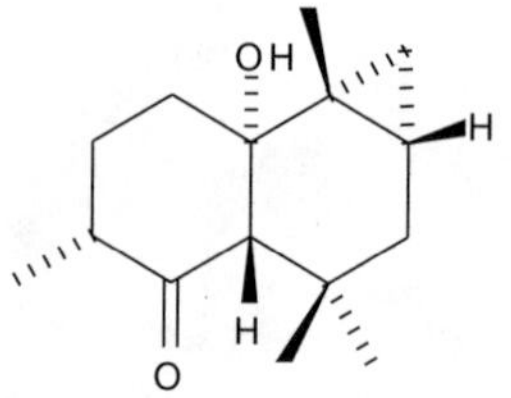

40

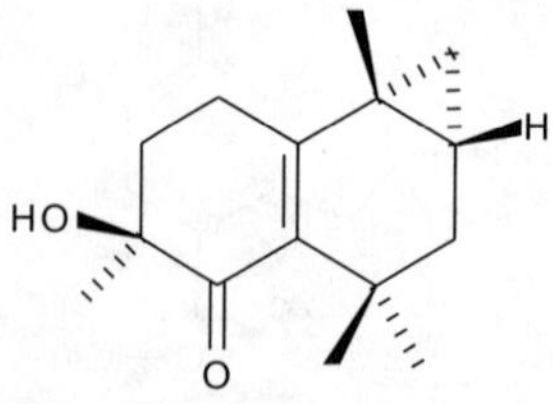

41

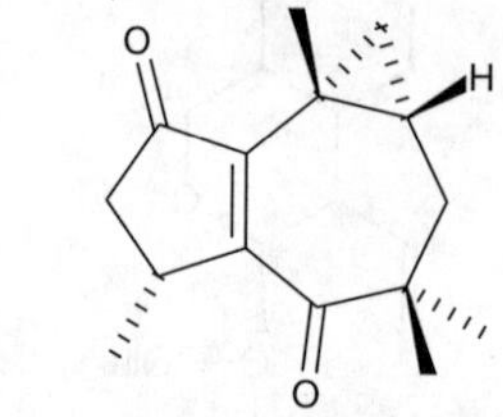

42

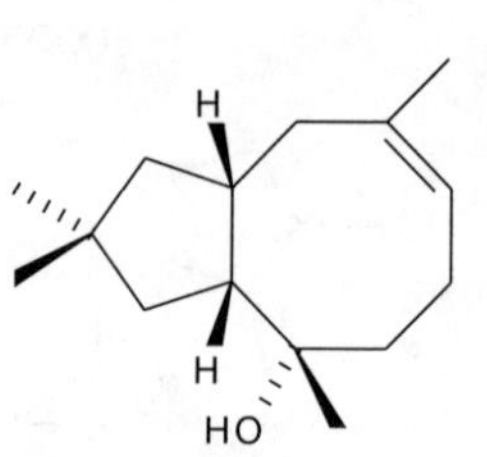

43

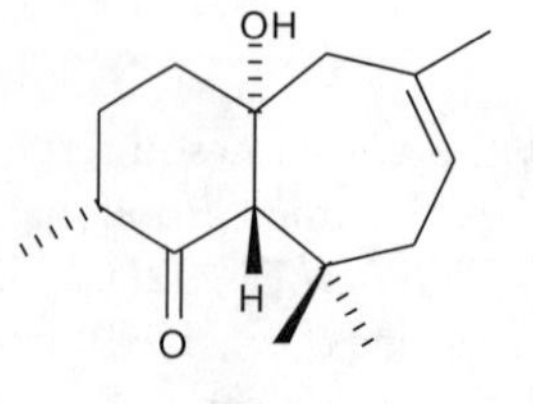

44

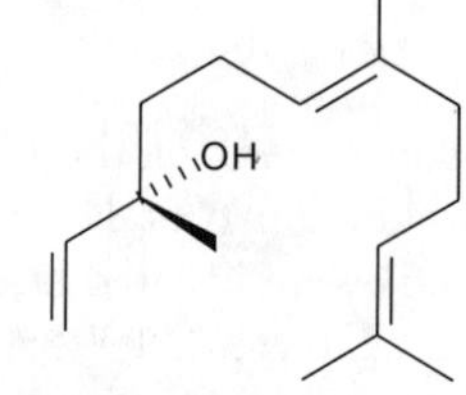

45

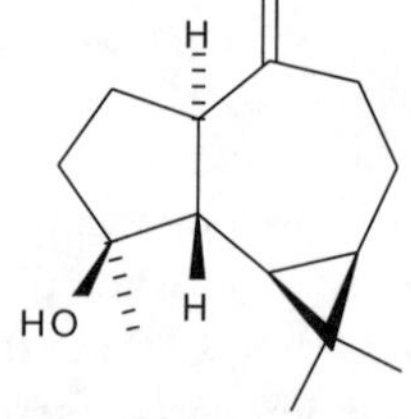

46

Figure 5.5 (Continued)

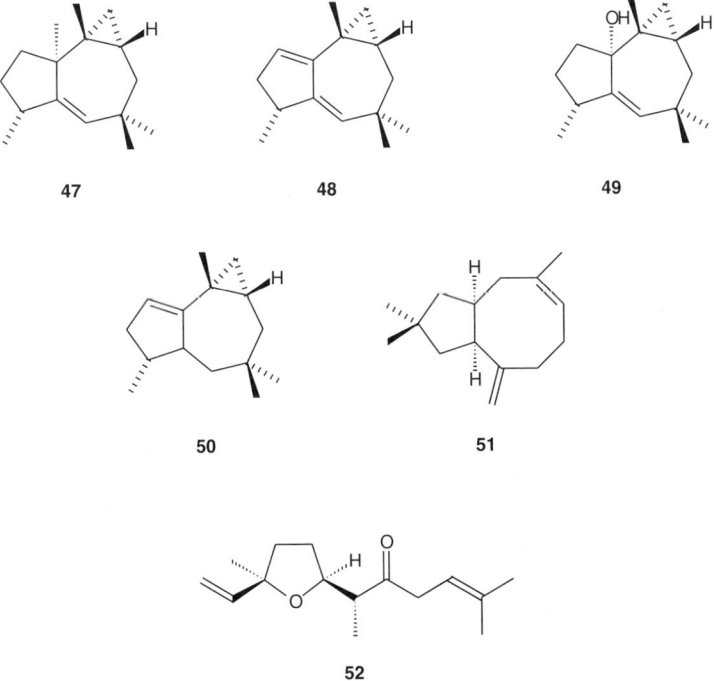

Figure 5.5 Novel and uncommon sesquiterpenoids isolated from *Lippia integrifolia*.

diene 51, (E)-β-caryophyllene, α-humulene, davanone (52) and lippifoli-1(6)-en-5-one (36) were isolated from plants collected in Córdoba, Argentina (Figure 5.5).

The biogenesis of the sesquiterpenes with unusual carbon skeletons found in *L. integrifolia* can be rationalized starting from α-humulene (53) (de Lampasona *et al.*, 1999), which is also a constituent of the oil (de Lampasona *et al.*, 1999; Fricke *et al.*, 1999). It is known that α-humulene is a key compound in sesquiterpene biosynthesis. Conformational studies of 53 showed that the molecule can exist in four stable conformers

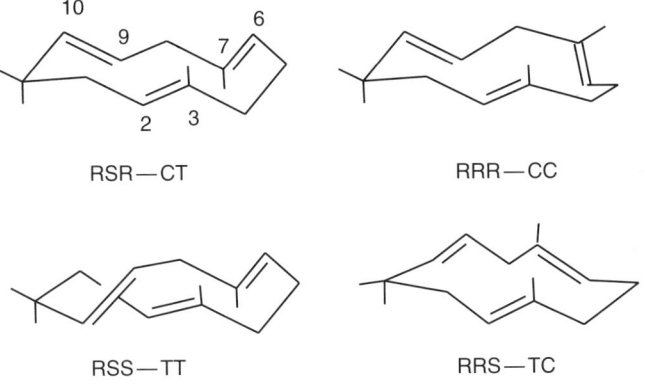

Figure 5.6 Stable conformers of α-humulene (53).

(Shirahama et al., 1980) named CT, CC, TT and TC. C and T indicate respectively a crossed and a parallel arrangement between the pair of double bonds $\Delta^{9,10}/\Delta^{2,3}$ and $\Delta^{2,3}/\Delta^{6,7}$ as shown in Figure 5.6. The highly stable CT conformer can produce a cis-fused cyclopropane as found in integrifolian-1,5-dione (35) (Figure 5.7). Intramolecular aldol condensation of 35 gives lippifoli-1(6)-en-5-one (36), the main constituent of the oil, whose hydration would afford 39 and 40. Alternatively, these ketoalcohols could also be formed by a nucleophilic attack of the C-6 anion of 35 to the C-1 carbonyl. However, it should be noted that 39 and 40 might well be artifacts formed by hydration of 36 during the hydrodistillation process (de Lampasona et al., 1999).

The CT conformer of 53 can also be precursor of the africanane derivatives found in L. integrifolia as shown in Figure 5.7 for 47.

The essential oils of wild and cultivated Lippia javanica Spreng. collected at several different places of Zimbabwe were analysed by GC-MS. The oil composition was variable (Chagonda et al., 2000) with one or more of the following being the major components: linalool (1.8–68.8 per cent), myrcene (0.5–54.0 per cent), limonene (0.4–39.9 per cent), 2,6-dimethylstyrene (trace-26.9 per cent), neral and geranial (trace-29.3 per cent) and geraniol (trace-15.8 per cent) while myrcenone (20.9–49.7 per cent), (Z)-tagetenone (24.9–39.9 per cent) and (E)-tagetenone (11.4–20.6 per cent) were found as the major components of plants collected in Kenya (Mwangi et al., 1991).

A number of papers on the essential oil of Lippia juneillana (Mold.) Tronc. collected at different places in the San Luis and Córdoba provinces of Argentina have appeared (Juliani et al., 1994; Zygadlo and Grosso, 1995; Zygadlo et al., 1995; Duschatzky et al., 1998; Juliani et al., 1998; Duschatzky et al., 1999). The yield and oil composition of leaves, stems and roots of L. juneillana collected at Merlo, San Luis province, were reported (Juliani et al., 1994) The highest oil content on dry weight, was found in leaves (1.5 per cent), in stems (0.04 per cent) and the lowest in roots (0.001 per cent). The content of oxygenated compounds was 76.5 per cent in stems, 65.8 per cent in roots and 58.5 per cent in leaves. The major constituents of the oil (leaves) were (Juliani

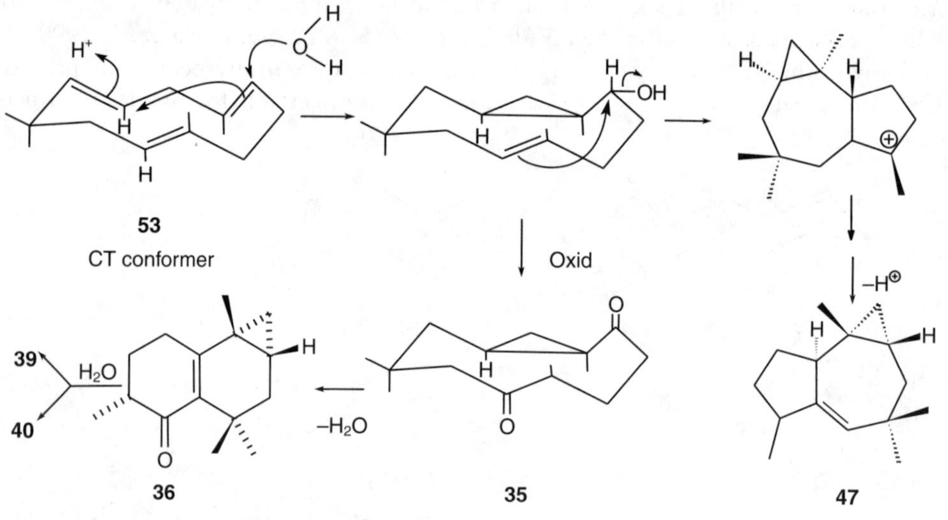

Figure 5.7 Plausible biogenetic pathway for integrifolian-1,5-dione (35), lippifolian-1(6)-en-5-one (36) and africanene 47.

et al., 1994): limonene (26.7 per cent), piperitenone (24.9 per cent), geranial (20.8 per cent), caryophyllene (5.3 per cent), camphor (4.2 per cent), lippione (2) (4.0 per cent), methyleugenol (3.0 per cent) and γ-bisabolene (3.1 per cent).

Plant material gathered at Lujan, also located in San Luis province contained (Duschatzky *et al.*, 1998) lippione (2) (36.5 per cent), limonene (23.1 per cent), camphor (7.9 per cent), spathulenol (6.5 per cent) and eucarvone (4.3 per cent) while plants collected at an undisclosed location of Córdoba province (Zygadlo *et al.*, 1995) afforded dihydrocarvone (17.9 per cent), myrcenone (17.2 per cent), camphor (10.2 per cent), limonene (10.1 per cent) and myrcene (7.0 per cent). The seasonal variation of the oil composition (Juliani *et al.*, 1998) as well as its antifungal activity (Zygadlo and Grosso, 1995) have been reported. The effect of harvesting period on the essential oil composition of a taxon determined as *Lippia aff. juneillana* (Mold.) Tronc. collected in San Luis, Argentina (Duschatzky *et al.*, 1999) has been examined. The percentage of the major constituents in the summer and winter harvest was respectively: limonene (26.8 and 19.9 per cent), *trans*-dihydrocarvone (16.0 and 0 per cent), lippione (2) (22.9 and 47.7 per cent), eucarvone (4.4 and 4.6 per cent), bicyclogermacrene (4.6 and 6.3 per cent) and camphor (8.1 and 0.2 per cent).

The flower oil of *Lippia lycioides* (Cham.) Steud. (syn. *Aloysia grattissima* (Gill. et Hook.) Tronc.) collected in Córdoba, Argentina showed a large amount of pulegone (Zygadlo *et al.*, 1995) (65.8 per cent) together with dihydrocarvone (2.1 per cent), menthone (2.0 per cent), limonene (3.6 per cent), α-thujone (1.1 per cent), β-thujone (0.8 per cent), camphor (0.7 per cent) and spathulenol (2.1 per cent).

The essential oil of *Lippia microphylla* Cham. collected at an undisclosed location in north-eastern Brazil (Lemos *et al.*, 1992) contained 1,8-cineole (36.4 per cent), terpinen-4-ol (10.2 per cent), α-terpineol (10.2 per cent), methylthymol (5.9 per cent), sabinene (4.8 per cent), γ-terpinene (4.5 per cent), thymol (4.3 per cent), β-caryophyllene (3.9 per cent), α-humulene (3.8 per cent), α-terpinolene (3.8 per cent), *p*-cymene (2.4 per cent), α-copaene (1.6 per cent) and myrcene (1.2 per cent) (Lemos *et al.*, 1992).

Lippia micromera Schauer in DC. is known as orégano, orégano poleo, petit thyme, false thyme, small thyme or Spanish thyme. It is native to Colombia, Venezuela and the Caribean Islands, being cultivated in many countries of Tropical America for seasoning foods. Until 1993, the chemical information on this taxon was sparse and fragmented (Terblanché *et al.*, 1996). Plants cultivated at Camden, Delaware using a stock from Saint Thomas, Virgin Islands (Tucker *et al.*, 1993), yielded an oil dominated by carvacrol (26.5 per cent), γ-terpinene (22.5 per cent), *p*-cimene (11.5 per cent) and thymol (7.1 per cent) together with carvacryl acetate (1.6 per cent), methylthymol (1.5 per cent), terpinen-4-ol (2.0 per cent), 1,8-cineole (2.2 per cent), α-terpinene (4.9 per cent), α-thujene (3.8 per cent) and myrcene (3.4 per cent) as minor significant components. Plants cultivated in Cuba (Pino *et al.*, 1998) contained a higher amount of carvacrol (42.2 per cent) and lower content of γ-terpinene (4.5 per cent) than the former as well as β-caryophyllene (12.1 per cent), *trans*-α-bergamotene (9.3 per cent), caryophyllene oxide (9.0 per cent), *p*-cymene (7.2 per cent), α-humulene (4.0 per cent), terpinen-4-ol (3.1 per cent), eugenol (0.9 per cent), humulene epoxide II (1.4 per cent) and caryophylla-1(12), 7-dien-9-ol (1.1 per cent). The Cuban *L. micromera* is similar to the high-carvacrol forms of Mexican oregano (*L. graveolans* H.B.K) which has (Terblanché *et al.*, 1996) 43.7–48.0 per cent carvacrol, 14.5–16.2 per cent *p*-cymene and 11.7–14.2 per cent γ-terpinene but a lower carvacrol content than high-carvacrol forms of Turkish oregano (*Origanum vulgare* L. subsp. *hirtum* (Link) Ietswaart) which has 58.7–60.7 per cent carvacrol, 11.5–13.0 per cent *p*-cymene and 8.1–14.2 per cent γ-terpinene (Tucker and Maciarello, 1993).

Lippia oatesii Rolfe is an undershrub growing wild in the semi-dry middle and western parts of Zimbabwe. So far, only one paper on the essential oil of this plant has been published (Chagonda *et al.*, 2000).The major constituents were: geranial (23.0–38.9 per cent), neral (18.1–28.5 per cent), *p*-cymene (6.8–25.7 per cent) and α-phellandrene (2.9–16.8) while limonene (1.7–3.7 per cent), β-phellandrene (1.8–3.8 per cent) and γ-terpinene (0.6–3.4 per cent) were notable minor components.

The composition (Zygadlo *et al.*, 1995) of the flower oil of *Lippia polystachia* (Gris.) Mold. and its antifungal activity (Zygadlo and Grosso, 1995) have been reported. Flowers of wild specimens collected at Colon Department, Córdoba province, Argentina yielded an oil containing α-thujone (41.1 per cent), carvone (10.0 per cent), limonene (5.6 per cent), camphor (3.0 per cent), myrcene (1.4 per cent), β-thujone (0.9 per cent), lippifoli-1(6)-en-5-one (36) (1.0 per cent), sabinene (3.0 per cent) and α-pinene (0.9 per cent).

Lippia rugosa A. Chev. collected in Burkina Faso yielded an essential oil containing a fairly high amount of oxygenated sesquiterpenes (Menut *et al.*, 1993). The main constituents were: β-caryophyllene (25.7 per cent), elemol (23.2 per cent), epoxy-caryophyllene (8.1 per cent), α-phellandrene (3.2 per cent), *trans*-β-ocimene (2.9 per cent), γ-eudesmol (2.8 per cent), α-eudesmol (2.2 per cent), β-eudesmol (2.1 per cent), 1,8-cineole (1.7 per cent) and camphor (1.0 per cent).

Lippia sarvoryi Meikle, also gathered in Burkina Faso, yielded an oil rich in sesquiterpene hydrocarbons (60 per cent) (Menut *et al.*, 1993). The most significant components were: β-caryophyllene (41.0 per cent), germacrene (isomer not identified) (14.1 per cent), 1,8-cineole (10.3 per cent), elemol (9.4 per cent), *trans*-β-ocimene (4.6 per cent), α-phellandrene (2.5 per cent), sabinene (2.1 per cent), α-humulene (2.0 per cent), α-eudesmol (1.3 per cent), β-eudesmol (0.9 per cent), camphor (1.6 per cent), limonene (1.2 per cent) and γ-eudesmol (0.8 per cent).

Lippia scaberrima Sond. is a shrub native to South Africa where it is used as a medicinal plant by various cultural groups. Only one paper on the composition of its essential oil has been published (Terblanché *et al.*, 1998). Less than 40 per cent of the oil was identified. Flowerheads contained 1,8-cineole (23.1–33.6 per cent), camphor (0.2–3.1 per cent), α-phellandrene (0.3–1.5 per cent), β-pinene (0.4–0.6 per cent), camphene (0–1.9 per cent) and linalool (0.2–0.3 per cent). The oil from leaves was void of 1,8-cineole and only a few minor constituents were identified: camphor (0.6–3.3 per cent), linalool (0.3–0.7 per cent), *endo*-borneol (1.6–4.8 per cent), camphene (0.3–1.4 per cent), β-pinene (0.2–0.7 per cent), α-terpineol (0.2–0.4 per cent), *trans*-carveol (0–0.2 per cent), δ-cadinene (0–0.2 per cent), limonene (0–6.1 per cent), α-pinene (0.1–0.6 per cent), myrcene (traces 0.2 per cent) and γ-terpinene (traces 0.4 per cent).

Lippia sidoides Cham. is a wild aromatic shrub native to north-eastern Brazil used by the local people to prepare antiseptic remedies. Leaves collected near Mossoró in Rio Grande do Norte (De Abreu Matos *et al.*, 1999) yielded 4.5 per cent oil very rich in thymol (73.1 per cent) with minor amounts of β-caryophyllene (7.3 per cent), *p*-cymene (6.0 per cent), caryophyllene oxide (4.4 per cent), methythymol (2.2 per cent), 1,8-cineole (1.1 per cent), terpinen-4-ol (1.0 per cent), δ-cadinene (0.6 per cent), aromadendrene (0.6 per cent), α-humulene (0.4 per cent), linalool (0.4 per cent) and myrcene (0.4 per cent).

Flowers of *Lippia turbinata* Griseb. collected at Colon Department, Córdoba province Argentina, gave an oil (Zygadlo *et al.*, 1995) containing α-thujone (30.2 per cent), carvone (10.1 per cent), limonene (6.1 per cent), camphor (4.0 per cent), lippifoli-1(6)-en-5-one (36) (3.3 per cent), spathulenol (4.4 per cent) and bornyl acetate (1.2 per cent) as the most significant constituents. This oil was similar to the one obtained from leaves (Velasco Negueruela *et al.*, 1993; Terblanché *et al.*, 1996). On the other hand, the aerial parts of plants collected at Comechingones mountains, San Luis province Argentina, yielded an oil which differed from the previous ones (Velasco Negueruela *et al.*, 1993; Zygadlo *et al.*, 1995); in this case the major components being (Duschatzky *et al.*, 1998) limonene (43.3 per cent), lippione (2) (24.8 per cent) and 1,8-cineole (14.7 per cent).

Leaves of *Lippia wilmsii* H.H.W. Pearson collected in Kenya gave 1.6 per cent essential oil containing piperitenone (12 per cent) in addition to the previously reported compounds (Terblanché *et al.*, 1996).

PHENYL-PROPANOIDS

Phenyl-propanoids are metabolites derived from *L*-phenylalanine or *L*-tyrosine where the starting aminoacid skeleton (C_6—C_3) remains intact. Phenylethanoids (C_6—C_2 skeleton) are produced biogenetically from tyrosine by loss of the side chain carboxyl group controlled by specific decarboxilases. Caffeic and ferulic acids and their derivatives are typical phenyl-propanoids found in plants while β-arylethylamines, arylacetaldehydes and acetophenones are well known representatives of phenyl-ethanoids.

Verbascoside (acteoside) (19) which contains both phenyl-propanoid and phenyl-ethanoid residues attached to a sugar moiety (Figure 5.3) was isolated as an analgesic principle (Nakumura *et al.*, 1997) from leaves of *Lippia citriodora* Kunth (*L. triphylla* (L'Hér.) Kuntze); 19 together with isoverbascoside (20) and derhamnosyl-verbascoside (21) have been found in *L. adoensis* Hochst. (Taoubi *et al.*, 1997) (= *L. multiflora* Mold.) (Figure 5.3).

Flavonoids are phenolic plant products based on a C_{15} skeleton of mixed biosynthetic origin. The benzene nucleus of ring A (Figure 5.8) is formed following the polyketide pathway and condensates with a phenylalanine derived C_6—C_3 precursor to produce the characteristic C_6—C_3—C_6 system of flavonoids. This skeleton is shown in formula 54. The flavonoids are thus 1,3-diarylpropanes. Flavonoids are divided into several classes according to the oxidation level of the central pyran ring C of 55: flavones (56), flavonols (57), flavanones (58) and dihydroflavonols (59).

The flavonoid molecule is commonly substituted by hydroxyl and methoxyl groups but also by sugars, methylenedioxy groups, isoprenyl chains, sulphate groups, etc. which leads to a countless array of compounds.

Reports on flavonoids of five *Lippia* species are available. Flavones (56) were found in *L. citriodora* (Tomas-Barberán *et al.*, 1987; Skaltsa and Shammas, 1988), *L. sidoides* (Macambira *et al.*, 1986), *L. nodiflora* (Nair *et al.*, 1973; Tomas-Barberán *et al.*, 1987) and *L. canescens* (Tomas-Barberán *et al.*, 1987) while flavanones (58) were isolated from *L. graveolans* (Dominguez *et al.*, 1989). Leaves of *L. citriodora* collected in Greece (Skaltsa and Shammas, 1988) contained the flavone aglicones salvigenin (70), eupatorin (71), eupafolin (68), 6-hydroxyluteolin (69), luteolin (63), hispidulin (64), cirsimaritin (72), diosmetin (61), chrysoeriol (62), apigenin

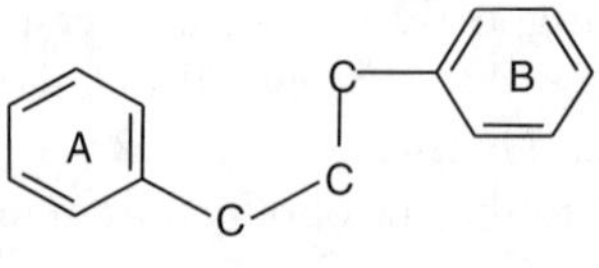

Figure 5.8 Structural formulae of the major flavonoids classes.

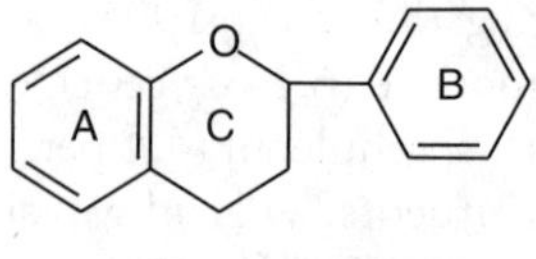

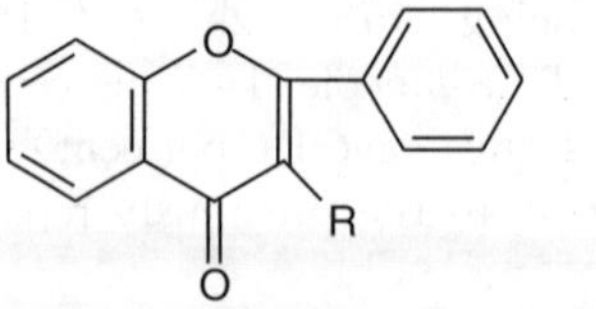

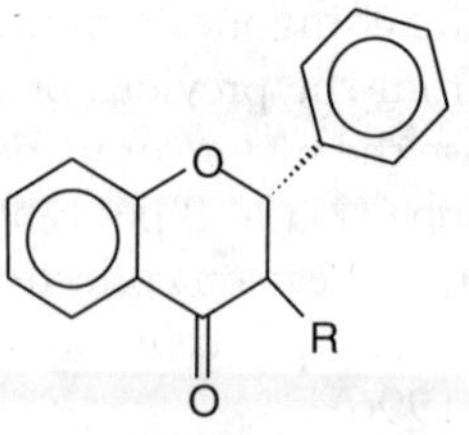

(60), pectolinaringenin (73) and cirsiliol (74) (Table 5.2) as well as the O-glycoside luteolin-7-O-β-glucoside (87) (Figure 5.9). On the other hand, aerial parts from a specimen growing in the Botanical Garden of Reading University, UK, gave three flavone glucuronides of luteolin, apigenin and diosmetin which were identified as the corresponding 7-O-glucuronylglucosides 90, 91 and 92 (Tomas-Barberán et al., 1987) (Figure 5.9).

A careful investigation of the aerial parts of a Spanish collection of *L. nodiflora* (Tomas-Barberán et al., 1987) afforded nepetin (65), jaceosidin (67), hispidulin (64) and the 12 new flavone sulphates: 75–86 (Table 5.2). No flavonoid glycoside was detected. Four additional samples of *L. nodiflora* collected in Israel, Egypt, Saudi Arabia and Malaysia showed almost identical flavonoid patterns to the Spanish sample

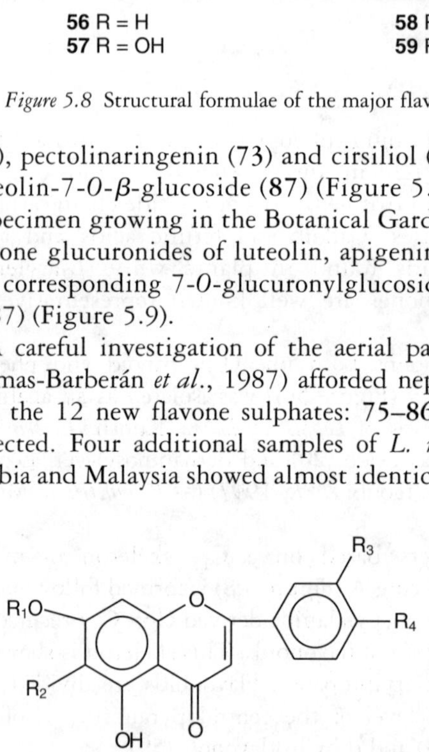

	R_1	R_2	R_3	R_4	References
87	β-glucoside	H	OH	OH	Skaltsa and Shammas, 1988
88	arabinoside	OH	OH	OH	Nair et al., 1973
89	arabinoside	OH	OH	O-rhamnoside	Nair et al., 1973
90	glucuronylglucoside	H	OH	OH	Tomas-Barberán et al., 1987
91	glucuronylglucoside	H	OH	OMe	Tomas-Barberán et al., 1987
92	glucuronylglucoside	H	H	OH	Tomas-Barberán et al., 1987

Figure 5.9 Flavonoid glycosides of *Lippia*.

Table 5.2 Flavonoids of *Lippia*

Compound	Substitution pattern			Trivial name	References
	OH	OMe	OSO$_3^-$		
60	5,7,4'	–	–	apigenin	Skaltsa ans Shammas, 1988
61	5,7,3'	4'	–	diosmetin	Skaltsa ans Shammas, 1988
62	5,7,4'	3'	–	chrysoeriol	Skaltsa ans Shammas, 1988
63	5,7,3',4'	–	–	luteolin	Skaltsa ans Shammas, 1988
64	5,7,4'	6	–	hispidulin	Skaltsa ans Shammas, 1988; Tomas-Barberán et al., 1987
65	5,7,3',4'	6	–	nepetin	Tomas-Barberán et al., 1987
66	5,6,7,4'	3'	–	nodifloretin	Tomas-Barberán et al., 1987
67	5,7,4'	6,3'	–	jaceosidin	Tomas-Barberán et al., 1987
68	5,7,3',4'	6	–	eupafolin	Skaltsa ans Shammas, 1988
69	5,6,7,3',4'	–	–	6-hydroxyluteolin	Skaltsa ans Shammas, 1988
70	5	6,7,4'	–	salvigenin	Skaltsa ans Shammas, 1988
71	5,3'	6,7,4'	–	eupatorin	Skaltsa ans Shammas, 1988
72	5,4'	6,7	–	cirsimaritin	Skaltsa ans Shammas, 1988; Macambira et al., 1986
73	5,7	6,4'	–	pectolinaringenin	Skaltsa ans Shammas, 1988
74	5,3',4'	6,7	–	cirsiliol	Skaltsa ans Shammas, 1988
75	5,4'	6	7	hispidulin-7-sulphate	Tomas-Barberán et al., 1987
76	5,7	6	4'	hispidulin-4'-sulphate	Tomas-Barberán et al., 1987
77	5,3',4'	6	7	nepetin-7-sulphate	Tomas-Barberán et al., 1987
78	5,4'	6,3'	7	jaceosidin-7-sulphate	Tomas-Barberán et al., 1987
79	5,6,3',4'	–	7	6-hydroxyluteolin-7-sulphate	Tomas-Barberán et al., 1987
80	5,7,3',4'	–	6	6-hydroxyluteolin-6-sulphate	Tomas-Barberán et al., 1987
81	5,6,4'	3'	7	nodifloretin-7-sulphate	Tomas-Barberán et al., 1987
82	5,3',4'	–	6,7	6-hydroxyluteolin-6,7-disulphate	Tomas-Barberán et al., 1987
83	5,4'	3'	6,7	nodifloretin-6,7-disulphate	Tomas-Barberán et al., 1987
84	5,7	6	3',4'	nepetin-3',4'-disulphate	Tomas-Barberán et al., 1987
85	5	6,3'	7,4'	jaceosidin-7,4'-disulphate	Tomas-Barberán et al., 1987
86	5	6	7,4'	hispidulin-7,4'-disulphate	Tomas-Barberán et al., 1987

(Tomas-Barberán et al., 1987). However, in an earlier work on *L. nodiflora* growing in India (Nair et al., 1973) the flavonoid glycosides **88** and **89** were characterized (Figure 5.9).

L. canescens collected in Spain showed a very similar flavone sulphate pattern to that of the Spanish sample of *L. nodiflora* (Tomas-Barberán et al., 1987). Cirsimaritin (72) was isolated from *L. sidoides* (Macambira et al., 1986) while *L. graveolans* collected in Mexico contained the flavanones naringenin (22) and pinocembrin (23).

NAPHTOQUINOIDS

Lapachenol (24) (Figure 5.4), isocatalponol (93) and the new naphtoquinoid (94) were isolated from *L. sidoides* (Macambira *et al.*, 1986). A biomimetic synthesis of 94 was carried out by epoxidation of isocatalponol (93) with *m*-chloroperbenzoic acid followed by reaction with BF_3, Et_2O (Macambira *et al.*, 1986). Lapachenol (24) has also been found in *L. graveolans* (Dominguez *et al.*, 1989).

93   94

MISCELLANEOUS

Several fatty acids, triterpenes and sterols have been reported in the genus. Palmitic, linoleic and linolenic acids along with ursolic acid, stigmasterol, phytol and camphene glycol were isolated from leaves of *L. ukambensis* (Chogo and Grank, 1982) while palmitic, stearic, behemic, arachidic and lignoceric acids, β-sitosterol and the napthoquinoids described above were characterized in *L. sidoides* (Macambira *et al.*, 1986).

CONCLUSION

It is clear that the chemistry of *Lippia* has been largely concentrated on essential oils, an outstanding feature of the genus being the remarkable qualitative differences observed in the composition reported for the same species from different geographic origins. However, it is worth to note that oils obtained from the same plant stock remain chemically constant for many years (Soler *et al.*, 1986; Souto-Bachiller *et al.*, 1996). For instance, sterile plantles obtained by *in vitro* germination of seeds from wild plants of *Lippia dulcis* have been cloned by single node culture; after repeated subculture for more than 5 years, it was found that the variations in the essential oil composition and (+)-hernandulcin content were negligible (Souto-Bachiller *et al.*, 1996).

The above facts indicate that the genus possesses a rich genetic diversity to biosynthesize mono- and sesquiterpenoids and also that the putative intraspecific variations are inherited. Consequently, breeding and selection of cultivars of wild species producing potentially valuable oils appears as a promising research area.

On the other hand, the literature on nonvolatile constituents of *Lippia* is scarce and fragmented. Flavones are the prevailing type of flavonoids in *Lippia*. Comparative studies on the flavonoid pattern of different species could be useful to establish relationships at subgeneric level. It can also be anticipated that investigations on iridoids, which are chemotaxonomic markers of the Verbenaceae, and naphthoquinoids as well as other non-volatile secondary metabolites will be rewarding to comprehend this very complex genus.

REFERENCES

Bellakhdar, J., Il Idrissi, A., Canigueral, S., Iglesias, J. and Vila, R. (1994) Composition of Lemon Verbena (*Aloysia triphylla* (L'Herit.) Britton) oil of Moroccan origin. *J. Essent. Oil Res.* 6, 523–526.

Bellakhdar, J., Il Idrisi, A., Canigüeral, S., Iglesias, J. and Vila, R. (1993) Analysis of the essential oil of the Odorant Vervain (*Lippia citriodora* H.B. and K.). *Planta Med. Phytother.* 26, 269–273.

Bubnov, Y.N. and Gurskii, M.E. (1986) Synthesis of (±)-Hernandulcin, a sweet substance of Lippia dulcis, using boron and silicon enolates. *Izv. Akad. Nauk SSSR, Ser. Khim.* 1448.

Cabrera, A.L. (1993) Flora de la Provincia de Jujuy. Tomo XIII, Parte IX. Colección Científica del INTA, Buenos Aires.

Caceres, A. (1996) Plantas de uso Medicinal en Guatemala. Editorial Universitaria. Universidad de San Carlos, ciudad de Guatemala, p. 287.

Catalán, C.A.N., de Lampasona, M.E.P., de Fenik, I.J.S., Cerda-Garcia-Rojas, C.M., Mora-Perez, Y. and Joseph-Nathan, P. (1994) Minor constituents of *Lippia integrifolia*. *J. Nat. Prod.* 57, 206–210.

Catalán, C.A.N., de Lampasona, M.E.P., de Fenik, I.J.S., Cerda-Garcia-Rojas, C.M. and Joseph-Nathan, P. (1993a) Structure and conformation of a Humulendione from *Lippia integrifolia*. *J. Nat. Prod.* 56, 381–385.

Catalán, C.A.N., de Fenik, I.J.S., Cerda-Garcia-Rojas, C.M., Mora-Perez, Y. and Joseph-Nathan, P. (1993) Total assignment of the ^{13}C NMR spectra of integrifolian-1,5-dione and derivatives by 2D spectroscopy. *Spectroscopy* 11, 1–8.

Catalán, C.A.N., de Fenik, I.J.S., de Arriazu, P.J. and Kokke, W.C.M.C. (1992) 4,5-Seco-African-4,5-dione from *Lippia integrifolia*. *Phytochemistry* 31, 4025–4026.

Catalán, C.A.N., de Lampasona, M.E.P., Cerda-Garcia-Rojas, C.M. and Joseph-Nathan, P. (1995) Trace constituents of *Lippia integrifolia*. *J. Nat. Prod.* 58, 1713–1717.

Catalán, C.A., de Iglesias, D.A., Retamar, J.A., Iturraspe, J., Dartayet, G.H. and Gros, E.G. (1983) A sesquiterpene diketone from *Lippia integrifolia*. *Phytochemistry* 22, 1507–1508.

Catalán, C.A., de Fenik, I.J.S., Dartayet, G.H. and Gros, E.G. (1991) Integrifolian-1,5-dione and a revised structure for 'Africanone', biogenetically related sesquiterpene ketones from *Lippia integrifolia*. *Phytochemistry* 30, 1323–1326.

Catalán, C.A.N., Merep, D.J. and Retamar, J.A. (1977) El Aceite Esencial de *Lippia alba* (Miller) N.E. Brown de la Provincia de Tucumán. *Riv. Ital. Essenze, Prof. Piante Off., Aromi, Sap., Cosmet., Aerosol* 59, 513–518.

Chagonda, L.S., Makanda, C.D. and Chalchat, J.C. (2000) Essential oils of wild and cultivated *Lippia javanica* (Spreng.) and *L. oatesii* (Rolfe) from Zimbabwe. *J. Essent. Oil Res.* 12, 1–6.

Chanh, P.H., Koffi, Y. and Chanh, A.P.H. (1998) Comparative hypotensive effects of compounds extracted from *Lippia multiflora* leaves. *Planta Médica* 54, 294–296.

Chogo, J. and Grank, G. (1982) Essential oil and leaf constituents of *Lippia ukambensis* from Tanzania. *J. Nat. Prod.* 45, 186–188.

Compadre, C.M., Hussain, R.A., de Compadre, R.L.L., Pezzuto, J.M. and Kinghorn, A.D. (1987) The intensely sweet sesquiterpene Hernandulcin: Isolation, synthesis, characterization and preliminary safety evaluation. *J. Agric. Food Chem.* 35, 273–279.

Compadre, C.M., Pezzuto, J.M., Kinghorn, A.D. and Kamath, S.K. (1985) Hernandulcin: an intensely sweet compound discovered by review of ancient literature. *Science* 227 (N° 4685), 417–419.

Compadre, C.M., Robbins, E.F. and Kinghorn, A.D. (1986) The intensely sweet herb *Lippia dulcis* Trev.: Historical uses, field inquires and constituents. *J. Ethnopharmacol.* 15, 89–106.

Cullmann, F. and Becker, H. (1998) Terpenoid constituents of *Pellia Epiphylla*. *Phytochemistry* 47, 237–245.

de Abreu Matos, F.J., Machado, M.I.L., Craveiro, A.A., Alencar, J.W. and de V. Silva, M.J. (1999) Medicinal plants of northeast Brazil containing Thymol and Carvacrol. *Lippia sidoides* Cham. and *L. gracillis* H.B.K. (Verbenaceae). *J. Essent. Oil Res.* 11, 666–668.

de Abreu Matos, F.J., Lacerda Machado, M.I., Craveiro, A.A. and Alencar, J.W. (1996) essential oil composition of two chemotypes of *Lippia alba* grown in northeast Brazil. *J. Essent. Oil Res.* 8, 695–698.

de Lampasona, M.E.P., de Fenik, I.S., Catalán, C.A.N., Dartayet, G.H., Gros, E.G., Cerda-Garcia-Rojas, C.M., Mora-Perez, Y. and Joseph-Nathan, P. (1999) Constituents of the essential oil of *Lippia integrifolia* (Griseb.) Hieron. (Verbenaceae). N. Caffini *et al.* (eds), Proc. WOCMAP-2, *Acta Horticulturae* 500, 81–88.

Demissew, S. (1993) A description of some essential oil bearing plants in Ethiopia and their indigenous uses. *J. Essent. Oil Res.* 5, 465–479.

Djerrari, A., Benjillali, B. and Crouzet, J. (1993) Effect of harvest period on the Morocco verbena volatile fraction composition. *Riv. Ital. EPPOS* 4 (Spec. Num.), 610–614.

Dominguez, X.A., Sanchez, V.H., Suarez, M., Baldas, J.H., Del R. and Gonzalez, M. (1989) Chemical constituents of *Lippia graveolans*. *Planta Med.* 55, 208–209.

Duschatzky, C., Bailac, P., Firpo, N. and Ponzi, M. (1998) Composición de los Aceites Esenciales de *Lippia juneillana*, *L. integrifolia* y *L. turbinata* de la Provincia de San Luis (Argentina). *Rev. Colombiana de Química* 27, 9–16.

Duschatzky, C., Bailac, P., Carrascull, A., Firpo, N. and Ponzi, M. (1999) Essential oil of *Lippia aff. juneliana* grown in San Luis, Argentina. effect of harvesting Period on the essential oil. *J. Essent. Oil Res.* 11, 104–106.

Ford, R.A., Api, A.M. and Letizia, C.S. (1992) Verbena oil. *Food Chem. Toxicol.* 30(Suppl.), 1375–1385.

Ford, R.A., Api, A.M. and Letizia, C.S. (1992a) Verbena absolute. *Food Chem. Toxicol.* 30(Suppl.), 1355.

Fricke, C., Hardt, I.H., König, W.A., Joulain, D., Zygadlo, J.A. and Guzman, C. (1999) Sesquiterpenes from *Lippia integrifolia* essential oil. *J. Nat. Prod.* 62, 694–696.

Frighetto, N., de Oliveira, J.G., Siani, A.C. and Calagodas Chagas, K. (1998) *Lippia alba* (Mill.) N.E.Br. (Verbenaceae) as a source of Linalool. *J. Essent. Oil Res.* 10, 581–584.

Garneau, F.X., Gagnon, H., Jean, F.I., Koumaglo, H.K., Moudachirou, M. and Addae-Mensah, I. (1996) Les Chemotypes de *Lippia multiflora*, *Maleleuca quinquenervia* et *Clausena anisata* du Togo, Benin et Ghana. Actes du Colloque de Saint-Jean-sur-Richelieu, 18–24 October 1995, pp. 125–135, Univ. Chicoutimi, Quebec.

Gomes, E.C., Ming, L.C., Moreira, E.A., Gomes, M.O., Dallarmi, M., Kerber, V.A., Conti, A. and Weiss Filho, A. (1993) Constituents of the essential oil from *Lippia alba* (Mill.) N.E. Brown (Verbenaceae). *Rev. Bras. Farm.* 74, 29–32.

Grundy, D.L. and Still, C.C. (1985a) Inhibition of acetylcholinesterase by pulegone-1, 2-epoxide. *Pestic. Biochem. Physiol.* 23, 383–388.

Grundy, D.L. and Still, C.C. (1985) Isolation and identification of the major insecticidal compound of Poleo (*Lippia stoechadifolia*). *Pestic. Biochem. Physiol.* 23, 378–382.

Herbert, J.M., Maffrand, J.P., Taoubi, K., Augereau, J.M., Fouraste, I. and Gleye, J. (1991) Verbascoside isolated from *Lantana camara*, an inhibitor of protein kinase C. *J. Nat. Prod.* 54, 1595–1600.

Juliani, H.R., Koroch, A.R., Trippi, V.S. and Juliani, H.R. (1998) Variación Estacional del Contenido y Composición del Aceite Esencial de *Lippia juneillana*. *An. Asoc. Quím. Argentina* 86, 193–196.

Juliani, Jr., H.R., Trippi, V.S., Juliani, H.R. and Ariza-Espinar, L. (1994) Essential oils from various plant parts of *Lippia juneillana*. *An. Asoc. Quím. Argentina* 82, 53–55.

Kaneda, N., Lee, I.S., Gupta, M.P., Soejarto, D.D. and Kinghorn, A.D. (1992) (+)-4-β-Hydroxyhernandulcin, a new sweet Sesquiterpene from leaves and flowers of *Lippia dulcis*. *J. Nat. Prod.* 55, 1136–1141.

Kanko, C., Koukoua, G., N'Guessan, Y.T., Lota, M.L., Tomi, F. and Casanova, J. (1999) Composition and intraspecific variability of the leaf oil of *Lippia multiflora* Mold. from the Ivory Coast. *J. Essent. Oil Res.* 11, 153–158.

Kanko, C., Koukoua, G., N'Guessan, Y.T., Tomi, F., Casanova, J. and Fournier, J. (1999) Composition de L'Huille Essentielle de *Lippia multiflora* (Verbenaceae). Comparaison des Huiles Essentielles de quelques Especes Africaines et Americaines de *Lippia* colles de *Lippia multiflora*. *J. Soc. Quest–Afr. Chin.* 1, 51–58.

Koumaglo, K.H., Akpagana, K., Glitho, A.I., Garneau, F.X., Gagnon, H., Jean, F.I., Moudachirou, M. and Addae-Mensah, I. (1996a) Geranial and neral, major constituents of *Lippia multiflora* Moldenke leaf oil. *J. Essent. Oil Res.* 8, 237–240.

Koumaglo, H.K., Dotse, K., Glitho, I.A., Garneau, F.X., Moudachirou, M. and Addae-Mensah, I. (1996) *Riv. Ital. EPPOS* 7 (Spec. Num.), 680–691.

Leclerq, P.A., Silva Delgado, H., Garcia, J., Hidalgo, J.E., Cerrutti, T., Mestanza, M., Rios, F., Nina, E., Nonato, L., Alvarado, R. and Menendez, R. (1999) Aromatic plants oils of Peruvian Amazon. Part 1. *Lippia* alba (Mill.)N.E.Br. and *Corniuta odorata* (Poeppig) Poeppig ex Schauer, Verbenaceae. *J. Essent. Oil Res.* 11, 753–756.

Lemos, T.L.G., Monte, F.J.Q., Matos, F.J.A., Alencar, J.W., Craveiro, A.A., Barbosa, R.C.S.B. and Lima, E.O. (1992) Chemical composition and antimicrobial activity of essential oils from Brazilian plants. *Fitoterapia* 63, 266–268.

Menut, C., Lamaty, G., Samate, D., Nacro, M. and Bessiere, J.M. (1993) Contribution a L'etude des Lippia Africaines: Constituants Volatils de Trois Especes du Burkina Faso. *Riv. Ital. EPPOS* 4, 23–29.

Menut, C., Lamaty, G., Sohounhloue, D.K., Dangou, J. and Bessiere, J.M. (1995) Aromatic plants of Tropical West Africa III. chemical composition of leaf essential oil of *Lippia multiflora* Moldenke from Benin. *J. Essent. Oil Res.* 7, 331–333.

Menut, C., Lamaty, G., Bessiere, J.M., Koudou, J. and Maidou, J. (1995a) Aromatic plants of Tropical Central Africa. Part XVII. 6,7-Epoxymyrcene, the major unusual constituent of *Lippia multiflora* s.l. Moldenke essential oil from the Central Africa Republic. *Flavour Fragrance J.* 10, 75–77.

Moldenke, H.N. (1959) A resume of the Verbenaceae, Avicenniaceae, Stilbaceae, Symphoremaceae and Eriocaulaceae of the World as to Valid Taxa. *Geographic Distribution and Synonymy*. U.S.A.

Macambira, L.M.A., Andrade, C.H.S., Matos, F.J.A., Craveiro, A.A. and Braz Filho, R. (1986) Naphtoquinoids from *Lippia sidoides*. *J. Nat. Prod.* 49, 310–312.

Mori, K. and Kato, M. (1986) Synthesis and absolute configuration of (+)-Hernandulcin, A new Sesquiterpene with intensely sweet taste. *Tetrahedron Lett.* 27, 981–982.

Mori, K. and Kato, M. (1986a) Synthesis of (6S, 1′)-(+)-Hernandulcin, a sweetener and its isomers. *Tetrahedron Lett.* 42, 5895–5900.

Mwangi, J.W., Muriuki, G.G., Addae-Mensah, I., Munavu, R.M., Lwande, W., Craveiro, A.A. and Alencar, J.W. (1989) Essential oil of *Lippia wilmsii* H.H.W. Pearson. *Rev. Latinoamericana Química* 20, 143–144.

Mwangi, J.W., Addae-Mensah, I., Munan, R.M. and Lwande, W. (1991) Essential oils of Kenyan *Lippia* Species. Part III. *Flavour Fragrance J.* 6, 221–224.

Nair, A.G.R., Ramesh, P., Nagarajan, S. and Subramanian, S.S. (1973) New flavone glycosides from *Lippia nodiflora*. *Indian J. Chem.* 11, 1316–1317.

Nakumura, T., Okuyama, E., Tsukada, A., Yamazaki, M., Satake, M., Nishibe, S., Deyama, T., Moriya, A., Maruno, M. and Nishimura, H. (1997) Acteoside as analgesic principle of Cedron (*Lippia triphylla*), a Peruvian medicinal plant. *Chem. Pharm. Bull.* 45, 499–504.

Naomesi, B.K., Adebayo, G.I. and Bamgbose, S.O.A. (1985) The vascular actions of aqueous extracts of *Lippia multiflora*. *Planta Medica* 51, 256–258.

Özek, T., Kirimer, N., Baser, K.H.C. and Tümen, G. (1996) Composition of the essential oil of *Aloysia triphylla* (L'Herit.) Britton grown in Turkey. *J. Essent. Oil Res.* 8, 581–583.

Pelissier, Y., Marion, C., Casadebaig, J., Milhau, M., Kone, D., Loukou, G., Nanga, Y. and Bessiere, J.M. (1994) A chemical, bacteriological, toxicological and clinical study of the Essential oil of *Lippia multiflora* Mold. (Verbenaceae). *J. Essent. Oil Res.* 6, 623–630.

Pettit, G.R., Numata, A., Takemura, T., Ode, R.H., Narula, A.S., Schmidt, J.M., Cragg, G.M. and Pase, C.P. (1990) Antineoplastic agents, 107. Isolation of Acteoside and Isoacteoside from *Castilleja liniariaefolia*. *J. Nat. Prod.* 53, 456–458.

Pino, J.A., Garcia, J. and Martinez, M. (1997) Solvent extraction and supercritical carbon dioxide extraction of *Lippia alba* (Mill.) N.E. Brown Leaf. *J. Essent. Oil Res.* 9, 341–343.

Pino, J.A., Ortega, A. and Rosado, A. (1996) Chemical Composition of the essential oil of *Lippia alba* (Mill.) N.E. Brown from Cuba. *J. Essent. Oil Res.* 8, 445–446.

Pino, J.A., Alea, L., Ortega Perez, A.G., Rosado, A. and Baluja, R. (1996a) Composition and antibacterial properties of the essential oil of *Lippia alba* (Mill.) N.E. Brown. *Rev. Cubana Farm.* 30, 29–35.

Pino, J.A., Rosado, A. and Menendez, R. (1998) Leaf oil of *Lippia micromera* Schauer in DC. from Cuba. *J. Essent. Oil Res.* 10, 189–190.

Rastrelli, L., Caceres, A., Morales, C., de Simone, F. and Aquino, R. (1998) Iridoids from *Lippia graveolans*. *Phytochemistry* 49, 1829–1832.

Retamar, J.A., Delfini, A.A., Juliani, H.R., Giussani, C.D. and Piagentini, R. (1981) Aceite Esencial de *Lippia grisebachiana*. *Essenze Deriv. Agrum* 51, 91–97.

Retamar, J.A. (1994) Variaciones Fitoquímicas de la Especie *Lippia alba (Salvia morada)* y sus Aplicaciones en la Química Fina. *Essenze Deriv. Agrum.* 64, 55–60.

Ricciardi, A.I.A., Pipet, N.V., Romero Fonseca, L., Veglia, J.F. and Lancelle, H.J. (1981) Sobre la Existencia de Lippiafenol (diosfenoleno) en el Aceite Esencial de *Lippia turbinata* Griseb., el "poleo" de Córdoba. *Facena* 4, 163–168.

Sauerwein, M., Yamazaki, T. and Shimomura, K. (1991a) Hernandulcin in Hairy root cultures of *Lippia dulcis*. *Plant Cell Rep.* 9, 579–581.

Sauerwein, M., Flores, H.E., Yamazaki, T. and Shimomura, K. (1991) *Lippia dulcis* shoot cultures as a source of the sweet sesquiterpene hernandulcin. *Plant Cell Rep.* 9, 663–669.

Shimizu, S., Katsuhara, J. and Inouye, Y. (1996) Optical rotatory dispersion of Rotundifolone. *Agr. Biol. chem.* 30, 89–93.

Shirahama, H., Osawra, E. and Matsumoto, T. (1980) Conformational studies on Humulene by means of empirical force field calculations. Role of stable conformers of humulene in biosynthetic and chemical reactions. *J. Am. Chem. Soc.* 102, 3208–3213.

Skaltsa, H. and Shammas, G. (1988) Flavonoids from *Lippia citriodora*. *Planta Medica* 54, 465.

Soler, E., Dellacassa, E. and Moyna, P. (1986) Composition of *Aloysia gratissima* flower essential oil. *Planta Medica* 52, 488–490.

Souto-Bachiller, F.A., de Jesus-Echevarria, M., Cárdenas-Gonzalez, O.E., Acuña-Rodriguez, M.F., Melendez, P.A. and Romero-Ramsey, L. (1997) Terpenoid composition of *Lippia dulcis*. *Phytochemistry* 44, 1077–1086.

Souto-Bachiller, F.A., de Jesus-Echevarria, M. and Cárdenas-Gonzalez, O. (1996) Hernandulcin is the major constituent of *Lippia dulcis* Trev. (Verbenaceae). *Nat. Prod. Lett.* 8, 151–158.

Taoubi, K., Fauvel, M.T., Gleye, J., Moulis, C. and Fourasté, I. (1997) Phenylpropanoid glucosides from *Lantana camara* and *Lippia multiflora*. *Planta Medica* 63, 192–193.

Terblanché, F.C. and Kornelius, G. (1996) Essential oil constituents of Genus *Lippia* (Vebenaceae)-A literature review. *J. Essent. Oil Res.* 8, 471–485 and references therein.

Terblanché, F.C., Kornelius, G., Hassett, A.G. and Rohwer, E.R. (1998) Composition of the essential oil of *Lippia scaberrima* Sond. from South Africa. *J. Essent. Oil Res.* 10, 213–215.

Tomas-Barberán, F.A., Harborne, J.B. and Self, R. (1987) Twelve 6-Oxygenated-Flavone Sulphates from *Lippia nodiflora* and *L. canescens*. *Phytochemistry* 26, 2281–2284.

Toursarkissian, M. (1980) Plantas Medicinales de la Argentina. Editorial Hemisferio Sur, Buenos Aires, pp. 135.

Tucker, A.O. and Maciarello, M.J. (1993) Oregano: Botany, Chemistry and Cultivation. In G. Charalambous (Ed.), *Herbs, Spices and Edible Fungi*. Amsterdam.

Tucker, A.O., Maciarello, M.J., Espaillat, J.R. and French, E.C. (1993) The essential oil of *Lippia micromera* Schauer in DC. (Verbenaceae). *J. Essent. Oil Res.* 5, 683–685.

Valentin, A., Pelissier, Y., Benoit, F., Marion, C., Kone, D., Mallie, M., Bastide, J.M. and Bessiere, J.M. (1995) Composition and Antimalarial activity *in vitro* of volatile components of *Lippia multiflora*. *Phytochemistry* 40, 1439–1442.

Velasco Negueruela, J.A., Perez-Alonso, M.J., Guzman, C.A., Zygadlo, J.A., Ariza Espinar, L., Sanz, J. and Garcia Vallejo, M.C. (1993) Volatile constituents of four *Lippia* species from Córdoba, Argentina. *J. Essent. Oil Res.* 5, 513–524.

Viana, G.S.B., Do Vale, T.J., Rao, V.S.N. and Matos, F.J.A. (1998) Analgesic and Antiinflammatory Effects of two chemotypes of *Lippia alba*: A comparative study. *Pharmaceutical Biol.* 36, 347–351.

von Roman Kaiser and Lamparsky, D. (1976) Constituents of Verbena oils. 1st Communication. Natural occurrence of Photocitrals and some of their derivatives. *Helv. Chim. Acta* 59, 1797–1802.

von Roman Kaiser and Lamparsky, D. (1976a) Constituents of Verbena oils. 2nd. Communication. Caryophyllane-2,6-β-oxide, a New Sesquiterpenoid compound from the oil of *Lippia citriodora* Kunth. *Helv. Chim. Acta* 59, 1803–1808.

Zoghbi, N.G.B., Andrade, E.H.A., Santos, A.S., Silva, M.H.L. and Maia, J.G.S. (1998) Essential oils of *Lippia alba* (Mill.) N.E. Brown growing wild in the Brazilian Amazon. *Flavour Fragrance J.* 13, 47–48.

Zrira, S. and Benjillali, B. (1998) L'Huile Essentielle de Verveine (*Lippia citriodora*) du Maroc. *Riv. Ital. EPPOS*, 9(Spec. Num.), 668–675.

Zygadlo, J.A., Lamarque, A.L., Maestri, D.M., Guzman, C.A., Lucini, E.I., Grosso, N.R. and Ariza-Espinar, L. (1994) Volatile constituents of *Aloysia triphylla* (L'Herit) Britton. *J. Essent. Oil Res.* 6, 407–409.

Zygadlo, J.A., Lamarque, A.L., Guzman, C.A. and Grosso, N.R. (1995) Composition of the flowers oils of some *Lippia* and *Aloysia* species from Argentina. *J. Essent. Oil Res.* 7, 593–595.

Zygadlo, J.A. and Grosso, N.R. (1995) Comparative study of the Antifungal activity of essential oils from Aromatic plants growing wild in Central Region of Argentina. *Flavour Fragrance J.* 10, 113–118.

Part 4
Cultivation and breeding

6 Cultivation of Oregano

Olga Makri

INTRODUCTION

The name oregano is derived from the Greek oros meaning 'mountain' and ganos meaning 'joy'. The plant grows wild in the mountains of Greece and is commonly called wild marjoram. The Greeks used it as a poultice for wounds, and Pliny recommended it for scorpion and spider bites. The colonists brought it to America, where it escaped into the wild. Records on the use of oregano date back thousands of years: the famous 'hyssop', mentioned in the bible, is believed to be an *Origanum syriakum* L. plant (Hepper, 1987).

Literature searchers have found at least 61 species of 17 genera belonging to six families mentioned under the name oregano. The family Lamiaceae (Labiatae) is considered to be the most important group containing the genus *Origanum* that provides the source of well known oregano spices – Turkish and Greek types. Two genera of the Verbenaceae family (*Lanata* and *Lippia*) are used for production of oregano herbs. The other families (Rubiaceae, Scrophulariaceae, Apiaceae and Asreraceae) have a restricted importance. However, we frequently encounter the herbs of the above mentioned families under the name of oregano in the market (Bernath, 1996).

Today, oregano plant parts and biochemical extracts (herb, leaf, essential oil, etc.) are commonly used in the food industry as a spice. Oregano can be considered one of the most important spices both on Mediterranean countries (Vokou *et al.*, 1988, 1993; Carmo *et al.*, 1989; Baser *et al.*, 1992, 1993) and elsewhere (Anonymous, 1985; International Organisation for Standardization, 1985; Bernath, 1993).

The popularity of oregano is increasingly growing as a result of scientific developments achieved in the area of its cultivation and utilisation. More and more new interesting varieties are being produced, thus contributing to broadening the horizon of its actual application.

Morphology

Oregano (Lamiaceae family) has square stems with opposite aromatic leaves. The flowers are arranged in clusters at the base of the uppermost leaves or in terminal spikes. The individual flowers have two lips, the upper ones two-lobed and the lower three-lobed. Each flower produces, when mature, four small seed like structures. The foliage is dotted with small glands containing the volatile or essential oil that gives to the plant its aroma and flavour (Simon *et al.*, 1984).

Oregano is a perennial species that grows spontaneously in areas across the Mediterranean region, particularly in high locations. In these areas oregano is harvested mainly from wild populations, once or twice a year, at flowering stage.

Origanum majorana L.

Marjoram, *Origanum majorana* L., is a tender perennial herb native to North Africa and south-west Asia and naturalised in southern Europe. Formerly classified as *Majorana hortensis* Moench. and also sweet or knotted marjoram, the plant reaches a height of 20–40 cm, has thin square, glabrous to tomentose, reddish stems, and has small, grey-green, ovate leaves, pink or purple flowers, and erect, stems. Marjoram is cultivated in France, Greece, Hungary, the United States, Egypt, and several other Mediterranean countries (Sarlis, 1994).

The reported life zone of marjoram is 6–28 °C with an annual precipitation of 0.5–2.7 m and a soil pH of 4.9–8.7. The plant is adapted to well-drained, fertile loam soils. The cold-sensitive plant cannot survive northern climates. For cultivation, marjoram is both seeded directly and transplanted into fields. Harvesting is generally accomplished at full bloom and can be done two or three times per year, depending upon the growing region. Plant material is often dried in drying sheets to avoid direct sunlight and thus preserve the green colour and the aroma (Sarlis, 1994).

There are a wide range of ecotypes and chemotypes of marjoram, and the plant is often confused with other *Origanum* species. Pot marjoram, *Origanum onites* L., is a short perennial with papillose, hirsute stems, ovate leaves, and white or purple flowers. Formerly classified as *M. onites* (L.) Benth., this plant is native to south-east Europe, Turkey, and Syria. Wild marjoram refers to several plants, generally of *Origanum* species that are collected and used as oregano. *Thymus mastichina* L., a native of Spain and North Africa, is the source of an essential oil known as Spanish wild marjoram oil.

Marjoram and pot marjoram are both generally recognised as safe for human consumption as natural flavourings/seasonings, and marjoram is generally recognised as safe as an extract/essential oil.

Origanum vulgare

Under the common name oregano, four different species are reported: *Origanum vulgare* L., and the subspecies *O. vulgare* subsp. *vulgare*, *O. vulgare* subsp. *viride*, and *O. onites* (Sarlis, 1994). Oregano originates from the Mediterranean and is closely related to marjoram. Its pungency is in direct proportion to the amount of sun it receives. It grows to a height of about 20 cm, with woody stems and dark green leaves around 2 cm long. Small, white flowers are borne on long spikes. The plant protects the inclined soils, and it is quite tolerant to cold and dryness. During the winter the aerial parts are destroyed, but the roots maintain their vitality for the revegetation in spring. Oregano grows in medium soils, and in areas with high elevation and cool summer (Sarlis, 1994).

Plant seeds in warm soil in late summer (August). Plants can be moved outdoors after three to four months (October–November). Oregano is best treated as an annual in cold climates where it will not over winter well. When grown as a perennial, roots should be divided every 3 years for best growth and flavour. Older plants will do well as a potted plant as long as they receive sufficient sunlight. As with most herbs, desiccated plant parts should be removed as frequently as necessary (Sarlis, 1994).

Harvesting the leaves and stem tips should start when plants are at the flowering stage, beginning at 10 cm from the ground. The flavour will improve after the flower buds form, just before flowering. To harvest, cut the stem tops down to the first two sets of leaves. New stems and shoots will grow, producing second and sometimes third crops. Leaves should be dried in a warm, dry, shaded place, and stored in an airtight container (Sarlis, 1994).

Wild oregano (wild marjoram, common oregano, *Origanum vulgare*)

Wild oregano is a herbaceous perennial, native to Asia, Europe and North Africa. It is a beautiful plant, flowering in heady corymbs, with reddish bracts and purple corollas. The plant is rangy and sprawling if not cut back. The foliage is finely textured and grey-green; the variety 'Aureum' (Golden Oregano) has yellow leaves. Flowers come in late summer, grow in spikes, and are purplish white. Height is 30–60 cm with comparable width.

Along with Greek Oregano, it is the source of highly antiseptic essential oils including carvacrol and thymol. There has been much commercial focus on carvacrol as a healing agent, but it is the whole herb that does the work, including all the essential oils as well as the tonifying tannins found in the plant. And, carvacrol itself occurs at therapeutic levels in many medicinal herbs besides Oregano.

Growing wild oregano is rather easy. Optimum pH is 6.8. Wild oregano grows well in shade, the cultivated sub-species *O. v. hirtum* does not. Pungency declines in rich soils, and after flowering. Oregano has a spreading root system. Propagation is usually by seed or cuttings.

Cuttings of new shoots (about 30 cm long) are removed in late spring once the leaves are firm enough to prevent wilting when placed in sand. Well-rooted cuttings are placed in the ground about 30 cm apart or planted outside in pots. If seeds are used, they should be sown in a seedbox in spring and planted outside when seedlings are 7.5 cm tall. Old wood that becomes leggy should be cut out at the end of winter and plants should be replaced every four years or so to prevent legginess.

Fertilisation, pest and insect control

Beside the soil preparation (ploughing), oregano cultivation demands fertilisation with ammonium phosphate during November to December, and efficient pest control (weeding out) (Sarlis, 1994).

Fortunately, the savoury herbs are not especially subject to serious damage by disease or insect pests, particularly when grown on a small scale. This may be due in part at least to the repellent or inhibitory action of their aromatic oils. When they are grown on a commercial scale, however, certain diseases and insect pests do cause damage under some conditions. In unusually dry weather the red spider mite may cause some damage to oregano by developing brown leaf spots, but since these diseases and insects are of infrequent occurrence and seldom cause serious damage the grower need not be greatly concerned about them. The aphids can be controlled easily with commercial dusts or spray solutions containing nicotine, rotenone, or pyrethrum. The red spider mite and the fungus diseases are more difficult to control, but they present no problem under normal conditions.

Oregano attracts honeybees, which pollinate other flowering plants. With its low compact growth, oregano makes a good border plant. Once in bloom, the plants may produce flowers throughout the growing season. Plants should be pinched back to encourage bushier growth.

Comparative methods of oregano cultivation

In central European countries, especially in Hungary, the cultivation of *O. majorana* has a long tradition. Commercial material of oregano (*O. vulgare*) is partially collected from wild plants even today. To avoid the disadvantages of exploiting oregano directly from the wild, efforts have been made in the area of its domestication and cultivation. The selection of new cultivars is underway and the material already selected is characterised by 0.5–1.5 per cent oil content containing carvacrol and thymol as the main compounds (Bernath, 1996). In Hungary, oregano is cultivated on light, dry and well-drained soils, which are somewhat alkaline. Propagation can be done vegetatively by separation of roots or by seed. Roots are planted in the field in September–October. The seed should be sown in an open-air nursery with inter-row distances of 25 cm in April. One gram of seed is usually sufficient for sowing 1 m^2 of nursery. The depth of sowing is 5–10 mm. Seedlings can be transplanted to the field in May, when they reach 10–12 cm in height. They are planted with a spacing of 50–60 cm between rows and 20–25 cm within rows. Irrigation is required at the time of planting and a few other times in the first year. In the following years, plants have developed an efficient root system and thus no further irrigation is usually needed. Plants are harvested at blooming stage and dried afterwards in shade or by artificial means to preserve the colour and fragrances of the herb. The lifespan of the crop is about 5–6 years and usually one harvest is done in the first year and two in the following years. On average, the yield ranges from 2.5 to 3.5 tons/ha (Bernath, 1996).

The cultivation of marjoram (*Origanum majorana* L.) has a long historical tradition in the eastern parts of Germany. For more than 100 years marjoram has been cultivated in the area around the town of Aschersleben (Heeger, 1956). Based on this tradition, marjoram is still grown today on 550 ha in Germany. More than 95 per cent of this cultivation is practised on the small area near Aschersleben in Saxony-Anhalt. This area produces about 8 per cent of the total medicinal and aromatic cultivation in Germany. The basic cultivation requirements for growing oregano are good climatic conditions and a highly fertile soil. The fields have to be without stones. Marjoram is mainly cultivated after legumes or potatoes and it comes in the crop rotation before wheat or barley. In autumn the field has to be well ploughed and levelled, and in spring needs to be ready for sowing. Mid-April is the best time to sow marjoram. This is sown directly in the field at 0.5–1 cm depth. In August, marjoram commences flowering and reaches the right stage for harvesting (Hammer and Junghanns, 1996).

Two species of oregano, *O. vulgare* L. and *O. syriacum* L., are grown commercially in Israel for use as fresh and dried herbs. The two species have been selected from wild populations originated from Israel and Greece. The selection of high quality cultivated varieties has been the result for the availability of large genetic diversity gathered during extensive germplasm-collecting missions targeting wild *Origanum* populations. Because of the very small size of oregano seeds, the species perennial habit and due to the fact that the plant is harvested more than once a year, the crop is propagated by stem cuttings planted directly in the field. In Israel oregano germplasm collections are conserved both as living plants and as seed (Putievsky *et al.*, 1996).

Scientists from Slovenia and the Federal Republic of Yugoslavia reported on experiences with oregano cultivation in different regions of former Yugoslavia (Cok and Kota, 1989; Kota and Cok, 1989; Macko and Cok, 1989). In the Istrian area, Greek oregano (*O. heracleoticum* L.) was introduced into cultivation in 1984, after preliminary ecological studies on species acclimation and on herb quality have been made. Cultivations have been set up on three sub-Mediterranean areas (Dvori-Isola, 260 m asl, limestone brown soils on calcareus flish, plant density 63 500 plants/ha, Smarje-Capodistria, 200 m asl, limestone brown soils on calcareus flish, plant density 57 000 plants/ha, Savudrija, sea level, terra rossa soil type, plant density 63 500 plants/ha). (Baricevic, 1996). These areas border with the Adriatic sea (45°31' lat., 300 days of growth period, average temperature in the growth period 15.2 °C, average winter temperature 5.5 °C, average rainfall in the growing period 912 mm and during winter time 139 mm, 2346 h of insulation) (Baricevic *et al.*, 1995). In all the localities oregano plants were manually planted in May. Plants were harvested at the beginning of the flowering period and naturally dried (dry, airy and shady place) (Baricevic, 1996).

Oregano has been used in the Anatolian region of Turkey since ancient times. Records in Turkey date its use back to the seventh century BC. In this region the crop has mainly been used as a spice and as a medicine to treat various health disorders. The natural occurrence of the 23 species of oregano, reported to be indigenous to Turkey, is recorded in the following floristic regions: the Euro–Siberian, the Mediterranean (including the Aegean part), and the Irano–Turanian. The cultivation of oregano is very popular in Turkey and a marked increase in the area devoted to the cultivation of this crop has been noticed in the last few years (Kitiki, 1996). Although Turkey is one of the main oregano exporters in the world, a small amount of this crop is also being imported into the country. The largest importing country of oregano from Turkey is United States, whose import corresponds to approximately 50 per cent of the total Turkish export. With regard to the planting pattern, a distance of 45 cm between rows is found the best and also the most suitable distance for mechanisation. The best harvest time to collect the highest amount of essential oil is when 50 per cent of the plants in the field have entered flowering. In relatively small fields, harvest is usually done manually, mechanical harvesting is being recommended only for large fields. After harvesting, plants are dried in the shade. A 25 cm stack height is preferred during drying operations in order to facilitate the accumulation of essential oil content. Although drying under natural conditions is a common procedure, drying ovens operating at 30–35 °C can also been used in commercial scale production. Moisture content of 7 per cent (min) to 12 per cent (max) is required (Kitiki, 1996).

The ecogeographical characteristics of Albania, particularly its soil and climate, represent ideal conditions for the spontaneous growth of oregano in many places throughout the country, especially in Lauretum–Castanetum plant association areas. Oregano is also being cultivated in the country, but on a limited area. Three species are mentioned in the Excursionist Flora of Albania, i.e. *O. vulgare* L or 'red type', which is widely dispersed in the northern part of Albania, *O. heracleoticum* L. or 'white type', which is widely dispersed in the southern part of Albania, and *O. majorana*. Both types grow as perennial plants. Normally, flowering takes place in July-September for the red type and in May–July for the white type. Oregano annual production in Albania (dried leaves with 13 per cent moisture content) ranges from 550 to 600 tons, of which about 500 tons are exported (mainly to Greece, Germany and Italy). The rest of the production is used for national consumption. Harvest, processing and trading of oregano follow official standards set out by National Authorities (Xhuveli and Lipe, 1996). Two types of oregano cultivation are practised in Albania: in home gardens and in open fields. The traditional home garden cultivation is practised more in the southern part of the country. In this case, plants are grown to meet family needs only. The cultivation of oregano in open fields is not done following any particular modern agronomic criteria. As for the home gardens, seeds are taken from spontaneous populations and also being used for field cultivation. Over the last few years, the Forest and Pasture Institute of Tirana has set up a small oregano experimental field.

In Egypt, *M. hortensis* (*O. majorana*) is cultivated for its leaves and essential oil. Plants have been planted in rows (40-cm spacing between plants, 50-cm row-spacing) on newly reclaimed land (formerly belonging to the Egyptian Armed Forces) near Cairo. The soil texture was sandy with a pH of 8.3, and calcium phosphate (475 kg/ha) was added before cultivation. Plants were watered as required by a sprinkler system. Nitrogen (NH_4SO_4, 350 kg/ha) and potassium (K_2SO_4, 120 kg/ha) were applied to the soil after planting, and after each harvest. Plants were harvested 3 times during the first growing season (July 7 and October 15, 1990, and February 20, 1991), and 3 times during the second growing season (May 10, August 15 and November 15, 1991). The yield of leaf essential oil and the composition of the oil were determined after all harvests. The best yields of herb, leaves and oil were obtained from the second and third harvest of each season. The yield of vegetative tissue and the mass of leaves per plant were greater in the second year compared with the first year at all harvests, but the oil content of the leaves was lower in the second year. Total oil yields were higher in the second year (103.8 kg/ha) than in the first year (79.8 kg/ha) due to increased vegetative yields. Nine components were identified from the essential oils, and although there were qualitative and quantitative differences, the main constituent of all oils was terpinen-4-ol (26.7–41.6 per cent) (Omer *et al.*, 1994).

Studies related to oregano cultivation

Kozlowski and Szczyglewska (1994) studied the biology of germination of *O. vulgare* L. seeds. Seeds of *O. vulgare*, obtained from the garden of the Institute of Medicinal Plants in Poznan, were stored in a non-heated room. Seeds were germinated under laboratory conditions on Petri dishes or in the Jacobsen apparatus. In the first year after harvest, germination capacity was between 60 and 75 per cent, and germination capacity was maintained for 4–5 years of storage. Germination capacity was reduced to approximately 12 per cent following storage for >5 years, seeds stored for 10 years were not viable.

The production of the culinary herbs, among them marjoram (*Origanum majorana*), in a system involving the use of conventional full-bed polyethylene mulch with furrow irrigation was studied by Csizinszky (1992) during different seasons: winter (December 1988–January 1989), spring (March–June 1989) and winter/spring (January–May 1990, planted after a tomato crop). Disease was not present but aphids (Aphididae), leafminers (*Lyriomiza* spp.), thrips (Thripidae) and whitefly (*Bemisia* spp.) were found in large numbers on marjoram when planted in March. In Florida, species originating from the temperate climatic zones are probably best grown during November–March, and those from the tropics during February–May or October–December.

Ecophysiological aspects of seed germination were investigated in the widely distributed Mediterranean-endemics. *O. vulgare* seeds had an absolute light requirement, seed germination could be promoted by green safelight or far-red light. The seeds were germinated over a relatively low temperature range, with an optimum temperature around 15–20 °C. Germination was also dependent upon the age of the seeds, old seeds germinated to a higher percentage than fresh ones. The seeds dispersed within the persistent fruiting calyces. Germination occurs within these calyces. Essential oils in the calyx strongly inhibit germination. This essential oil-induced dormancy is overcome under natural conditions by leaching of the inhibitors by rainwater. It is suggested that this dormancy operates as an adaptation strategy that prevents germination and subsequent seedling establishment during the early phase of the rainy period, which is usually interrupted by drought spells in the Mediterranean climate (Thanos et al., 1995).

An experiment was conducted to find out how humic substances affect nutrient uptake of plants. Oregano (*Origanum vulgare*) grown in nutrient film technique at 2 pH levels (4.5 and 6.5) in 2 substrates (peat and perlite) and at 3 levels of humic substance (0, 20 and 50 per cent v/v solutions of a peat extract). At low pH, the high concentration of humic substance resulted in a low shoot FW, perhaps caused by a toxicity of the humic substance at low pH. This was less pronounced at high pH (Kreij and Basar, 1995).

The rate and application method of bentazon(e) and terbacil for use with sweet marjoram (*Origanum majorana*) were evaluated in field studies at the South El-Tahrir Horticulture Research Station in Egypt during 1991–1992. Terbacil at 0.96 kg active ingredient (a.i.)/ha provided satisfactory weed control for both annual and perennial (mainly *Cynodon dactylon*) weeds, while bentazon was less effective. Terbacil application between the rows resulted in the highest yield of sweet marjoram. Neither the herbicides nor application methods (direct and between rows) altered essential oil yield or composition in sweet marjoram (El-Masry et al., 1995).

Trials were conducted at Monreale (400 m asl) and Villalba (650 m asl) in 1991–1993 on a local ecotype of *Origanum vulgare* which was planted in rows 100 cm apart at a plant spacing of 25, 50, 75 or 100 cm. Fresh herbage yields were highest at both sites in 1992. Plant density had no significant effect on yield or essential oil content. Essential oil content did not differ significantly between years (Leto et al., 1994).

In a field trial in Germany in 1995, *O. vulgare* was given a basal application of 40 kg N/ha and one or two top dressings of 40 kg N on 16 May or 16 June, and no supplementary irrigation or three irrigations (20 mm each) on 16 May, 16 June or 16 July. Increase in N rate decreased DM yield from 369 to 456.5 g/m^2, seed yields from 16.95 to 28.81 g/m^2 and ether extract from 1.92 to 2.06 ml/100 g. Irrigation increased DM yield from 371.7 to 435 g/m^2 and seed yields by 12 per cent. Germination averaged 84.05 per cent and while the effects of N rate and irrigation could not be established, application of 80 kg N and 60 mm irrigation increased germination in the light to 94.66 per cent in laboratory tests. Germination capacity of seed covered with 1–10 mm depth of coarse sand or soil decreased from 62 to 8 per cent and from 84 to 15 per cent, respectively, when assessed 21 days after sowing (Kadner, 1996).

The effects of varying nutrient concentrations on essential oil yield and composition, and growth of *O. majorana* plants grown in a greenhouse during spring (from 13 April to 5 June 1998) and summer (from 4 June to 27 July 1998) were investigated. In spring, plants were treated with N and P, while in summer, N, P and K were supplied. In both seasons, nutrients were applied at 0, 0.3, 1.0 and 3.0 mM, but N was also applied at 10 mM. In the spring treatment, the total yield of volatile oil extracted from the leaves increased by 50 per cent as N and P levels were increased from 0 to 10 mm and 0 to 3.0 mm, respectively, and sabinene hydrate acetates and terpinene components of the oil increased with increasing N levels, but not with increasing P levels. Essential oil yield and composition were similar in all treatments during summer. Effects on growth (measured by plant weight, number of leaves and plant height) were greater in spring, where a 3-fold increase in plant fresh weight was observed when N at 10 mM was applied and a 2.5-fold increase when P at 3.0 mM was applied. In summer, increases in plant weight at 3.0 mM levels of N, P and K were 1.5-fold, under 2-fold and 2.5-fold, respectively; a slightly inhibitory effect on growth was observed at 10 mM N. Day length and temperature may have been the factors that contributed to the differences in essential oil yield and composition and plant growth in *O. majorana* in the different seasons (Guerrero Trivino and Johnson, 2000).

Experiments were conducted at Rio Primero, Province of Cordoba, Argentina, during 1996 and 1997, to determine efficiency of chemical control for annual weeds in oregano fields. The assays were performed on preplanted crops grown in loam sandy soil. The predominant weeds were *Sisymbrium irio* and *Chenopodium album*, less prominent were *Raphanus sativus*, *Coronopus didymus*, *Polygonum aviculare*, *Lamium amplexicaule*, *Apium leptophyllum*, *Stellaria media*, *Bowlesia incana* and *Carduus nutans*. The following herbicides and rates were used: preplanting, trifluralin 0.96 kg i.a./ha and pendimethalin 0.99 kg/ha, postplanting, prometryn 1.5 kg/ha, linuron 1.00 kg/ha, lenacil 0.8 kg/ha and pyridate 1.125 kg/ha. The assays were done in random in blocks with four repetitions and control plots were untreated. The plots were formed by five 10-m length rows separated at 0.70 m. Recounting weed numbers and visual control percentage evaluated the control efficiency at days 21 and 49 postplanting. Recounting sprouted plants and estimating damage symptoms according to a 0–100 per cent scale evaluated the phytotoxic effect on the crop. The results indicated that at preplanting, trifluralin and pendimethalin gave a reasonable weed control, trifluralin performed better. At postplanting, prometryn was better than linuron, lenacil, and pyridate during both years in controlling weeds. None of the herbicides significantly reduced the number of plants produced at transplanting; moreover, the number of oregano plants was similar or higher than in the control. No symptoms of phytotoxicity or yield decrease were observed (Zumelzu et al., 1997).

The response of Egyptian oregano (*Origanum syriacum*) to N fertiliser application (0, 1, 2, 4 or 8 g $((NH_4)_2SO_4$/pot) in sandy soil was studied in pot experiments in 1994–1995 and 1995–1996. Nitrogen fertiliser significantly increased plant height, fresh weight and dry weight in all cuts in both years. Plant biomass increased linearly and significantly as N application rate increased up to 4 g/pot, but increases from applying up to 8 g/pot were insignificant. Oil content increased significantly up to 2 g $((NH_4)_2SO_4$/pot), above which it tended to decrease. N application increased the essential oil content compared to the control in all cuts in both years, but 2 g $((NH_4)_2SO_4$/pot) was generally sufficient to satisfy the demands of *O. syriacum*. Thymol and carvacrol were the main essential oil components in all treatments, but nitrogen fertiliser application tended to increase their biosynthesis at the expense of γ-terpinene and *p*-cymene (Omer, 1999).

Oregano and marjoram are crops for which genetic improvement is most necessary because of their high chemical and physiological heterogeneity. Crop improvement is highly recommended considering their widespread use and the great difficulties that non-uniform material may cause to the commercial sector. Taking into consideration both producers' and users' needs, efforts of any oregano breeding programme should be directed to the improvement of the following targets: yield-related parameters, e.g. growth habit, leaf/stem ratio, stress (salt, cold) tolerance, resistance to diseases and quality-related parameters, e.g. better aromatic characteristics, essential oil content and composition, antioxidant and antimicrobial properties. To achieve these goals, selection and hybridisation methods, combined with analytical controls on the variability encountered in the material, are the most appropriate tools for crop improvement. Local strains of *Origanum vulgare* subspecies and *O. majorana* (*M. hortensis*), as well as spontaneous hybrids (*O. majoricum*, *O.* × *intercedens*), are traditionally cultivated in many countries. In addition, several ornamental varieties are also present in the market. Breeding of oregano has started in relatively recent times. Breeding work has focused mainly on *O. majorana*, *O. syriacum*, *O. virens*, *O. vulgare* subsp. *hirtum* and some hybrids, by using chemotaxonomy results and male sterility as tools for controlled crossings. Results so far are promising, as shown by the good results obtained in trials made with some new varieties (Franz and Novak, 1996).

REFERENCES

Anonymous (1985) Spices and condiments. Dried marjoram (*Origanum vulgare*). Whole cut or ground leaves. Specifications French-Standard, NF V 32-170, pp. 5.

Baricevic, D., Raspor, P., Spanring, T., Prus, T. and Gomboc, S. (1995) Funding cannot match Slovenia's intense interest in biodiversirty conservation and research. *Diversity* 11(1–2), 105–106.

Baricevic, D. (1996) Experiences with oregano (*Origanum* spp.) in Slovenia. In *Proceedings of the IPGRI International Workshop on Oregano*, 8–12 May 1996, CIHEAM, Valenzano (Bari), Italy, pp. 110–121.

Baser, K.H.C., Ozek, T., Kurkcuoglu, M. and Tumen, T. (1992) Composition of the essential oil of *Origanum sipyleum* of Turkish origin. *J. Essential Oil Res.* 4(2), 139–142.

Baser, K.H.C., Ozek, Tumen, T. and Sezik, E. (1993) Composition of the essential oils of Turkish *Origanum* species with commercial importance. *J. Essential Oil Res.* 5(6), 619–623.

Bernath, J. (1993) Vadon termo es termesztett gyogynovenyek. Mezogazda kiado, Budapest.

Bernath, J. (1996) Some scientific and practical aspects of production and utilisation of oregano in central Europe. In *Proceedings of the IPGRI International Workshop on Oregano*, 8–12 May 1996, CIHEAM, Valenzano (Bari), Italy, pp. 76–93.

Carmo, M.M., Frazao, S. and Vanancio, F. (1989) The chemical composition of Portuguese *Origanum vulgare* oils. *J. Essential Oil Res.* 1(2), 69–71.

Cok, H. and Kota, E. (1989) Experiences with cultivation of Greek oregano (*Origanum heracleoticum* L.) in Yugoslavia. *Sodobno kmetijstvo* 22(7/8), 434–437.

Csizinszky, A.A. (1992) The potential for aromatic plant production with plastic mulch culture in Florida. First world congress on medicinal and aromatic plants for human welfare (WOCMAP), Maastricht, Netherlands, 19–25 July.

El-Masry, M.H., Charles, D.J. and Simon, J.E. (1995) Bentazon and terbacil as postemergent herbicides for sweet basil and sweet marjoram. *J. Herbs, Spices and Med Plants* 3(3), 19–26.

Franz, C. and Novak, J. (1996) Breeding of *Origanum* species. Oregano. In *Proceedings of the IPGRI International Workshop on Oregano*, 8–12 May 1996, CIHEAM, Valenzano (Bari), Italy.

Guerrero Trivino, M. and Johnson, C. B. (2000) Season has a major effect on the essential oil yield response to nutrient supply in *Origanum majorana*. *J. Horticultural Sci. Biotechnol.* 75(5), 520–527.

Hammer, K. and Junghanns, W. (1996) *Origanum majorana* L. – some experiences from Eastern Germany. In *Proceedings of the IPGRI International Workshop on Oregano*, 8–12 May 1996, CIHEAM, Valenzano (Bari), Italy, pp. 100–102.

Heeger, E.F. (1956) Handbuch des Arznei- und Gewurzpflanzenanbaus. Berlin.

Hepper, F.N. (1987) Planting a Bible Garden. HMSO, London, UK.
International Organizaton for Standardization (1985) Dried oregano (*Origanum vulgare* Linnaeus) – Whole or ground leaves specification. International Standard (ISO), pp. 7925–1985.
Kadner, R. (1996) Do nitrogen and supplementary irrigation affect seed yield. *TASPO Gartenbaumagazin* 5(11), 22–23.
Kitiki, A. (1996) Status of cultivation and use of oregano in Turkey. In *Proceedings of the IPGRI International Workshop on Oregano*, 8–12 May 1996, CIHEAM, Valenzano (Bari), Italy, pp. 122–132.
Kota, E. and Cok, H. (1989) Content of some biogenic elements in oregano plants (*Origanum heracleoticum* L.) grown in different localities in Istria and Vojvodina (Yugoslavia). *Sodobnokmetijstvo* 22(11), 466–470.
Kozlowski, D. and Szczyglewska, D. (1994) Biology of germination of medicinal plant seeds. Vd. *Origanum vulgare* L. seeds. *Herba Polonica* 40(3), 79–82.
Kreij, C. and Basar, H. (1995) Effect of humic substances in nutrient film technique on nutrient uptake. *J. Plant Nut.* 18(4), 793–802.
Leto, C., Carrubba, A. and Trapani, P. (1994) Effects of planting density on the cultivation of oregano (*Origanum heracleoticum* L.) in two Sicilian environments. Italy, 2–3/7 569–577.
Macko, V.H. and Cok, H. (1989) Efficiency and selection of herbicides in oregano (*Origanum heracleoticum* L.). In L.J. Vasiljevic (ed.), *Proceedings of Yugoslav Conference on the Application of Pesticides*, Opatija, Yugoslavia, 4–7 December 1989, pp. 193–196.
Omer, E.A., Ouda, H.E. and Ahmed, S.S. (1994) Cultivation of sweet marjoram, *Marjorana hortensis*, in newly reclaimed lands of Egypt. *J. Herbs, Spices & Med. Plants* 2(2), 9–16.
Omer, E.A. (1999) Response of wild Egyptian oregano to nitrogen fertilization in a sandy soil. *J. Plant Nutr.* 22(1) 103–114.
Putievsky, E., Nativ, D. and Uzi, R. (1996) Cultivation, selection and conservation of oregano species in Israel. In *Proceedings of the IPGRI International Workshop on Oregano*, 8–12 May 1996, CIHEAM, Valenzano (Bari), Italy, pp. 103–110.
Sarlis, G. (1994) Aromatic and Pharmaceutical Plants. Agricultural University of Athens, Greece.
Simon, J.E., Chadwick, A.F. and Craker, L.E. (1984) Herbs: An Indexed Bibliography. 1971–1980. The Scientific Literature on Selected Herbs, and Aromatic and Medicinal Plants of the Temperate Zone. Archon Books, 770pp., Hamden, CT.
Thanos, C.A., Kadis, C.C. and Skarou, F. (1995) Ecophysiology of germination in the aromatic plants thyme, savory and oregano (Labiatae). *Seed Sci. Res.* 5(3), 161–170.
Vokou, D., Kokkini, S. and Bessiere, J.M. (1988) *Origanum onites* (Lamiaceae) in Greece: distribution, volatile oil yield, and composition. *Econ. Bot.* 42(3), 407–412.
Vokou, D., Kokkini, S. and Bessiere, J.M. (1993) Geographic variation of Greek oregano (*Origanum vulgare* ssp. *hirtum*) essential oils. *Biochem. Systematics Ecol.* 21(2), 287–295.
Xhuveli, L. and Lipe, Q. (1996) Oregano (*Origanum vulgare* L.) in Albania. In *Proceedings of the IPGRI International Workshop on Oregano*, 8–12 May 1996, CIHEAM, Valenzano (Bari), Italy, pp. 133–137.
Zumelzu, G., Darre, C., Novo, R. J. and Bracamonte, R. E. (1997) Preemergent control of annual weeds in oregano (*Origanum vulgare* L.). In *Proceedings of the Second World Congress on Medicinal and Aromatic Plants for Human Welfare*, WOCMAP-2. Agricultural production, post-harvest techniques and biotechnology, Mendoza, Argentina, 10–15 November.

7 Breeding of Oregano

Chlodwig Franz and Johannes Novak

INTRODUCTION

Due to its multipurpose use as kitchen herb, in the food as well as in the flavour industry the demand on oregano has grown tremendously within recent years and is still increasing (Verlet, 1994; Olivier, 1997). Nevertheless, most of the so called "Greek Oregano" (oregano from *Origanum* sp.) is still gathered from the wild exceeding 10 000 tons crude drug per year. To avoid a threatening overexploitation even of a species as widely distributed in the Mediterranean as oregano and to improve the market supply, efforts have been made to introduce oregano into systematic cultivation. This was successfully done during the last decade in Turkey, Greece, France and Italy, and large scale field production was established furthermore in Israel, Chile, Argentina and China. Since *Origanum* sp. shows a high biodiversity especially also regarding phytochemical characters, the question arises which oregano is grown for which purpose based on which specifications.

Phenotypic diversity of aromatic plants was detected very early because of their striking sensorial properties, and this is true also for oregano. For instance it is known in the herb trade that oregano from either side of the same mountain may be substantially different (Olivier, 1997). The main reasons for the above heterogeneity are

1. the individual genetic diversity,
2. the morpho- and ontogenetic variability, and
3. modifications due to environment (including cultivation practices).

representing contemporarily the main factors influencing the quality. It is therefore important to separate genetic from any other sources of variation, for example, by growing plants under identical conditions and harvesting specific plant parts at a well defined development stage, before conclusions on e.g. phytochemical breeding values and on the breeding progress can be drawn.

Inter- and infraspecific variation

The natural chemical polymorphism within the genus *Origanum* is very high with respect to the essential oil, showing an increasing number of chemotypes identified with ongoing investigations of further species and populations. It is obvious that *O. vulgare* and its subspecies, especially the most frequently analysed ssp. *hirtum* are characterised by carvacrol, thymol, *p*-cymene and γ-terpinene as main compounds, although some exceptions also

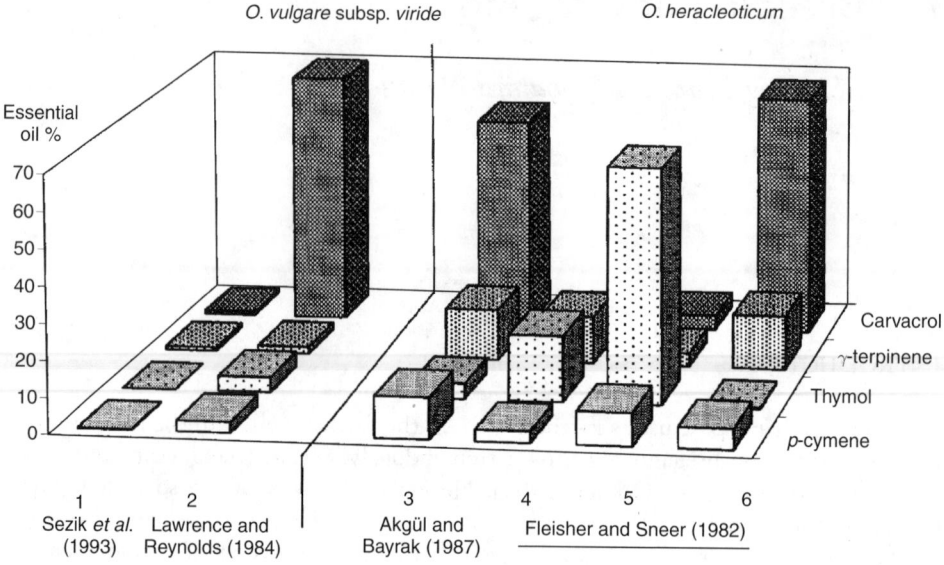

Figure 7.1 Classification of *Origanum* spp. based on chemical compounds (according to Bernath, 1997).

have been reported. Other *Origanum* species as e.g. *O. onites*, *O. syriacum* or *O. dictamnus* may contain – besides of the mentioned substances – linalool, terpinene-4-ol, sabinene or even caryophyllene and borneol as major compounds (Figure 7.1).

Compiled data on this subject are given by Bernàth (1997) and more recently by Skoula *et al.* (1999). For *O. majorana*, in addition, it is not yet clear if the cultivated marjoram represents a *cis*-sabinene-hydrate chemotype only (Novak and Franz, 2000), since investigations of wild populations show also oil compositions with mainly carvacrol, thymol and linalool (Figure 7.2). Also the essential oil content in total shows a wide range from e.g. 0.034 per cent up to 7.6 per cent in *Origanum laevigatum* (Baser *et al.*, 1996; Özgüven *et al.*, 1996) or even up to 9.9 per cent in *O. vulgare* ssp. *hirtum* (Pasquier, 1997).

Remarkable chemical variations have not only been observed between but also within populations and accessions. Single plant investigations of a grouping of *O. vulgare* ssp. and their offsprings resulted in an unexpected differentiation into chemotypes including one with a marjoram-like profile, but growth characteristics and winter hardiness of a *O. vulgare* ssp. (Marn *et al.*, 1999, Figure 7.3 and 7.4). A detailed assessment of 54 *O. vulgare* ssp. *hirtum* and *O. onites* clones has also shown clearly distinguished chemical groups being the basis of a selection program (Pasquier, 1997). Carvacrol contents ranging from traces up to 95 per cent of the essential oil are described (Kokkini *et al.*, 1991), and almost the same goes for the other compounds mentioned. Just recently, some individuals producing essential oils consisting of more than 95 per cent of enantiomer pure single substances could be detected (Novak, unpublished).

The large inter- and infraspecific variability is, however, not only true for chemical quality characters as it is the essential oil and its compounds, but also valid for leaf and flower colour, trichome density (besides essential oil bearing glands), yield, leaf:stem-ratio a.s.o., or in general for morphological characters.

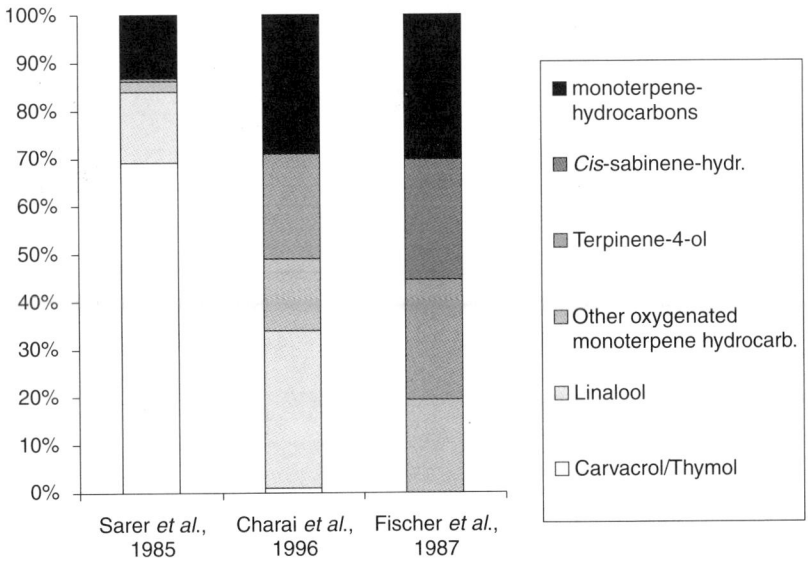

Figure 7.2 Essential oil composition of *Origanum majorana* L. according to different authors.

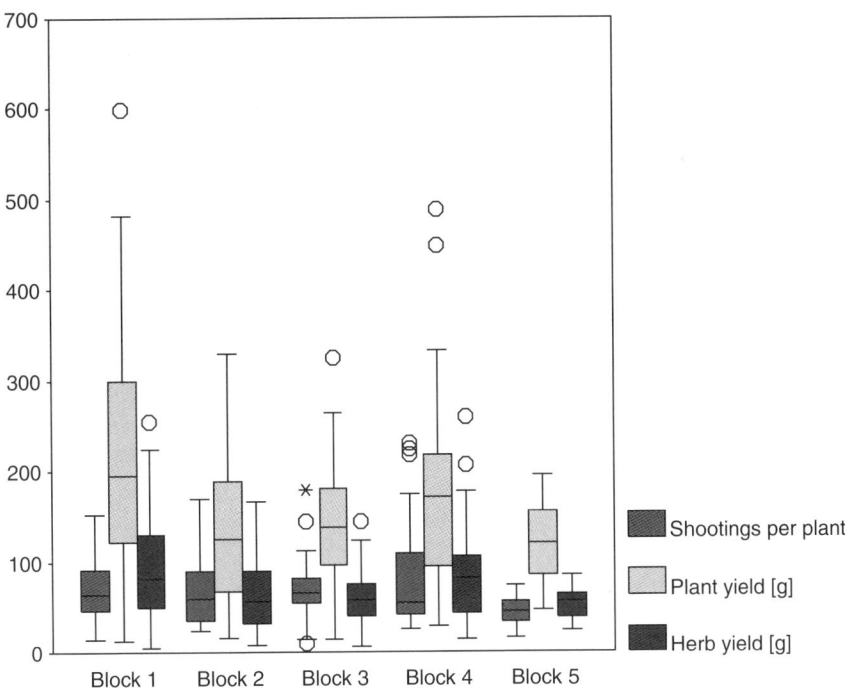

Figure 7.3 Yield related parameters of five single plant offsprings ("blocks") of *Origanum vulgare* ssp.

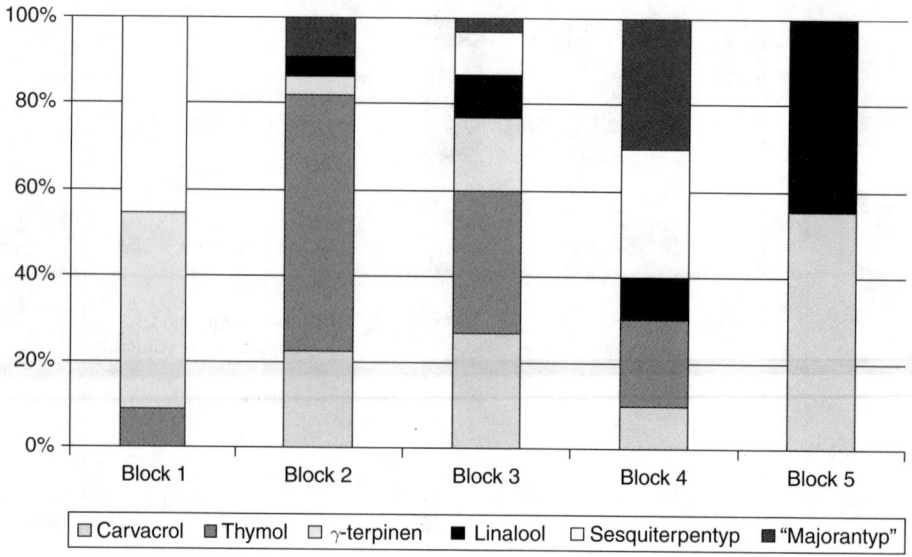

Figure 7.4 Chemotypical segregation of five single plant offsprings of *Origanum vulgare* ssp. crossings (Marn et al., 1999).

Biosynthesis

Knowledge on the biosynthesis of the essential oil compounds and their inheritance is useful for a more effective selection and establishment of a targeted breeding programme (Figure 7.5). In contrast to e.g. *Mentha* sp. where the biosynthesis and inheritance of the monoterpenes is known in detail (Croteau and Gershenzon, 1994), only some key enzymes have been identified so far for carvacrol, thymol and linalool synthesis in *Origanum* (Croteau and Karp, 1991). Perhaps some parallelism could be seen with the biogenetic sequence in *Thymus vulgaris* (Gouyon and Vernet, 1982) since thymol, carvacrol and linalool are also the main compounds. As regards *O. majorana*, Novak et al. (2001) studied *cis*- and *trans*-sabinene hydrate in several occassions (Figure 7.6).

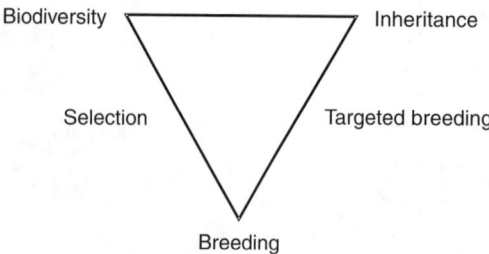

Figure 7.5 "A plant breeding program ... requires a basic knowledge of the inheritance of at least the major essential oil compounds" (Murray and Reitsema, 1954).

1...*cis*-sabinene hydrate
2...*trans*-sabinene hydrate
3...*cis*-sabinene hydrate acetate
4...sabinene

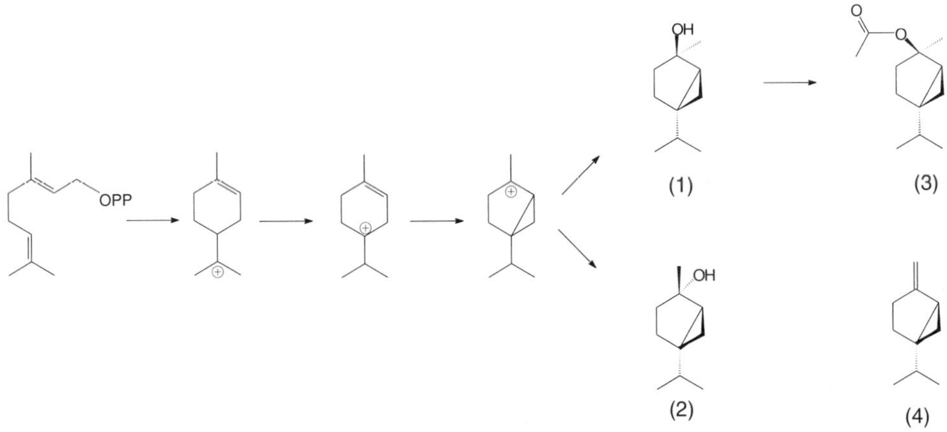

Figure 7.6 Biosynthetic "route" to the formation of *trans*- and *cis*-sabinene hydrate.

BREEDING TARGETS

Owing to the extremely large morphological and chemical variety encountered in *Origanum* sp. and taking into account the demand for homogenous raw material, wild collection accompanied with quality as well as species maintaining assurance systems (sustainability, GHP) and/or field production of reliable genotypes are the future methods of choice for quality products. As to the latter, selection and breeding represent an important part of the quality assurance system. The mentioned large variability represents at the same time a challenge for the breeder in search of homogeneity and an excellent basis for selection activities.

Generally, there are three major uses of oregano:

1 as ornamental in gardens, especially rockeries,
2 as kitchen herb responsible for the classical "pizza" oregano flavour,
3 for industrial purposes as source of antimicrobial essential oils and/or antioxidant extracts.

Ornamentals

Many oregano species and varieties offered as ornamentals are selected on their visual appearance only, and therefore without relevance for herb or essential oil production due to their low essential oil content and atypical composition. Some of them would still not be interesting for crude herb production even if containing much essential oil since the unusual leaf colour would result in an "off-colour" impression.

Quality characteristics

Regarding the use of oregano as crude drug or as source of essential oils and extracts, the following parameters are of particular interest:

- Composition of essential oils (in oregano high carvacrol content, in marjoram high *cis*-sabinene hydrate content);
- Quantity of essential oils (in marjoram more than 2 per cent is desired);
- Colour of the dried herb (green is preferred over grey);

Essential oil composition

There is now a very clear quality profile for "oregano" to be observed: carvacrol is regarded as the valuable sensorial as well as important antimicrobial compound. Breeding will focus therefore mainly on the optimisation of the carvacrol content with as less other compounds present as possible to not interfere with sensorial/antimicrobial properties of the end product. Selecting the right chemotype means at first of course selecting for the presence of carvacrol, but on the second hand also selecting against other compounds like e.g. *cis*- and *trans*-sabinene hydrate, linalool or even thymol. A first rough selection can be based on simple organoleptical testings for the characteristic phenolic smell. It can easily be distinguished between the carvacrol and thymol taste, carvacrol being rather pungent than thymol.

Sampling for the determination of the carvacrol content is crucial due to the many changes in the essential oil profile during the ontogenesis. So it was observed that phenolic compounds were increasing during the vegetation period (Putievsky *et al.*, 1985). But these compounds are also increasing from the top to the bottom of a plant, which has been proven for thymol by Werker *et al.* (1985), a fact that has to be considered in sampling, especially if using minor quantities of the herb for chemical analyses to not damage the plant before seed ripening if the selection is based on single plant evaluations. In addition, studies concerning the influence of environmental factors have shown the effect of temperature and daylength on the essential oil of oregano. As the day becomes longer, the essential oil content as well as the phenolic compounds increase, whereas at short day conditions (spring and autumn) the precursor *p*-cymene prevails (Dudai *et al.*, 1992). Similar results have also been obtained with *O. majorana* (Circella *et al.*, 1995).

Apart from the typical "oregano" profile correlated with the carvacrol content, the enormous inter- and infraspecific chemical polymorphism of *Oregano* sp. offers a wide range for selection towards the production of specific monoterpenes as fine chemicals, new odour and flavour profiles a.s.o. In *O. syriacum*, for instance, a certain geraniol and geranyl-esters as well as ethylcinnamate content is responsible for "a tender desert note" of the herb (Fleisher and Fleisher, 1991). And some of the minor compounds of *O. vulgare* ssp. *hirtum* have shown a very high antioxidant effect (Deans *et al.*, 2000).

Essential oil content

Regarding the essential oil content, the breeding targets will differ significantly between genotypes to be selected for herb production or for essential oil production. Due to the enormous possible range of the essential oil content (see above) an optimisation has to be considered in the context of use. While oregano used as herb in cooking contains in average between 1 and 3 per cent essential oil, genotypes having more than 3 per cent oil would make it necessary to change recipes. But by using less herb on a pizza, for example, the visual impression of herbs would almost be absent. Therefore, the essential oil content should for that purpose not exceed approx. 3 per cent. For the production of the essential oil, on the other side, the situation is quite different, since by selecting for essential oil contents of more than 7 per cent the productivity is dramatically increasing.

It is very likely, that in the future there will be separate genotypes selected for herb and essential oil production, respectively. This will of course not exclude using herb genotypes for essential oil production in combined systems, where e.g. the first cut is sold as herb, the consecutive cuttings in a year may be used for essential oil production.

Leaf colour and density of non-essential oil producing trichomes

Selection for favourable leaf colour and against non-essential oil producing trichomes are breeding targets for the sensorial impression of the herb. These are of course characters only to be observed for the herb use and not for essential oil production. The importance of leaf colour in herb production is not equal but depends on the requested quality features of the customer, on necessary post-harvest processes and selling as fresh, frozen or dried material. For marjoram, a green leaf colour is preferred, whilst greyish is more typical for oregano. Leaf colour has of course a higher impact on fresh and frozen than on dried material.

Agronomical Characters

Agronomical characters are of great importance for a higher productivity of the crop but in some cases also prerequisites for certain cultivation techniques (like e.g. seed weight). The most significant features are:

- Yield of dry matter;
- Upright growth (to avoid soil contamination and spoilage of leaves);
- Ratio of leaves to stem (of special interest for herb-processing companies);
- Quick development of young plants (especially for marjoram which is a slowly established crop and is rather weak when facing weed problems);
- Resistance to pathogens (for example, *O. majorana* is severely affected by *Alternaria* and *Fusarium*);
- Salt and drought tolerance (a much desired trait in Mediterranean areas);
- Winterhardiness (desired for biennial/perennial production in Central Europe).

The variability stated for the essential oil content and composition continues in the agronomical characters. Out of this group, yield is of course one of the most important parameters securing the necessary productivity for being competitive on the market. The yield can be expressed as weight per area unit in herb production or as essential oil yield per area unit. The variation between single plants can range between approx. 10 g dried leaf/flower-fraction per plant up to 250 g (Marn, 1999). Therefore also in yield high progress can be obtained within a relatively short time of breeding/selecting.

Of somewhat minor importance than yield are the leaf-stem ratio and machinability. However, having the opportunity to observe also these characters in breeding will result in better production techniques and qualities.

Oregano (*Origanum* sp.) belongs to the species with very small fruits (nutlets) weighing only approx. 60 µg per seed (thousand seed mass = 0.06 g) (Thanos, 1995). Also due to this reason direct sowing of oregano is difficult and until now plantlet production under protected conditions and planting is preferred. Selecting for higher seed weight will be a first step to enhance the production technique of direct sowing, since seed quality, germinability and vigour depend on it.

In a situation like oregano, where the selection starts from wild collected populations, non-resistant genotypes should of course be very strictly eliminated to avoid susceptibility in selected high-productive genotypes.

Breeding methods

Since breeding of oregano will start principally from wild populations, simple positive mass selection will still be the method of choice for a first progress. It has to be obeyed, that *Origanum* species are open-pollinated species. Therefore characters evaluated and selected before flowering will guarantee a by far better selection success than characters selected after flowering.

If valuable character expressions are scattered over several populations crossings are necessary to combine them in one genotype. Prerequisite of such breeding programmes is the knowledge of the floral and reproduction biology. Artificial pollination will be difficult due to the small flower size and the high number of flowers within an inflorescence. Here, the in Lamiaceae often occurring male sterility could be used to control crossings efficiently (Lewis and Crowe, 1952; Kheyr-Pour, 1980, 1981). Such a system of pollination control like male sterility could also be used to produce hybrids and to use advantages of hybrid systems like heterosis effects and uniformity as has been already established for marjoram (*Origanum majorana*) (Circella *et al.*, 1995; Novak *et al.*, 1999). Beyond "hybrids" in hybrid breeding also naturally occurring interspecific hybrids do exist in a vast number (Appl, 1928; Ietswaart, 1980) indicating the principal possibility to artificially create interspecific hybrids. In the case of the

Table 7.1 Ornamental Origanum cultivars used mainly in rockeries (Facciola, 1990)

Name	Description
White Anniversary	Bright green leaves, broadly margined in white. Spring growth in a white ground hugging mat, changing to a pale cream by fall.
Golden Oregano (*O. vulgare* ssp. *vulgare* "Aureum"; Golden Creeping Oregano)	Compact, creeping habit, to 6 inches high. Attractive golden coloured foliage. Good ground cover for rock gardens and edges of flower beds. Mild, thyme-like oregano flavour.
Silver (Silver Oregano)	Ornamental silver leaves. Mild oregano flavour. Can be used in cooking. Tender perennial.
Variegated Oregano	Attractively streaked with golden variegation that contrasts prominently against the deep green background. Mildly flavoured. Excellent for edging or in the rock garden.

spontaneous *Origanum × intercedens* hybrid (Kokkini and Vokou, 1993), an interesting increase in the biomass yield has been detected, thus suggesting a particular economic significance of this plant. In some cases, however, the hybridisation is accompanied by complete sterility and therefore further genetic improvement of the material is not possible. On the other hand, if the hybrid is an exceptionally interesting plant, its

Table 7.2 Recently developed cultivars/varieties of *Origanum* sp. (1995–2000)

Species	Cultivar/ Variety	Country	Year of Reg.	Breeding method	Specific characters
O. dictamnus		GR		Selection	High biomass
O. × intercedens		GR		Selection	Carvacrol-rich, high ess. oil, high biomass
O. majorana	Erfo	D	1997		
	Marietta	D	1997		
	Max	D	1997		
	Tetrata	D	1997		
	G1	F	1998	Polycross	% essential oil = XX, CSH 40%
	G2	F	1998	Polycross	% essential oil = XX, CSH 20%
	G3	F	1998	Polycross	% essential oil = XX, CSH 26%
	G4	F	1998	Polycross	% essential oil = XX, CSH 16%
O. microphyllum		GR		Selection	*Cis*-sabinene hydrate rich
O. onites		GR		Selection/ Self pollination	Carvacrol rich, High ess. oil yield
O. sp. *syriacum*	Carmeli	ISR	1999	Selection	Carvacrol type
	Tavor	ISR	1999	Selection	Thymol type
O. virens	9.103	F		Clonal selection	% essential oil = 2–3%, L 40 %, T 20%
	9.112	F		Clonal selection	% essential oil = 2–3%, L 20 %, T 30%
	9.105	F		Clonal selection	% essential oil = 2–3%, T 35%
	9.136	F		Clonal selection	% essential oil = 2%, C 40%
O. vulgare	Oren	ISR	1998	Selection	dry and fresh marked-high yield
	O-1	DK	2000	Recurrent Pedigree	1, 75% rosmarinic acid, high intensity of flavour, ozone/white fly tolerant
O. vulgare ssp. *hirtum*	Hirtum	GR		Selection/ self pollination	Carvacrol-rich, high ess. oil yield

Disposal of the above data is geatly acknowledged to:
D,F – Pank, BAZ
DK – B.F. Knudsen, Bot. Operations ApS
F – A. Panel, ITEIPMAI
GR – M. Skoula, MAICH
ISR – E. Putievsky, Neve Ya'ar
Notes
C, Carvacrol; T, Thymol; L, Linalool; CSH, *cis*-sabinene hydrate.

economic cultivation could still be possible by using vegetative propagation methods (stem cuttings or *in vitro* techniques), as shown by several clonal varieties.

Breeding of essential oil crops as e.g. *Origanum* sp. needs also several analytical methods. For the determination of the essential oil content usually distillation techniques are employed. Apart from microdistillation they have the disadvantages of using too much material and being time-consuming. A possible future method for a simple and fast determination of the essential oil content on only minor quantities (or even better in the field on living plants) could be by NIRS, a method successfully used to determine essential oil content in marjoram (Schulz, 1999).

The most common analytical methods used today for determining the essential oil composition are GC/MS (gas chromatography coupled with mass spectrometry), but also GC coupled with olfactorial tests (scent detectors), and head-space-techniques for the determination of genuine oils. The application of all these analytical methods makes breeding more time consuming and expensive and has to be taken in consideration when designing a breeding program for a new and up to now under-utilised crop.

Breeding results: Oregano varieties

A number of commercial varieties of oregano exists on the market, many of them used as horticultural crops, particularly ornamentals suited for rockeries (Table 7.1). Systematic genetic improvement of oregano and establishment of new cultivars has started, in addition, in the context of domestication and introduction as new crop. Systematic breeding programmes are known in the meantime from several mostly mediterranean countries using indigenous wild species as starting material. In Turkey, for instance, *Origanum onites* is the species mainly used, in Greece *O. vulgare* ssp. *hirtum* and *O.* × *intercedens*, whereas in Israel besides of *O. vulgare* ssp. *hirtum* especially *O. syriacum* ("Za'atar") is preferred and in France *O. virens* (Ceylan *et al.*, 1996; Pasquier, 1997; Putievsky *et al.*, 1997; Skoula and Kamenopoulos, 1997). According to an actual survey a number of varieties and cultivars of *Origanum* sp. have been bred and registered within the last years, compiled in Table 7.2.

CONCLUSION

Due to the high biodiversity of *Origanum* sp., especially the high chemical and physiological heterogeneity, crop improvement is highly recommended in consideration of the widespread use of oregano and of the difficulties caused by non-uniform material to supplier as well as customer. Taking into account both producers' and user's needs, efforts of any oregano breeding should be directed towards the improvement of yield-related as well as quality-related targets. To achieve these goals, selection and hybridisation methods combined with analytical controls are the most appropriate tools for breeding of oregano. Up to now, local strains of *O. vulgare* ssp. and *O. majorana* as well as spontaneous hybrids are traditionally cultivated in many countries. In addition, several ornamental varieties are present on the seed and plantlet market. Systematic breeding of oregano has started, however, just in relatively recent times, focused mainly on *O. majorana*, *O. onites*, *O. syriacum*, *O. virens* and *O. vulgare* ssp. *hirtum*. Chemogenetic results and male sterility have been used up to now as main tools for controlled crossings. Results so far are promising, but further research is needed, particularly for investigating the species genetic background and possible applications of biotechnology. Investments in such new varieties are rather high since agronomic as well as chemical targets have to be considered. Due to the relatively unfavourable cost: benefit-ratio to be expected with specialist minor crops as e.g. oregano, a strong variety protection is needed. For choosing the appropriate type of protection, a case by case decision will be necessary: plant variety rights if the plant material meets the DUS (distinctness–uniformity–stability) requirements, plant patents, on the contrary, if the prerequisites novelty, inventive step and industrial applicability are satisfied (Franz, 2001). Breeding of *Origanum* sp. is, however, a challenge to obtain herbs, essential oils, natural antioxidants or antimicrobial agents of high quality with an economic significance not to be neglected.

REFERENCES

Appl, J. (1928) Über einen Bastard von *Origanum majorana* und *Origanum vulgare* und dessen Aufspaltung in der F_2-Generation. *Preslia* 6, 3–13.

Baser, K. H. C., Ozek, T., Kurkcuoglu, M. and Tümen, G. (1996) Essential oil of *Origanum laevigatum* Boiss. *J. Essential Oil Res.* 8, 185.

Bernáth, J. (1997) Some scientific and practical aspects of production and utilisation of oregano in central Europe. Oregano. Promotion the conservation and use of under-utilised and neglected crops 14. S. Padulosi (ed.), *IPGRI* pp. 76–93.

Ceylan, A., Bayram, E., Schneider, M., Weinbrenner, G. and Marquard, R. (1996) Selective breeding with *Origanum onites* based on collections from the wild flora in West-Turkey. *Beiträge zur Züchtungsforschung* 2(1), 167–170.

Circella, G., Franz, C.H., Novak, J. and Resch, H. (1995) Influence of day length and leaf insertion on the composition of Marjoram essential oil. *Flavour Frag. J.* bd. 10, 371–374.

Croteau, R. and Gershenzon, J. (1994) Genetic control of monoterpene biosynthesis in mints (*Mentha*: Lamiaceae). In B.E. Elli, G.W. Kuroki and H.A. Stafford (eds), *Genetic Engineering of Plant Secondary Metabolism Plenum Press*, New York.

Croteau, R. and Karp, F. (1991) Origin of natural odorants. In P.M. Müller and D. Lamparsky (eds), *Perfumes – Art, Science and technology*, Elsevier Applied Science, London and New York.

Deans, S.G., Dorman, H.J.D., Vender, C., Novak, J. and Skoula, M. (2000) Antioxidant properties of the volatile oil and phytoconstituents of *Origanum vulgare* ssp. *hirtum*. In *Proceedings of 2nd International Symposium on Breeding Research on Medicinal and Aromatic Plants*, Chania, Crete, Greece.

Dudai, N., Putievsky, E., Ravid, U., Palevitch, D. and Halevy, A.H. (1992) Monoterpene content in *Origanum syriacum* as affected by environmental conditions and flowering. *Physiologia Plantarum* 84(3), 453–459.

Facciola, S. (1990) Cornucopia – A Source Book of Edible Plants, Vista, *CA:Kampong Publications*.

Fleisher, A. and Fleisher, Z. (1991) Chemical composition of *Origanum syriacum* L. essential oil. Aromatic plants of the holy land and the sinai, Part V. *J. Essent. Oil Res.* 3, 121–123.

Franz, Ch. (2001) Plant Variety Rights and Specialised Plants. *Proc. PIPWEG Conf.*, Angers (F), January 2001, pp. 131–138, Sheffi Ander Press.

Gouyon, P.H. and Vernet, P.H. (1982) The consegences of gynodioecy in natural populations of *Thymus vulgaris* L. *Ther. Appl. Genet.* 61, 315–320.

Kheyr-Pour, A. (1980) Nucleo-cytoplasmic polymorphism for male sterility in *Origanum vulgare* L. *J. Hered.* 71, 253–260.

Kheyr-Pour, A. (1981) Wide nucleo-cytoplasmic polymorphism for male sterility in *Origanum vulgare* L. *J. Hered.* 72, 45–51.

Kokkini, S., Vokou, D. and Karousou, R. (1991) Morphological and chemical variation of *Origanum vulgare* L. in Greece. *Botanica. Chronica.* 10, 337–346.

Kokkini, S. and Vokou, D. (1993) The hybrid *Origanum* × *intercedens* from the island of Nisyros (SE Greece) and its parental taxa: Comparative study of essential oils and distribution. *Biochem. Syst. Ecol.* 21(3), 397–403.

Ietswaart, J.H. (1980) *A Taxonomic Revision of the Genus Origanum (Labiatae)*, Leiden: Leiden University Press.

Lewis, D. and Crowe, L.K. (1952) Male sterility as an outbreeding mechanism in *Origanum vulgare. Heredity abst.* 6, 136–136.

Marn, M., Novak, J. and Franz, Ch. (1999) Evaluierung von Nachkommenschaften von *Origanum vulgare. Z Arzn Gew pfl* 4, 171–176.

Novak, J., Bitsch, C., Marn, M., Langbehn, J., Pank, F., Blüthner, W.D., Marchart, R., Junghanns, W. and Franz, Ch. (1999) Hybridsortensystem bei Majoran (*Origanum-majorana* L.). Vereinigung österreichischer Pflanzenzüchter. *BAL Gumpenstein*. pp. 77–82.

Novak, J. and Franz, Ch. (2001) Composition of essential oil compounds of a historic sample of Majoram (*Origanum majorana* L.) *Flavour Fragr. J.* (submitted).

Novak, J., Bitsch, C., Langbehn, J., Pank, F., Skoula, M., Gotsiou, Y. and Franz, Ch. (2000). Ratios of *cis*- and *trans*-Sabinene hydrate in *Origanum majorana* L. and *Origanum microphyllum* (Bentham) Vogel. *Biochem. Systematics Ecol.* 28, 697–704.

Olivier, G.W. (1997) The world market of oregano. Oregano, promotion of the conservation and use of under-utilised and neglected crops 14. S. Padulosi (ed.), *IPGRI*, 142–146.

Özgüven, M., Schneider, M. and Marquard, R. (1996) Yield and quality aspects of *Origanum* wild species collected in the cukorova region of Turkey. *Beiträge zur Züchtungsforschung* 2(1), 21–24.

Pasquier, A. (1997) Selection work on *Origanum vulgare* in France. Oregano, promotion of the conservation and use of under-utilised and neglected crops 14. S. Padulosi (ed.), *IPGRI*, pp. 94–99.

Putievsky, E., Ravid, U. and Husain, S.Z. (1985) Differences in the yield of plant material, essential oils and their main components during the life cycle of *Origanum vulgare* L. In A. Baerheim-Svendsen and J.J.C. Scheffer (eds), *Essential oils and Aromatic plants*, Dordrecht, The Netherlands:Martinus Nijhoff/Dr W. Junk Publishers, pp. 185–189.

Putievsky, E., Dudai, N. and Ravid, U. (1997) Cultivation, selection and conservation of oregano species in Israel. Oregano, promotion of the conservation and use of under-utilised and neglected crops 14. S. Padulosi (ed.), *IPGRI*, pp. 103–110.

Schulz, H., Krüger, H., Steuer, B. and Pank, F. (1999) Bestimmung von Inhaltsstoffen des Majoran (*Origanum majorana* L.) mittels Nah-Infrarot-Spektroskopie. *Z Arzn Gew pfl* 4, 62–67.

Skoula, M. and Kamenopoulos, S. (1997) *Origanum dictamnus* L. and *Origanum vulgare* L. subsp. *hirtum* (Link) Ietswaart: Traditional uses and production in Greece. Oregano, promotion of the conservation and use of under-utilised and neglected crops 14. S. Padulosi (ed.), *IPGRI*, pp. 26–32.

Skoula, M., Gotsiou, P., Naxakis, G. and Johnson, C.B. (1999) A chemosystematic investigation on the mono- and sesquiterpenoids in the genus *Origanum* (*Labiatae*). *Phytochemistry* 52(4), 649–657.

Thanos, C.A., Kadis, C.C. and Skarou, F. (1995) Ecophysiology of germination in the aromatic plants thyme, savory and oregano (*Labiatae*). *Seed Sci. Res.* 5(3), 161–170.

Verlet, N. (1994) An Overview of the Medicinal and Aromatic Plants Industry. ATTI Convegno Internazionale "Coltivazione e Miglioramento di Piante Officinali", Trento, 2–3 giugno 1994, ISAFA, 251–264.

Werker, E., Ravid, U. and Putievsky, E. (1985) Structure of glandular hairs and identification of the main components of their secreted material in some species of the Labiatae. *Isr. J. Bot.* 34(1), 31–45.

Part 5
Pharmacology

8 The biological/pharmacological activity of the *Origanum* Genus

Dea Baričevič and Tomaž Bartol

INTRODUCTION

In the past, several classifications were made within the morphologically and chemically diverse *Origanum* (*Lamiaceae* family) genus. According to different taxonomists, this genus comprises a different number of sections, a wide range of species and subspecies or botanical varieties (Melegari *et al.*, 1995; Kokkini, 1997). Respecting Ietswaart taxonomic revision (Tucker, 1986; Bernath, 1997) there exist as a whole 49 *Origanum* taxa within ten sections (*Amaracus* Bentham, *Anatolicon* Bentham, *Brevifilamentum* Ietswaart, *Longitubus* Ietswaart, *Chilocalyx* Ietswaart, *Majorana* Bentham, *Campanulaticalyx* Ietswaart, *Elongatispica* Ietswaart, *Origanum* Ietswaart, *Prolaticorolla* Ietswaart) the majority of which are distributed over the Mediterranean. Also, 17 hybrids between different species have been described, some of which are known only from artificial crosses (Kokkini, 1997). Very complex in their taxonomy, *Origanum* biotypes vary in respect of either the content of essential oil in the aerial parts of the plant or essential oil composition. Essential oil 'rich' taxa with an essential oil content of more than 2 per cent (most commercially known oregano plants), is mainly characterised either by the dominant occurrence of carvacrol and/or thymol (together with considerable amounts of γ-terpinene and p-cymene) or by linalool, terpinene-4-ol and sabinene hydrate as main components (Akgül and Bayrak, 1987; Tümen and Başer, 1993; Kokkini, 1997).

The *Origanum* species, which are rich in essential oils, have been used for thousands of years as spices and as local medicines in traditional medicine. The name hyssop (the Greek form of the Hebrew word 'ezov'), that is called '*za'atar*' in Arabic and *origanum* in Latin, was first mentioned in the Bible (Exodus 12: 22 description of the Passover ritual) (Fleisher and Fleisher, 1988). A comparative study of the traditional use of oregano-like herbs in the Mediterranean region, made by Fleisher and Fleisher (1988), established that the hyssop of the Bible is the carvacrol chemotype of the plant *Majorana syriaca* (L.) Feinbr. (syn.: *Origanum maru* L., *Origanum syriacum* L.). This plant, having a curative value in hypoglycaemic treatments (Yaniv *et al.*, 1987), was an important part of the purification rites and was used as a medicine and as a condiment (Fleisher and Fleisher, 1988). A number of plants were found to have a similar flavour to that of hyssop, such as *Coridothymus capitatus* (L.) Reichenb. (Spanish oregano), *Satureja thymbra* L., *Thymbra spicata* L. and *Origanum vulgare* L. After the destruction of the Temple of Jerusalem at the beginning of the Christian era, the ritual use of hyssop ceased, while the tradition of using hyssop as a flavouring persisted, giving rise to two cultures of condiments, characterised by the high content of carvacrol in essential oils: '*za'atar*' in the Middle East and oregano in Europe.

Carvacrol-rich essential oils of *Origanum heracleoticum* L. (syn.: *O. hirtum* L., *O. creticum* Sieber × Bentham, *O. vulgare* L. subsp. *hirtum* (Link) Ietswaart), *O. maru* L. and *O. smyrnaeum* L. (syn.: *O. onites* L., *Majorana smyrnaea* (L.) Kostel, *M. onites* (L.) Bentham), that are native to Turkey and other East Mediterranean countries, have been used as a seasoning and in local materia medica in Turkey.

Aerial flowering parts of *O. vulgare* ssp. *viride* (Boiss.) Hayek (*O. viride* (Boiss.) Halacsy) are used in Iranian traditional medicine as diuretic, stomachic, antineuralgic, antitussive and expectorant. Afsharypuor et al. (1997) report on composition of essential oil of *O. vulgare* ssp. *viride* (Boiss.) Hayek, that grows wild in northern parts of Iran (with linalyl acetate, sabinene, β-caryophyllene as main components) and differs substantially from the composition of essential oil of the same species, growing wild in the Balkan area (Bulgaria, Albania, Turkey, Greece, Yugoslavia) (carvacrol chemotypes) or cultivated in Israel (thymol chemotype).

O. dubium Boiss., an endemic Mediterranean shrub, is widely spread on Cyprus (Paphos forest), in Greece and in Southern Turkey. In Cyprus it is locally known as 'rigani' and widely used in the preparation of local foods like 'souvla' and 'suovlakia'. In the traditional Cyprus medicine, an infusion of the leaves, flowering stems and flowers is used as a digestive and carminative, while carvacrol-rich essential oil is used externally as an antirheumatic (Arnold et al., 1993). In Greece, similar local names ('righani', 'aroani', 'rhoani') and traditional uses are found to be used for *Origanum onites* L. (= *O. smyrnaeum* L.). The essential oil of *Origanum majorana* L. var. *tenuifolium* Weston, which is endemic to Cyprus and locally named 'sampsishia', is traditionally used against common cold and fever or as a spasmolytic in gastro-intestinal disorders when applied internally or as an antirheumatic after external administration. *Cis*-sabinene hydrate (leaves: 33.3 per cent, flowers: 24.0 per cent, stems: 7.4 per cent), terpinene-4-ol (leaves: 21.6 per cent, flowers: 16.6 per cent, stems: 19.0 per cent), γ-terpinene (leaves: 8.3 per cent, flowers: 10.6 per cent, stems: 11.1 per cent) and α-terpineol (leaves: 7.3 per cent, flowers: 12.4 per cent, stems: 14.2 per cent) were found to be the leading compounds in 'sampsishia' essential oil, obtained after hydrodistillation of the above-ground plant parts. *Cis*-sabinene hydrate was also found to be the major component (21.5 per cent) of the essential oil of *Origanum rotundifolium* Boiss., aerial parts of which are used as a flavoured herbal tea in Turkey (Başer et al., 1995).

Aerial parts of *Origanum hypericifolium* O. Schwarz et P.H. Davis, an endemic Turkish species, collected before flowering, are used in Turkey (Burdur, Gölhisar) as a condiment, as meat flavouring and as herbal tea for treatment of common cold, stomach complaints and debility (Başer and Tümen, 1994). It was found, that aerial parts of this herb, when obtained in pre-flowering stage, contained relatively high amounts of essential oil (2.5 per cent), rich in carvacrol (64.3 per cent).

Aerial parts of *Origanum sipyleum* L. (syn.: *Majorana sipylea* (L.) Kostel, *Amaracus sipyleus* (L.) Rafin), a species of the eastern Mediterranean area, is widely used as a spice (central Anatolia) and against gastrointestinal disorders and cough (west Anatolia). The essential oil of this species was found to be rich in γ-terpinene and aromatic monoterpenes (Başer and Tümen, 1992). Dry leaves and flowering tips of *O. majorana* L. (syn. *Majorana hortensis* Moench.) and their tincture are used in the formulation of vermouths and bitters. The essential oil is used in the formulation of compounded oils for flavouring sauces, condiments, canned meats and other products (de Vincenzi and Mancini, 1997). In India this plant is traditionally used as an astringent, diuretic, antihysterial, antiasthmatic and antiparalytic drug (Yadava and Khare, 1995).

Based on ethnobotanical investigation in Lavras Novas (Brazil), where a rich folklore in the use of medicinal plants has been observed due to the remoteness of villages from modern medical facilities and the lack of a health care system, *O. majorana* L. is used for treatment of earache and influenza in children (Stehmann and Brandão, 1995).

Origanum vulgare L., commonly known as oregano or wild marjoram, is a well known flavouring for many international dishes and has antioxidant applications in human health (Dorofeev *et al.*, 1989; Deighton *et al.*, 1993). Antimicrobial action is reported for *O. vulgare* extracts, which contained phenolcarboxylic acids (cinamic, caffeic, *p*-hydroxybenzoic, syringic, protocatechuic, vanillic acid) as presumably active substances (Mirovich *et al.*, 1989). Also, the fumigant toxicity of oregano essential oils for storeroom insects has been established (Shaaya *et al.*, 1991; Baricevic *et al.*, 2001). Traditionally, *Origani vulgaris herba* was used in respiratory tract disorders such as cough or bronchial catarrh (as expectorant and spasmolitic agents), in gastrointestinal disorders (as choleretic, digestive, eupeptic and spasmolitic agents) as oral antiseptic, in urinary tract disorders (as diuretic and antiseptic) and in dermatological affections (alleviation of itching, healing crusts, insect stings) (Blumenthal, 1998; Bruneton, 1999). Although, the monograph documentation of drug plant *O. vulgare* was submitted to the German Ministry of Health, the staff responsible for phytotherapeutic medicinal domain – Commission E – evaluated *O. vulgaris herba* negatively (Banz. Nr. 122 from 6th July 1988), because of lack of scientific proofs for the above mentioned indication areas (Blumenthal, 1998). Nevertheless, many of the studies confirmed the beneficial effects of oregano for human health. In view of the folk medicinal usage of New Zealand plants, the bioactivity of commercial essential oils of *O. vulgare* L. and of *O. majorana* L. was studied *in vitro* for their antibacterial, antifungal, antioxidative and spasmolytic activities. Oregano and marjoram were found to be effective antimicrobial agents and had a significant spasmolytic action on smooth muscle (Lis-Balchin *et al.*, 1996). Extracts of *O. vulgare* exhibited a high capacity to compete with progesterone binding to intracellular receptors for progesterone, and showed considerable progestine activity *in vitro* (Zava *et al.*, 1998).

Origanum vulgare, *O. majorana*, *O. dubium* and *O. dictamnus* contain in their leaves flavonoids (flavanone group – naringin, flavone group – apigenin and luteolin, flavonol group – quercetin) and flavonoid–glycosides (luteoline-7-glucoside, apigenin-7-glucoside, eriodictyol-7-glucoside) (Harvala and Skaltsa, 1986; Skaltsa and Harvala, 1987; Bohm, 1988; Soulèlès, 1990), some of which are known to possess spasmolitic activity. The antioxidative effect of plants belonging to the *Origanum* genus is probably the consequence of content of polar hydroxycinnamic derivatives and flavonoid glycosydes (Nakatani and Kikuzaki, 1987; Lamaison *et al.*, 1993; Sawabe and Okamoto, 1994; Takácsová *et al.*, 1995). These are found prevalently in essential oil 'poor' *origanum* taxa (less than 0.5 per cent of essential oil), such as *O. calcaratum* Jussieu or *O. vulgare* L. ssp. *vulgare*. Also non-polar phenolic compounds like thymol and carvacrol, which are major components of essential oil 'rich' *Origanum* taxa, possess remarkable antioxidant properties (Lagouri *et al.*, 1993).

Interesting results were obtained in Poland, where Skwarek *et al.* (1994) discovered that *O. vulgare* extracts, when applied to $ECHO_9$ Hill virus, cultured in monkey kidney cells, induced the formation of a substance with interferon-like activity. The findings of relatively new investigations on bioactive compounds from spices show oregano's potential as a source of pesticidal and cancer preventive chemicals (Anonymus, 1997; Craig, 1999). Essential oils from *O. vulgare* ssp. *hirtum* (Greek oregano), *Origanum onites* (Turkish oregano and *O. dictamnus* L. (syn. *Amaracus dictamnus* Benth., Cretan dittany), that are rich in phenolic compounds (carvacrol, thymol), possess antibacterial (Collier and Nitta, 1930; Kellner and Kober, 1954; Maruzzella and Sicurella, 1960; Aureli *et al.*, 1992; Biondi *et al.*, 1993; Vokou *et al.*, 1993) and antifungal properties against pathogenic or nonpathogenic fungi (Maruzzella and Liguori, 1958; Arras and Picci, 1984; Guérin and Réveillère, 1985; Colin *et al.*, 1989; Paster *et al.*, 1993).

In the West Indies wild oregano (Lamiaceae) is the common name for a variety of botanical names such as *Coleus amboinicus* Launert., *C. aromaticus* Benth., *Plectranthus aromaticus* Roxb. and *P. amboinicus* Launert (French Origanum). It is cultivated for its sweat-inducing and insecticidal properties and is used also as a culinary flavour, against stings from scorpions and poisonous centipedes, for cleaning textiles, and as a shampoo (Chatterjee *et al.*, 1958; Kuebel and Tucker, 1988; Prudent *et al.*, 1995). In India, the plant and its preparations are used topically against ulcers and inflammation of the mouth and internally as a digestive. In Vietnam, *C. aromaticus* is employed in the preparation of cough mixtures (Prudent *et al.*, 1995). According to Buznego *et al.* (1991) it possess antiepileptic properties. The essential oil of *C. aromaticus* is mainly composed of phenolic compounds, the major components depending on the geographical origin. *C. aromaticus* essential oil (Martinique origin), which contained carvacrol (60.96 per cent) and β-caryophyllene (13.26 per cent) as major components, showed both fungistatic (at MIC of 0.25 mg/ml towards human pathogens: *Aspergillus niger*, *Candida albicans* and towards plant pathogens: *Botrytis cinerea*, *Cylindrocarpon mali*, *Sclerotinia sclerotiorum*) and bacteriostatic activities *in vitro* (at MIC of 0.125 mg/ml towards *Mycobacterium smegmatis* and *Vibrio cholerae*, at MIC of 0.5 mg/ml towards *Staphylococcus aureus* and *Escherichia coli*) (Prudent *et al.*, 1995). Stiles and co-workers (1995) observed similar results. They found, that *O. vulgare* essential oil was an effective anti-*C. albicans* agent (MIC against 3 different Candida strains <0.1 µg/ml), and that this activity might be due to the carvacrol content.

The world trends toward increasing usage of spices indicate an overall change in food habits and tastes. The appetiser effect of culinary herb mixtures with oregano was studied in animals and in human experiments. The authors generally agree that herb mixtures, when added to the diet in a moderate (about 0.3 per cent) quantity cause higher liveweight gains of experimental animal groups (pigs, calves) than the controls (Gunther, 1991; Stenzel *et al.*, 1998), and act as appetisers in human diet (Yeomans, 1996). An improved feed conversion as well as protein and energy utilisation in fish that were fed an oregano supplemented (3 per cent) diet, was observed also by El-Maksoud *et al.* (1999). However, aversive postingestive feeding effects after initially increased preference for oregano flavoured feed as well as generalisation of aversion from familiar to novel feeds with a similar flavour were observed in animal (lambs) experiments (Launchbaugh and Provenza, 1994; Villalba and Provenza, 1996), causing lower preference for long-term feeding with oregano flavoured straw. Aversions to sauces flavoured with oregano have been reported also in human diet – in pregnant women (Hook, 1978).

ANTIFUNGAL ACTIVITY

When assessing the food-preservative and health promoting potential of spices, more attention in recent literature has been placed on the studies of inhibitory effects of oregano essential oil or its components to the fungal growth and/or sporulation than on that of oregano crude drug or its extracts. The antifungal activity is strongly correlated with the type of essential oil (that depends on plant species and origin), its concentration and pH of the testing medium *in vitro* (Deans and Svoboda, 1990; Thompson, 1990; Biondi *et al.*, 1993). Authors generally agree that there is a relationship between the chemical structure of the most abundant essential oil components and their antifungal and anti-aflatoxigenic potency. Phenols are believed to be the most potent antimicrobials followed by alcohols, ketones, ethers and hydrocarbons (Bullerman, 1977; Hitokoto *et al.*, 1980; Hussein, 1990; Daw *et al.*, 1994; Charai *et al.*, 1996). These presumptions are in accordance with the findings of Biondi *et al.* (1993), who reported that carvacrol – rich *O. onites* L. essential oil showed more potent antifungal activity against *A. niger*, *Aspergillus terreus*, *C. albicans* and against *Fusarium* spp. than the oregano oil, composed prevalently of terpinene-4-ol and of γ-terpinene. A potent antifungal effect of *O. vulgare* essential oil (rich in carvacrol and thymol) at concentration of 1 µl/ml against the common spoilage fungus *A. niger* was observed also by Baratta *et al.* (1998a).

However, there are many differences among genera of fungi with regard to sensitivity to the antifungal effects of oregano and of its essential oils. Also, the concentration of an essential oil and its origin may significantly influence their antifungal activity. When published results on the antimicrobial activity of oregano essential oils are compared, significant differences among results obtained by different research groups may be observed. These are due to different methods used in investigations, and also due to the incomplete specification (common name instead of botanical one) of the oregano species, especially, when no data are given about the chemical composition of the essential oil.

The ground *O. vulgare* (2 per cent) was found to possess a strong antifungal potential against several food-contaminating moulds, like *Trichoderma harzianum* Rifai, *Alternaria alternata* Keissler, *Fusarium oxysporum* Schlecht, *Mucor circinelloides f. griseo-cyanus* Schipper, *Cladosporium cladosporioides* de Vries, *Fusarium culmorum* Saac., *Aspergillus versicolor* Tiraboschi, but allowed selective growth of *Rhizopus stolonifer* Lind and *Penicillium citrinum* Thom in potato dextrose agar (Schmitz *et al.*, 1993). A moderate antifungal activity of oregano crude drug, when applied in a concentration of 2 per cent (the upper level, that is most often used in food industry), against *Penicillium citrinum*, *P. roqueforti*, *P. patulum* and against three mycotoxigenic fungi (*Aspergillus flavus*, *A. parasiticus* and *A. ochraceus*) was reported also by Azzouz and Bullerman (1982). Deans and Svoboda (1990) observed a significant potency of *O. majorana* essential oil against filamentous fungi, in particular mycotoxigenic strains.

An interesting comparative study between the antimicrobial activity of crude drug (ground dried plant), of aqueous extracts and essential oils of *Origanum compactum* Bentham (section *Prolaticorolla* Ietswaart) and of *O. majorana* L. of Morocco origin was made, in order to ascertain their antimicrobial potential against yeasts (*Saccharomyces cerevisiae*, *Candida utilis*, *Candida tropicalis*, *Candida lipolytica*), moulds (*Penicillium parasiticus*, *Geotrichum candidum*, *A. niger*), bacteria (*Pseudomonas fluorescens*, *E. coli*, *S. aureus*, *Bacillus cereus*) and lactic acid bacteria (*Lactobacillus plantarum*, *L. mesenteroides*) (Charai *et al.*, 1996). Sensitive microorganisms were generally more susceptible to *O. compactum* than to *O. majorana* regardless of the form of applied antimicrobial (crude drug, water extracts, essential oil). When considering the activity of the whole plant (at concentration of 1 per cent in solid media *in vitro*), yeasts proved to be the microorganisms most sensitive to the tested plant material, followed by lactic bacteria and other bacteria tested. Plants showed practically no inhibition potential towards moulds. By contrast, when considering water extracts of both plant species, moulds were the most sensitive, bacteria showed only a scarce sensibility (especially towards *O. majorana* extracts) and yeasts were not affected. Moulds were completely inhibited (*O. compactum*) or partly inhibited (*O. majorana*) by water extracts at 40 per cent concentration. Inhibition of microbial growth in the case of water extracts was probably due to the presence of water-soluble polyphenols and tannins.

Essential oil of *O. compactum* was characterised by carvacrol (49.5 per cent), *p*-cymene (21.2 per cent) and γ-terpinene (14.2 per cent) as major components, while *O. majorana* was found to be a linalool chemotype (linalool 32.6 per cent, terpinene-4-ol 22.3 per cent and *p*-cymene 8.1 per cent). It was found that both essential oils have inhibitory potential against microbial growth, *O. compactum* being the more potent inhibitor, probably due to the presence of a highly active phenolic moiety (carvacrol) in its essential oil. The following concentrations of essential oils were found to completely inhibit growth of tested microorganisms: bacteria at 4 ppm (lactic bacteria in the case of *O. compactum* at 1 ppm), yeasts at 1.6 ppm (*O. compactum*) or at 5 ppm (*O. majorana*), moulds at 1 ppm (*O. compactum*) or at 5 ppm (*O. majorana*).

Phenolic compounds, characteristic major constituents of essential oils of *O. vulgare* L. (section *Origanum* Ietswaart) or of *O. dictamnus* L. (section *Amaracus* Bentham), are probably responsible for the high inhibitory activity of carvacrol/thymol chemotypes of oregano against fungal growth, conidial germination and production of *Penicillium digitatum* (at essential oil concentration of 250–400 ppm) (Daferera *et al.*, 2000). Moreover, monoterpene components, which are present in essential oils in different proportions, seem to have more than an additive effect in fungal inhibition. Phenolic derivatives, present in essential oils, may also be involved in inhibition of yeast sporulation through depletion of cellular energy by reduction of respiration. It has been shown that oregano essential oil (at 100 ppm) significantly impaired the respiratory activity of *Saccharomyces cerevisiae* as evidenced by a reduction in CO_2 and ethanol production (Conner *et al.*, 1984). The growth of variety of yeasts was significantly inhibited by oregano essential oils (at concentration of 200 ppm: *Brettanomyces anomalus*, *G. candidum*, *Kluyveromyces fragilis*, *Lodderomyces elongisporus*, *Metchnikowia pulcherrima*, *Pichia membranaefaciens*, *Saccharomyces cerevisiae*, *Torulopis glabrata*, at 100 ppm: *Candida lypolytica*, *Hansenula anomala*, *Kloeckera apiculata*, *Rhodotorula rubra*, *Debaryomyces hansenii*), that were rich in phenolic compounds (Conner and Beuchat, 1984a). When exposed to sublethal heat treatments (48–54 °C), which correspond to those during food processing, the increased sensitivity of yeasts to oregano essential oil (at levels 100–200 ppm) or to oregano oleoresins (at levels 250–500 ppm) was observed. This might be due to the synergistic or additive effect of oregano oil or oleoresin (in combination with heat) on inactivation of yeasts and due to the affected recovery of heat stressed yeast cells after treatment with oleoresin, resulting in low viable populations (Conner and Beuchat, 1984b; Conner and Beuchat, 1985). Karanika *et al.* (2001) studied the inhibitory effect of aqueous extract (at concentration of 5 g/l) of *O. dubium* on *Yarrowia lipolytica*. It was found that the increased lag time and suppressed specific growth rate of this yeast were not attributed to the direct effect of the extract on yeast cells, but to the *O. dubium*-induced chelation of metal ions, which are essential for microbial growth.

Biological assays showed strong fungitoxic activity of essential oil (at 1000 ppm) from Turkey native populations of *Origanum minutiflorum* Schwarz Davis against *Fusarium moniliforme*, *Rhizoctonia solani*, *S. sclerotiorum* and *Phytophthora capsici*. It was found that the antifungal effect was due to the presence of phenolic components in the essential oil, like carvacrol and/or thymol, which are well known for their antifungal potency (Kurita et al., 1981; Farag et al., 1989; Curtis et al., 1996). Carvacrol or thymol, when applied in concentrations of more than 100 ppm led to a complete inhibition of growth of the above mentioned fungi *in vitro*. These phenolic compounds showed higher inhibition against *P. capsici* than the soil-applied systemic fungicide Previcur N (Muller-Riebau et al., 1995). Oregano essential oil, thymol and carvacrol (at 0.025 per cent or 0.05 per cent) were also found to be strong growth-inhibitory compounds against *Penicillium roqueforti, G. candidum* and *Mucor* spp. (Akgül and Kivanç, 1988; Akgül and Kivanç, 1989). Thompson (1990) established that carvacrol exhibited a fungicidal effect against *Aspergillus* spp., that was pH-dependent, and at pH 4 the fungicidal effect was more potent than at pH 6.

Adam et al. (1998) found a valuable therapeutic potency of essential oil of *O. vulgare* L. subsp. *hirtum* against experimentally induced dermatophytosis in rats (infection with *Trichophyton rubrum*). They report that carvacrol and thymol showed much higher antifungal activities against human pathogens (keratinophilic fungus *T. rubrum*, yeasts: *Malassezia furfur* and *Trichosporon beigelii*) than their biosynthetic precursors γ-terpinene and *p*-cymene.

Dose-dependent antiaflatoxinogenic effects of oregano (*O. vulgare* L.) essential oil and of carvacrol after exposure to toxigenic strains of *A. flavus* and *A. parasiticus* were observed in laboratory conditions (Llewellyn et al., 1981; Akgül et al., 1991; Özcan, 1998). Aflatoxins of *A. flavus* and *A. parasiticus* have been shown to be both toxic and carcinogenic in test animals and are also involved in the etiology of human liver cancer. Carvacrol, if used in sufficient amounts, can be an effective inhibitor of the growth and toxin production by *A. flavus* and *A. parasiticus*. At 500 ppm (28 °C) carvacrol completely inhibited growth and toxin production by both strains, aflatoxins (B_1, B_2, G_1) production being inhibited at lower concentrations than fungal growth. *A. flavus* was more sensitive to carvacrol than *A. parasiticus* (Akgül et al., 1991). Oregano (*O. vulgare* L.) essential oil, when applied in concentrations 1000 ppm or 750 ppm to a liquid culture medium, completely inhibited *A. ochraceus* growth and ochratoxin A production for up to 21 days and 14 days, respectively (Basilico and Basilico, 1999). When applied at 500 ppm, this essential oil was not effective. According to the results of Paster et al. (1990) higher concentrations of oregano essential oil are needed for total inhibition of spore germination (600 ppm) than for prevention of mycelial growth (400 ppm) of *A. flavus, A. ochraceus* or *A. niger*. The data obtained in the study of Hammer et al. (1998) showed inhibitory effect of *O. vulgare* essential oil (MIC = 0.12 per cent) against *C. albicans in vitro* (agar dilution method).

Oregano, its essential oil or isolated compounds, was studied also from the aspect of possible applications in plant protection, in post-harvest crop/fruit protection or in apiculture, where species specific fungi endanger the production systems.

The *in vitro* inhibitory effect of *O. vulgare* leaf extracts was observed in rice pathogen *Drechslera oryzae* (*Cochliobolus miyabeanus*) (Bisht and Khulbe, 1995). The strong antifungal effect of *O. vulgare* L. essential oil in relatively low concentrations (200 ppm) was observed against *Penicillium italicum* Wehm., *Botrytis cinerea* Pers., *Alternatia citri* Ell. et Pierce, that frequently infect fruits after harvesting and are difficult to control with synthetic phytopharmaceuticals due to the appearance of resistant strains (Arras and Picci, 1984). *Origanum compactum* Bentham essential oil (with carvacrol content of 37.7 per cent, Morocco origin) completely inhibited all three asexual reproduction steps (spore germination, mycelial elongation and sporulation) of common food contaminating fungi at concentrations of 1 per cent (*A. niger*, *Penicillium italicum*) or 0.1 per cent (*Zygorrhynchus* spp.) (Tantaoui-Elaraki *et al.*, 1993). The antifungal activity of essential oils, at all stages of asexual reproduction, was related to concentration, i.e. the higher the concentration the greater the inhibition. The fungistatic activity of essential oils of *O. syriacum* L. (syn.: *M. syriaca*) against mycelia of soil-borne pathogens (*Fusarium oxysporum*, *Macrophomina phaseolina*) and against foliar pathogens (*Botritis cinerea*, *Exserohilum turcicum*) was observed by Shimoni *et al.* (1993). Based on remarkable antifungal activity, these oils might serve as effective agents in control of *Fusarium oxysporum* and *Exserohilum turcicum* spread, although the authors are critical with regard to the toxicity of essential oils, which is comparably in the range of some synthetic fungicides.

Carvacrol rich essential oil of *O. vulgare* L. (Drôme origin) showed fungistatic as well as fungicide (minimal fungicidal concentration at 0.05 per cent) activity when assayed *in vitro* against bee pathogen *Ascosphaera apis*, and promises new possibilities for effective treatment of apiaries (in 0.1 per cent concentration) (Colin *et al.*, 1989). Similar results were obtained by Calderone *et al.* (1994), who studied the sensibility of this chalkbrood causing pathogen towards Spanish *Origanum* (*Thymus capitatus*) oil, rich in thymol. Essential oil of *Thymus capitatus* and thymol completely inhibited *in vitro* growth of *Ascosphaera apis* for 72 h at concentrations of 100 ppm and 10 ppm, respectively. The relevance of carvacrol in essential oils of *O. syriacum* L. (*za'atar*) for their antifungal activity was stressed in the experiments of Daouk *et al.* (1995). They have considered *O. syriacum* as a potent mould inhibitor, which could be used as a food preservative in small amounts without changing much the odour or taste of the stored food. They found that essential oil of Lebanese '*za'atar*' completely inhibited growth of *Penicillium* spp., *A. niger* and *Fusarium oxysporum* at 100 ppm, and had a pronounced effect already when applied to the culture media in concentrations as low as 0.05 per cent (50 ppm). The fumigant toxicity of *O. vulgare* L. essential oils against fungi (*A. flavus*, *A. ochraceus* and *A. niger*) attached to stored wheat grain was studied in Israel. The minimal inhibitory concentrations (MIC) of oregano oil needed to inhibit the mycelial growth and sporulation of the fungi were 2.0 and 2.5 ppm, respectively (Paster *et al.*, 1993; Paster *et al.*, 1995). Better inhibitory effect was achieved when the grain contained a relatively high moisture content (15–20 per cent). The results of this study indicate the possibility of using the oregano-based fumigants as an alternative to chemicals for preserving grains destined for human or animal consumption. Contrary to this, oregano essential oils could not be used in seed preservation, because the treatments considerably affected wheat germination. No phytotoxic effect on germination and corn growth was detected when maize grain was treated with *O. vulgare* essential oils against *A. flavus* contamination, but the effective essential oil concentration (>10 per cent) was too high for practical applications (Montes-Belmont and Carvajal, 1998). Similar results were obtained in Greece, where essential oils of *Origanum* spp. were studied for their antifungal effects against *Botrytis cinerea* (Thanassoulopoulos and Yanna, 1997). Although, oregano essential oils (at 500 ppm) were fungicidal towards *Botrytis cinerea* culture *in vitro*, the small reduction of Botryris-rot in kiwifruit was meaningless for the control of storage rot. Hence, the essential oils destroyed the qualitative characteristics of the fruits (flavour, flesh colour, taste) and were unmarketable.

High antifungal activities of isolated components of essential oils, like carvacrol and thymol against food-storage fungi (Thompson, 1989) and of carvone against *Penicillium hirsutum*, responsible for post-harvest *Penicillium* rot on tulip bulbs, were reported (Smid *et al.*, 1995).

Moulds, most frequently spread on bread (*Aspergillus glaucus* spp., *Eurotium repens* de Bary, *Cladosporium herbarum* Link ex Gray, *Penicillium expansum* Link ex Gray), can be effectively suppressed by oregano or its essential oils and, according to the authors (Kunz, 1994; Kunz *et al.*, 1995), offer new possibilities for the use of oregano in bakery and bakery products. On the other hand, only a weak fungistatic activity of oregano oleoresin was observed by Nielsen and Rios (2000), who studied antifungal effects towards bread spoilage fungi *Penicillium commune, P. roqueforti, A. flavus* and *Endomyces fibuliger*.

ANTIBACTERIAL ACTIVITY

Based on a broad spectra of antibacterial activity, oregano seems to be one of the most inhibitory spices tested. However, when considering the reported data, questions on the accuracy of the interpreted information often arise. When assessing the antimicrobial potency of herbs, spices and medicinal plants, that are used traditionally in gastronomy for preservative reasons or in prevention of human–plant diseases, one has to consider many factors which influence the efficacy of an extract or essential oil. These include the plant species used and its origin, concentration and composition of an extract/essential oil, the mode of dispersal of extract/oil to the medium, the concentration of tested organism in the growing medium, susceptibility of bacterial strains, the method used – *in vitro* or *in vivo* – pH, temperature . . .). Moreover, Skandamis *et al.* (2000) have stressed the importance of the fluidity of the culture medium for accurate estimation of the potency of essential oils against microorganisms. Different culture media (liquid culture or gel matrix) were found to influence the rate of consumption of glucose, thus influencing the growth of bacteria as well as their susceptibility toward extracts/essential oils.

Altogether, 52 essential oils (including *O. vulgare* and *O. majorana* essential oils) and extracts of different plant genera have been investigated (Hammer *et al.*, 1999) for their activity against *Acinetobacter baumanii, Aeromonas veronii* biogroup *sobria, C. albicans, Streptococcus faecalis, E. coli, Klebsiella pneumoniae, Pseudomonas aeruginosa, Salmonella enterica* subsp. *enterica* serotype *typhimurium, Serratia marcescens* and *S. aureus*. It was found that *O. vulgare* (Australian origin) yielded one of the most potent antibacterial agents, which considerably inhibited the growth of all tested microorganisms. The lowest minimum inhibitory concentration of *O. vulgare* essential oil was 0.12 per cent (v/v) and 0.25 per cent (v/v) of *O. majorana*. Among the tested bacteria, the most resistant was *Pseudomonas aeruginosa*, that was inhibited by *O. vulgare* essential oil at 2 per cent (v/v), but not by *O. majorana* oil.

Screening of Italian Medicinal Plants for their antibacterial activity using the *in vitro* paper disk diffusion method (paper disks Whatman No. 1) revealed the strong activity of *O. vulgare* L. dimethylsulphoxide (DMSO) extracts against Gram-positive (*Bacillus subtilis, S. aureus, Streptococcus haemolyticus*) and Gram-negative bacteria (*E. coli* 7075, *Klebsiella pneumoniae, Proteus mirabilis, Salmonella typhi* H) (Izzo *et al.*, 1995). The minimal inhibitory concentration (MIC) of applied extracts towards all tested bacteria was less than 4 µg/disk, with the exception of *Pseudomonas aeruginosa*, which was not susceptible. The essential oils of *Origanum onites* L. of Sicilian origin and of commercial oregano (which was found to be a mixture of two species, i.e. *O. vulgare* L. and *O. majorana* L.) exhibited bactericidal or bacteriostatic activity against a variety of Gram-positive

(*S. aureus*, *Streptococcus faecalis*, *Micrococcus luteus* and *Bacillus subtilis*) and Gram-negative bacteria (*Proteus vulgaris*, *E. coli*, *Hafnia alvei*) (Biondi et al., 1993). GC analysis revealed the carvacrol (61.68 per cent) as the main component of *O. onites* essential oil, while the commercial oregano sample consisted of terpinene-4-ol (24.87 per cent), γ-terpinene (15.91 per cent) and thymol (11.61 per cent) as leading compounds. When comparing the two species, *O. onites* essential oil was found to be more effective, inducing bactericidal effects against all G(−) tested organisms and against *S. aureus* and bacteriostatic effects against *Micrococcus luteus* and *Bacillus subtilis* at dilutions (in absolute ethanol) of 1:10 (final concentration of essential oil 1 µl/disk). The bactericidal effect of essential oil, distilled from commercial oregano sample showed bactericidal effects only against *Streptococcus faecalis* (dilution 1:2, final conc. of EO 5 µl/disk), *E. coli* and *Proteus vulgaris* (dilution 1:5, final conc. of EO 2 µl/disk), whilst bacteriostatic effects (dilution 1:10) were observed against other tested bacteria. *Pseudomonas aeruginosa* was not susceptible to any of the tested oils or concentrations. Similar results were obtained by Paster et al. (1990), who found that *Pseudomonas aeruginosa* was not affected by oregano (*O. vulgare* L.) essential oil at concentrations of up to 500 µg/ml. Under aerobic conditions the *O. vulgare* oil was very effective against *Campylobacter jejuni* (microaerophile) and *Clostridium sporogenes* (anaerobe) at 250 µg/ml (Paster et al., 1990). Also, good bacteriostatic (at concentration of 225 mg/l) and bactericidal (at concentration of 900 mg/l) effects *in vitro* against *Erwinia amylovora* were observed with *O. vulgare* essential oil (Scortichini and Rossi, 1989; Scortichini and Rossi, 1993).

Origanum vulgare essential oil (agar dilution method: 10 µl oil/Petri dish), characterised by high thymol (32.4 per cent) and carvacrol (16.7 per cent) content, showed a strong inhibitory (inhibition zone >20 mm) effect against a broad spectrum of tested bacteria, that were both G(+) or G(−) (*Alcaligenes faecalis*, *Bacillus subtilis*, *Beneckea natriegens*, *Brevibacterium linens*, *Brocothrix thermosphacta*, *Citrobacter freundii*, *Clostridium perfringens*, *Enterobacter aerogenes*, *Erwinia carotovora*, *Klebsiella pneumoniae*, *L. plantarum*, *Leuconostoc cremoris*, *Moraxella* spp., *Proteus vulgaris*, *Salmonella pullorum*, *Serratia marcescens*, *S. aureus*, *Streptococcus faecalis*, *Yersinia enterolitica*) (Baratta et al., 1998a). Good inhibitory activity (inhibition zone >10 mm <20 mm) was observed also against *E. coli*, *Flavobacterium suaveolens*, *Micrococcus luteus*, *Pseudomonas aeruginosa*, whilst two bacteria (*Acinetobacter calcoaceticus*, *Aeromonas hydrophila*) were not susceptible to oregano essential oil. In affected organisms, the origin of the bacterial strain or the fact of being G(+) or G(−) did not influence their susceptibility towards the oil. This is in agreement with the observations of Deans (Deans and Ritchie, 1987; Deans et al., 1992), who found that volatile oils of *O. vulgare* ssp. *hirtum* were equally effective against both G(+) and G(−) microorganisms.

Using the same bacterial strains and test conditions, the essential oils of *O. majorana* – consisting prevalently of terpinen-4-ol (20.8 per cent), γ-terpinene (14.1 per cent) and α-terpinene (14.1 per cent) – showed similar, but less potent antibacterial activity than *O. vulgare*. However, *O. majorana* exhibited a strong inhibitory effect against *Acinetobacter calcoaceticus* and *Aeromonas hydrophila* (Baratta et al., 1998b), against *Beneckea natriegens*, *Erwinia carotovora* and *Moraxella* spp. (Deans and Svoboda, 1990) but was not active against *Brevibacterium linens* and *Leuconostoc cremoris*.

Essential oils of *O. vulgare* of Turkish origin exhibited still more potent antibacterial activity as observed by Kivanç and Akgül (1986), who have studied the bactericidal effects of essential oils by both the agar diffusion and the serial dilution methods. Essential oils of *O. vulgare* showed pronounced bacteriostatic (dilution levels of

1:20–1:160) and bactericidal effects (dilution level of 1:20) against seven bacteria (*Aerobacter aerogenes*, *Bacillus subtilis*, *E. coli*, *Proteus vulgaris*, *Pseudomonas aeruginosa*, *Staphylococcus albus*, *Staphylococcus aureus*).

The growth of a wide range of bacteria (*Clostridium sporogenes*, *Enterobacter*, *E. coli*, *Klebsiella pneumoniae*, *Proteus vulgaris*, *Pseudomonas aeruginosa*, *Salmonella pullorum*, *S. aureus*, *Streptococcus faecalis*, *Yersinia enterolitica*) was strongly inhibited (zone inhibition >31.5 mm <71.2 mm), when grown in an agar medium supplemented with essential oil of *Origanum officinalis* (dilution 1:10), a special breed from Israel, that was selected for elevated oil yields (Deans et al., 1992).

Origanum majorana L. essential oil (Vojvodina origin), at concentration of 0.15 per cent showed moderate inhibitory activity against four bacteria (*E. coli*, *Proteus vulgaris*, *Salmonella enteritidis*, *Pseudomonas fluorescens*), that are frequently present as undesirable flora in the meat-processing industry (Sirnik and Gorišek, 1983). Deans and Ritchie (1987) reported, that *Origanum majorana* L. essential oil had a broad spectrum of antibacterial activities (at dilution level of 1:10) against bacteria of animal or human origin (*E. coli*, *Salmonella pullorum*, *Streptococcus faecalis*, *S. aureus*, *Clostridium sporogenes*) against soil bacteria (*Bacillus subtilis*, *Serratia marcescens*), plant pathogen (*Erwinia carotovora*) and aquatic bacteria (*Beneckea natriegens*). At dilution level of 1:5 of *O. majorana* essential oil, a remarkable effect was detected against *Yersinia enterocolica* and *Pseudomonas aeruginosa* (agar dilution method).

In a way comparable to antifungal activity, the antibacterial effects of oregano essential oils, as experienced in *O. vulgare* ssp. *hirtum*, *O. dictamnus* and commercial Greek *Origanum* oil, were mainly due to the presence of phenolic constituents of essential oils (carvacrol and/or thymol), whilst their biosynthetic precursors γ-terpinene and *p*-cymene were inactive (Pellecuer et al., 1980; Gergis et al., 1990; Panizzi et al., 1993; Sivropoulou et al., 1996; Adam et al., 1998). Hence, synergistic antibacterial activities of carvacrol and thymol were reported (Didry et al., 1993; Sivropoulou et al., 1996). According to Sivropoulou and co-workers (1996) *P. aeruginosa* exhibited resistance to all three tested essential oils as well as towards the compounds tested (carvacrol, thymol, γ-terpinene, *p*-cymene), although later findings of Dorman and Deans (2000) confirmed good inhibitory effects of *O. vulgare* essential oils and of carvacrol against this G(−) bacteria.

The essential oils of *O. vulgare* ssp. *hirtum* and *O. dictamnus* were extremely bactericidal (in *S. aureus*) at 1:4000 dilution, and even at dilutions as high as 1:50000 caused considerable decrease in bacterial growth rates. Essential oils of *O. vulgare*, rich in carvacrol (49.1 per cent) and of *Thymus vulgaris* L., rich in thymol (67.3 per cent), showed approximately the same range of antibacterial efficiency (MIC: 1:2000–1:3000) against *E. coli*, *S. aureus*, *Bacillus megaterium*, *Salmonella hadar* (Remmal et al., 1993). Dorman and Deans (2000) have studied the effects of essential oils of *O. vulgare* ssp. *hirtum* and of *Thymus vulgaris* toward a range of G(+) and G(−) bacteria. When comparing the relative efficacy of the two species, they found that thyme was generally more effective against the majority of G(+) bacteria (with the exception of *Acinetobacter calcoacetica* and *Yersinia enterolitica*) and against all G(−) bacteria (especially toward *Alcaligenes faecalis*, *Flavobacterium suaveolens*, *Klebsiella pneumoniae*, *Proteus vulgaris*, *Salmonella pullorum* and *Serratia marcescens*), but not toward *Pseudomonas aeruginosa*. This was more sensitive to essential oil of *O. vulgare*. The same results have been obtained by isolated phenolic compounds, showing a more pronounced effect of thymol against G(+) bacteria (with the exception of *Clostridium sporogenes*, that was well inhibited by *O. vulgare* and carvacrol) or G(−) bacteria (with the exception of *Pseudomonas aeruginosa*). This might indicate that the relative position of the hydroxyl group in the phenolic structure might contribute to the antibacterial potency of essential oil components (Dorman and Deans, 2000).

While most of the reported studies on the antimicrobial activity of oregano involved pathogenic bacteria, only a limited number of authors studied the inhibitory effects of oregano on lactic bacteria. Authors generally agree that, like all susceptible bacteria, lactic bacteria are also inhibited by *Origanum* essential oil in a concentration-dependent manner. High sensitivity of *Vibrio parahaemolyticus* was observed in media containing oregano (Beuchat, 1976). Zaika and Kissinger (1981) have found that oregano extracts were bactericidal toward lactic acid bacteria (at 8 g/l against *L. plantarum*, 4 g/l against *Pediococcus cerevisiae*), but these organisms became resistant toward the toxic effects of oregano, when sublethal concentrations (3 g/l in *L. plantarum*, 2 g/l in *P. cerevisiae*) were applied to the starter culture of bacteria. Instead of inhibition, low concentrations of oregano in culture medium stimulated the growth and production of acid production in resistant bacteria. Moreover, a phenomenon of cross-resistance of bacteria against different herb species was observed (Zaika *et al.*, 1983). This means that bacteria which had acquired a resistance to one herb species (oregano) were also resistant to other herbs (sage, rosemary, thyme). So far, the mechanism by which the starter cultures acquire their resistance is not known. Kivanç and co-workers (1991) studied the effects of *Origanum onites* leaves and essential oil on growth and acid production of *L. plantarum* and *L. mesenteroides*. They found that oregano leaves (0.5 per cent, 1.0 per cent or 2.0 per cent) had no significant influence on the growth of *L. plantarum*, but after 2 days of fermentation *in vitro* they stimulated its acid production. By contrast, the growth and acid production of *L. plantarum* were strongly inhibited by *O. onites* essential oil (150, 300 or 600 ppm). When considering *Leuconostoc mesenteroides*, both the leaves and the essential oil of *O. onites* at all tested concentrations inhibited the growth and acid production.

The stimulative effects of extracts of *O. majorana* on the growth of non-lactic acid bacteria have been reported by Adlova *et al.* (1998). They observed that the phenolic fraction of water extracts of *O. majorana*, present in media at low concentration (0.0001 per cent), exhibited a stimulating effect on *E. coli* and *Streptococcus pyogenes*, but had no influence on the growth of *Corynebacterium xerosis*. Vokou and Liotiri (1999) established that essential oil of *O. vulgare* L. ssp. *hirtum*, when added to soil samples (0.1 ml per 150 g of soil) of Mediterranean ecosystems, could be used by soil bacteria as a carbon and energy source, and that it stimulated soil microbial activity. Dose-dependent bactericidal (at ≥ 1 mmol/l) or bacteriostatic (MIC = 0.75 mmol/l) effects of isolated carvacrol on the foodborne pathogen *Bacillus cereus* were detected *in vitro* (Ultee *et al.*, 1998). The pH of the medium and the growth temperatures (8 °C or 30 °C) considerably influenced the bactericidal activity. Sensitivities recorded at pH 5.5 and 8.0 were two- and six-fold higher than those at pH 7.0, where the lowest sensitivity of *B. cereus* was detected. The study of the mechanism of action showed that the inhibitory effect of carvacrol was due to the interaction with membranes of *B. cereus*, changing their permeability for cations (K^+, H^+). The dissipation of the ion gradient influenced the membrane transport and led to impairment of cell essential processes and finally to cell death (Ultee *et al.*, 1999). However, a considerable decrease in sensitivity of *B. cereus* against carvacrol was observed after growth in the presence of sublethal concentrations of carvacrol (0.4 mmol/l). Concomitantly, a lower membrane fluidity of adapted cells was detected, indicating the changes in the fatty acid composition and rearrangement of the phospholipid bilayer in bacterial cell membranes during adaptation process (Ultee *et al.*, 2000). Pol and Smid (1999) report on the synergistic effects of carvacrol and nisin (bactericidal peptide, used as a biopreservative in certain foods) against *Bacillus cereus* and *Listeria monocytogenes in vitro*, which resulted in a much higher sensitivity of both pathogens (at 20 °C) to combined exposure (*B. cereus*: $MIC_{combination}$ = 0.63 mmol/l carvacrol and 1.25 µg/ml nisin, *L. monocytogenes*: $MIC_{combination}$ = 1.25 mmol/l carvacrol and 0.63 µg/ml nisin) than to individual compounds (*B. cereus*: $MIC_{carvacrol}$ = 1.25 mmol/l or MIC_{nisin} = 10 µg/ml, *L. monocytogenes*: $MIC_{carvacrol}$ = 2.50 mmol/l or MIC_{nisin} = 10 µg/ml). This means that lower concentrations of both carvacrol and nisin are needed for effective decrease in the number of colony-forming units of food-borne pathogens.

Oregano essential oils have been considered as an alternative natural additive in gastronomy and in the food processing industries. It was found that *O. vulgare* essential oil was effective in inactivation of *E. coli* in concentrations as low as 0.7 per cent, acting synergistically with the pH and storage temperatures, thus contributing to the intrinsic safety of home-made eggplant salad (Skandamis and Nychas, 2000). Also, Dorman and Deans (2000) believe that volatile oils (but not spices as integral ingredients) may have the greatest potential use as food preservatives. Due to their high antimicrobial potency they could be added to foodstuffs in small quantities and would cause no loss of organoleptic properties of the food. However, when assessing the potential use of essential oils in food and the food industries, one has to consider the whole system (food, essential oil, processing, storage temperatures, marketable/sensory characteristics of the processed food) to be able accurately to judge the usable value of plant essential oils/extracts. A considerable reduction in antimicrobial activity, when plant essential oils or extracts were evaluated in food systems/actual foods, was found by several authors (Shelef *et al.*, 1984; Ismaiel and Pierson, 1990b; Hao *et al.*, 1998). Similar observations were reported by Aureli *et al.* (1992), who found that the high antilisteric activities (4 strains of *L. monocytogenes*, one strain of *Listeria innocua*) of *Origanum* or *Thymus* essential oil (dilution 1:5 and 1:50 in absolute ethanol) *in vitro* were significantly reduced when essential oils were tested in a meat matrix (minced pork meat). Accordingly, the leaf extract of *O. majorana* exhibited no remarkable antilisteric effect, when assayed in cooked chicken breasts (Hao *et al.*, 1998). By contrast, a significant inhibitory activity of 0.8 per cent (v/w) oregano essential oil against *L. monocytogenes* on naturally contaminated beef meat fillets was observed under various packaging conditions at 5 °C (Tsigarida *et al.*, 2000). Based on the findings on the sporostatic and growth inhibiting activities of *Origanum* essential oil (at 150 and 200 ppm) against food contaminant *Clostridium botulinum in vitro* (TYG: thiotone yeast extract glucose medium) (Ismaiel and Pierson, 1990a), the same authors examined the potential antibotulinal effect of *Origanum* oil in minced pork. They observed considerable diminishing of antibotulinal activity of *Origanum* oil in the meat system. *Origanum* essential oil was effective only when used at 400 ppm and in combination with 50–100 ppm of sodium nitrite, depending on the spore inocula. The absence of inhibition by oregano oil in the meat system, in contrast to TYG medium, could be due to the high solubility of the oil components in the lipid fraction of the meat. Such concentrations ($\geq$400 ppm) of *Origanum* essential oil are questionable with regard to the possible effects on marketable characteristics of cured meat (flavour, colour, taste, and structure), although these were not assessed in the study of Ismaiel and Pierson (1990b). In view of practical implications in the preservation of food products, where the antimicrobial efficacy and sensory attributes of food have to be considered, the approach of combining different food-preservative compounds, proposed by Pol and Smid (1999), seems the most appropriate.

ANTIOXIDANT ACTIVITY

A large number of reports on the antioxidant effects of *Origanum* species have been published. A survey of the potential use of *Origanum* or oregano based preparations, that would replace synthetic substances such as BHT, as protectors of highly unsaturated lipids in foodstuffs has been made by numerous research groups. However, limited industrial applications are often ascribed to the characteristic oregano aroma and flavour, that influence the sensorial characteristics of processed food, so deodorization steps would be required (Nguyen *et al.*, 1991; Moure *et al.*, 2001). Dietary supplies of antioxidants from *Origanum* species were also considered as effective scavengers of the free radicals that are generated by metabolic pathways in the body, and in sufficient amounts prevent cellular damages and human diseases.

In these studies different methods, test/model systems (lard, bulk oils, emulsions, meat products, human cells as oxidation substrates), different plant preparations (whole spices, essential oils, hydrophilic or hydrophobic extracts, isolated phenolic compounds, etc.) have been used in the quantification of antioxidative potential (Madsen and Bertelsen, 1995; Madsen *et al.*, 1997; Pearson *et al.*, 1997; Moure *et al.*, 2001). Comparison between the different reports, which is often difficult due to a high variability in experimental design, mostly refer to the different potency of antioxidant activity of tested *Origanum* species or their compounds. This is why several authors (Laughton *et al.*, 1989; Frankel *et al.*, 1994; Pearson *et al.*, 1997) claim that a variety of testing systems is required when assessing the antioxidant potential of a substance, since a substance exhibiting high antioxidant activity in one system may have a prooxidant effect in another system. In the current literature, relatively little information is available on mechanisms of the antioxidative action, although phenolic compounds are most frequently cited as active ingredients, responsible for the antioxidant effect (Madsen *et al.*, 1997; Moure *et al.*, 2001). Comparison of the antioxidant activity of model phenolic compounds has shown that polymeric phenolic compounds are generally more potent antioxidants than simple monomeric phenolics (Moure *et al.*, 2001). Yamaguchi *et al.* (1999) observed that the degree of polymerisation of flavanols correlates with the superoxide-scavenging capacity. The antioxidant activity relies also on the polarity of tested compounds, depending on the type and polarity of the extracting solvent. It was found that hydrophilic antioxidants are generally more effective in bulk oil, whereas lipophilic antioxidants exhibit more potent effects in emulsions ('polar paradox' phenomenon) (Moure *et al.*, 2001).

Investigations on the antioxidative activity of herbs and spices date about 40 years ago, when Chipault with co-workers (Chipault *et al.*, 1952; Chipault *et al.*, 1955; Chipault *et al.*, 1956) screened the effects of 32 different spices in various model systems/substrates, measuring their persistence with antioxidant index (AI). This was defined as a ratio between substrate, containing herb ingredient, and substrate without herb addition. In order to investigate the stabilising capacity of *O. vulgare* in different substrates, these were exposed to autooxidation at substrate-specific temperature regimes: lard (at 99 °C), egg yolk (at 63 °C), oil in water emulsion (at 40 °C), minced pork (at −5 °C) and mayonnaise (at 20 °C). In all tested substrates dry oregano (at concentration of 0.1 per cent in o/w emulsion, at 0.25 per cent in minced pork and at 0.2 per cent in all other substrates) displayed high antioxidant activity, the AI being between 2.7 (egg yolk) and 8.5 (mayonnaise). The length of the induction period in autooxidation of lard, which was used as an indicator of antioxidative potency, showed a higher antioxidant potency of dry oregano

(*O. vulgare*) when compared with that of marjoram (*O. majorana*), although Saito *et al.* (1976) had found marjoram to be a more potent antioxidant herb than oregano at the same concentration tested (Gerhardt and Schröter, 1983).

Methanol extracts of *O. vulgare* and of *O. majorana* exhibited strong hydroxyl radical-scavenging activity, inhibiting the oxidation of 2-deoxyribose by more than 50 per cent at the 1 µg/ml concentration, however, the scavenging effect was much less evident when the antioxidant potential was measured by benzoic acid hydroxylation method (Chung *et al.*, 1997). By contrast with the high antioxidant activity of *O. vulgare* methanol extracts, as observed in the lard test system (Herrmann *et al.*, 1981; Banias *et al.*, 1992), oregano and marjoram extracts showed only scarce antioxidant effect in the β-carotene model system (Dapkevicius *et al.*, 1998).

Essential oil of *O. vulgare* showed the ability to form stable free radicals upon reaction with potassium superoxide (Deighton *et al.*, 1993). The essential oil monophenols, carvacrol and thymol, were identified as molecules which react with the superoxide anion (O_2^-), probably through hydrogen atom donation, and form stable paramagnetic species (free radicals) as found by EPR spectroscopy.

Origanum vulgare essential oil (Italian origin) demonstrated protective antioxidant properties in an egg yolk assay. At high concentrations (1000 ppm and 750 ppm), the antioxidative potency of oregano oil was higher than those of butylated hydroxytoluene (BHT) or of α-tocopherol. However, in a rat liver assay the antioxidative effect of oregano essential oil was much lower, at 750 ppm exhibiting the same range of antioxidant activity as α-tocopherol at 250 ppm (Baratta *et al.*, 1998a). Similar results were observed by Lagouri *et al.* (1993), who studied the antioxidative potency of *O. vulgare* L. ssp. *hirtum* Ietswaart and *Origanum onites* L. essential oils (at 1000 ppm) by measuring the autooxidation rate (peroxide value) of lard stored at 35 °C. The antioxidant activities of essential oils were attributed to their high phenol moiety (carvacrol and thymol) and were comparable to BHT (at 200 ppm). Still more potent activity was observed with *O. majorana* essential oil, the antioxidant activity being much higher than that of α-tocopherol and comparable to that of BHT at all concentration levels (100 ppm, 250 ppm, 500 ppm, 750 ppm, 1000 ppm) (Baratta *et al.*, 1998b).

A survey of the antioxidant activity of methanol extracts of *Origanum* species of Greek origin showed that *O. vulgare*, *O. dictamnus* and *O. majorana* at concentration of 0.02 per cent significantly prolonged the induction period of lard autooxidation at 75 °C and slightly decreased the rate of peroxide formation. However, their relative antioxidant efficiencies (oregano > dittany > marjoram) in comparison to those of BHT or rosemary extracts were much lower (Economou *et al.*, 1991).

Turkish oregano (*O. vulgare* L.) and Chilean oregano (*Origanum onites* L.) showed high antioxidative effects, measured by the oxygen depletion method as well as by EPR spectroscopy. The study of oregano water extracts, using both methods, allowed Madsen and co-workers (1996) to find that the antioxidant activity of oregano was due to at least two different antioxidative mechanisms. The activity might be due both to the non-phenolic group of compounds and to the phenolic group. The group of non-phenolic compounds act as scavengers of free radicals and are effective in early stages of oxidation. The group of phenolic compounds were effective in interrupting the chain processes responsible for oxygen consumption by a mechanism similar to that for tocopherols. Among the phenolic compounds that have been isolated from oregano, there were isolated at least five different groups of substances which were highly antioxidative active (rosmarinic acid, water soluble phenolic glycosides, flavonoids, carvacrol and thymol) (Madsen and Bertelsen, 1995). The ESR spin-trapping assay showed that the free radical scavenging capacity of the *O. dictamnus* water and methanol extracts correlated with the content of phenolics. When compared to acetone or ethanol extracts (poor in phenolics), the aqueous and methanol extracts (rich in phenolic compounds) were also the most effective in reducing oxygen consumption and thus had high chain-breaking properties, as evidenced by the oxygen depletion assay (Møller *et al.*, 1999). The ability of aqueous extracts of *O. dictamnus* to inhibit development of secondary lipid oxidation products was also confirmed in model food systems (turkey thigh meat homogenate), where dittany dose-dependently (from 0.0018 mg dittany/g meat upwards) inhibited development of thiobarbituric reactive substances.

Vekiari *et al.* (1993a) have studied the extracts of *O. vulgare* of different polarity, to find the active compounds responsible for the antioxidative effect of oregano. The main antioxidant factor of the non-polar hexane extract was isolated by repeated fractionations, and consisted mainly of terpene derivatives. Among polar compounds that were extracted from *O. vulgare* leaves, the most effective in stabilising lard against oxidation, with potency equal to BHT, were flavonoids (flavanone eriodictyol, the dihydroflavonols dihydrokaempferol and dihydroquercetin and flavone apigenine) (Vekiari *et al.*, 1993b). These compounds also showed marked antioxidant activity when tested on vegetable oils (corn, soybean and olive) under storage or frying conditions.

Nakatani (1997) reports that both the polar and non-polar fractions of oregano leaves significantly retarded oxidation of linoleic acid, measured by the ferric thiocianate (FTC) and thiobarbituric acid (TBA) methods. From a water-soluble fraction of methanol extracts, phenolic compounds with high antioxidant activities have been purified, the most potent being derivative of rosmarinic acid (2-caffeoyloxy-3-[2-(4-hydroxybenzyl)-4,5-dihydroxy] phenylpropionic acid and a new glycoside of protocatechuic acid ester, identified as 4-(3,4-dihydroxybenzoyl-oxymethyl)phenyl-β-D-glucopyranoside (Nakatani and Kikuzaki, 1987; Kikuzaki and Nakatani, 1989; Nakatani, 1992). These were more effective against linoleic acid oxidation than the natural α-tocopherol, and are comparable to the synthetic antioxidants, BHA (butylated hydroxyanisole) or BHT (butylated hydroxytoluene). Similar polyphenolic compounds were found in *O. majorana* leaves, but these also contain compounds such as 6-O-4-hydoxybenzoyl arbutin and 2-hydroxy-3-(3,4-dihydroxyphenyl)propionic acid, which possess moderate antioxidant activity (Nakatani, 1997). These findings are in agreement with Herrmann (1994), who observed that antioxidant active plant phenols often possess a 3,4-dihydroxybenzoyl- or 3-methoxy-4-hydroxybenzoyl group in their structure. Lamaison and co-workers (Lamaison *et al.*, 1990; Lamaison *et al.*, 1991; Lamaison *et al.*, 1993), who extensively studied the antioxidant activity of members of Lamiaceae, reported that the content of rosmarinic acid and of total hydroxycinnamic derivatives in hydroalcoholic extracts of *Origanum* taxa (*O. onites*, *O. tytthantum*, *O. vulgare* ssp. *hirtum*) was only partly correlated with their antioxidant effect, estimated by measuring the free radical scavenger effect on DPPH (1,1-diphenyl-2-picrylhydrazyl). They stressed the importance of flavonoid content for oregano antioxidant activity.

The antioxidant activity of isolated carvacrol and thymol in liposomal systems was confirmed by Aeschbach *et al.* (1994), and in biological systems (human aortic endothelial cells, HAEC) by Pearson *et al.* (1997). The potency of antioxidant activity of thymol ($ID_{50} = 4.02$ µM), measured as per cent of inhibition of HAEC-mediated human LDL oxidation, was significantly higher than that of carvacrol ($ID_{50} = 5.53$ µM), but it was found that both monophenols (thymol or carvacrol) had much lower antioxidant activities than rosmarinic acid ($ID_{50} = 0.74$ µM) (Pearson et al., 1997).

It was also found that combinations of spices or compounds with high antioxidant activities exhibited synergistic antioxidant effects, which would practically result in a better protection of foods from oxidation (Madsen et al., 1996). However, no practically important synergistic effects were observed, when *O. vulgare* methanol extracts – which were highly effective in stabilising lard (at concentration 0.02 per cent) stored at 75 °C – were mixed in lard with less potent antioxidants (methanol extracts of thyme, marjoram, spearmint, basil) (Economou et al., 1991).

In the study of Banias et al. (1992), combinations of methanol extracts of oregano, dittany or marjoram with primary antioxidants were used in the lard autooxidation process. The results showed that significant positive synergism in antioxidant activity existed in combinations of oregano or marjoram (0.1 per cent) with BHT (0.005 per cent) and in a combination of dittany (0.1 per cent) with ascorbyl palmitate (0.01 per cent). By contrast, high negative synergism was observed in combinations of oregano (0.1 per cent) or marjoram (0.1 per cent) with propyl gallate (PG) (0.01 per cent) and in a combination of oregano (0.1 per cent) or marjoram (0.1 per cent) with α-tocopherol. Milos et al. (2000) have studied the antioxidant activity of volatile aglycons, that are bound glycosidically in dry *O. vulgare* plants, in comparison to that of essential oil. Although, the total content of volatile aglycons in plant material (0.002 per cent) was significantly lower than the content of essential oil (2.9 per cent), a mixture of volatile aglycons (thymoquinone, benzyl alcohol, eugenol, thymol, carvacrol) showed similar antioxidant activity to that of essential oil. They inhibited hydroperoxide formation in lard stored at 60 °C even after 80 days and were significantly more effective than α-tocopherol. Thymoquinone, which was found to be a potent inhibitor of membrane lipid peroxidation (Houghton et al., 1995; Jerkovic et al., 2001) and the major component (40.2 per cent) among aglycons (Milos et al., 2000), as well as pure thymol as the major component (40.4 per cent) of essential oil, were much less active than a mixture of aglycons or essential oil of *O. vulgare*. These results indicated the importance of mixtures and their synergistic power in the antioxidant activity of *O. vulgare*.

In addition to numerous studies, where potent or moderate antioxidant effects of oregano were established in theoretic model systems, practical considerations on the use of oregano as stabilisers of edible oils (vegetable or fish oils) or of finished meat products have been made by several research groups. Generally, authors confirm the protective role of different *Origanum* taxa (*O. vulgare, O. compactum, O. majorana*) against the autooxidation process over time, although the potencies of oregano antioxidative effects are lower than those reported for rosemary or sage (Özcan and Akgül, 1995; Antoun and Tsimidou, 1997). Özcan and Akgül (1995) studied the antioxidant effects of methanol extracts and essential oils of numerous Turkish spices on sunflower oil, stored at 70 °C, and found that methanol extracts (including *O. vulgare* and *O. majorana*) exhibited higher antioxidant activity compared with essential oils. The increased delay in the onset of autooxidation might be due to the improved preservation of α-tocopherol, an internal antioxidative microcomponent principle of sunflower oil (Yanishlieva and Marinova, 1998; Beddows et al., 2000). When compared to hexane and ethyl acetate extracts, the ethanol extracts of several species of the *Lamiaceae* family were the most active in retarding the autooxidation process of sunflower oil exposed to 100 °C. However, *O. vulgare* ethanol extracts (at 0.08 per cent) showed only low antioxidant effect comparable to that of 0.02 per cent BHT (Yanishlieva and Marinova,

1995), or else did not improve the oxidation stability of sunflower oil, as is evident from the later study by Marinova and Yanishlieva (1997). Only a moderate stabilising effect of O. vulgare leaves in sunflower oil, exposed to autooxidation at room temperature, was observed also by de Felice et al. (1993), who measured the quality characteristics of the oil in the time period of 16 weeks.

By contrast, ground oregano (O. vulgare) inhibited lipid oxidation of fish/mackerel (Scomber scombrus) oil stored at 40° in dark at concentrations of 0.5 per cent and at 1 per cent as effectively as 200 ppm BHA and 200 ppm TBHQ (tertiary butylhydroquinone), respectively (Tsimidou et al., 1995). Dry leaves of O. vulgare ssp. hirtum (at concentration of 2 per cent) or essential oil of O. compactum (at concentrations 0.05 per cent and 0.1 per cent) also showed a high antioxidant activity in olive oil and, besides their stabilising effect, the organoleptic quality of the olive oil was significantly improved by addition of oregano, as assessed by Mediterranean consumer acceptability studies (Antoun and Tsimidou, 1997; Charai et al., 1999). A significant increase in the oxidative stability of fried chips, measured as the rate of peroxide formation during storage at 63 °C, was achieved both by addition of ground O. vulgare (1 per cent, after frying) or its petroleum ether extracts (1.1 per cent, before frying) (Lolos et al., 1999). The oregano antioxidant activity was almost as effective as that of TBHQ up to 6 days of observation, although the peroxide value of cottonseed oil, extracted from oregano-treated potato chips increased after one week. However, the results of this study indicated that ground oregano or its extract might be used to extend the storage life of potato chips as they decrease the oxidative deterioration of the oil absorbed into the chips.

The importance of the testing substrate (bulk oil or o/w emulsion) in the evaluation of oregano antioxidant potency has been shown in the study of Abdalla and Roozen (1999), who found that acetone extract of O. vulgare effectively inhibited the autooxidation process of sunflower oil at both 600 ppm and 1200 ppm, but exhibited only moderate antioxidant activity when tested in a 20 per cent sunflower o/w emulsion. It has been also shown that oregano extracts acted as pro-oxidants in both oil and emulsions, when exposed to light (Abdalla et al., 1999).

When assessing the antioxidative potential of O. majorana or O. vulgare in preventing rancidity of meat products, only limited practical significance has been documented. El-Alim et al. (1999) report that O. majorana and O. vulgare, and especially their ethanol extracts, had a strong antioxidant activity, inhibiting lipid peroxidation both in fresh chicken meat as well as in heat-treated pork. Because of their shelf time prolonging properties, oregano and oregano-based preparations have been recommended for use in semi-prepared meat products. However, other studies show less optimistic results. Ground O. majorana was added at a concentration of 0.2 per cent to the laboratory and industrial prepared sausage model systems, that were exposed to ripening. Only a low antioxidative effect was observed on the basis of the redox potential reduction of the marjoram-supplemented model compared to the control (Palic et al., 1993). Korczak and co-workers (1988) have found relative low antioxidant efficacy of O. majorana (at 0.5 per cent) in minced meat model systems when compared to those of rosemary or sage. Hence, the pro-oxidising activity of marjoram, which is probably influenced by elevated temperature, diminishes its practical value as a natural additive in meat processing (Korczak et al., 1988).

The antioxidative effects of O. vulgare drug plant (Origani herba) have been studied in the light of both direct use as stabilisers of fat and, indirectly, as feed additives in order to improve the shelf-life of meat and fat-containing food (Vichi et al., 2001).

ANALGESIC, ANTIINFLAMMATORY AND ANTISPASMODIC ACTIVITY

Carvacrol-rich (67 per cent) essential oil of *Origanum onites*, collected at the Izmir locality (Turkey), showed a marked analgesic activity as assessed by the tail-flick method in male albino mice. The analgesic activity of *O. onites* essential oil was dose-dependent. When applied at 0.33 ml/kg, the activity of *O. onites* oil was comparable to that of morphine (applied at 1 mg/kg), but at 0.03 ml/kg more potent than the analgesic activity of fenoprofen (at 8 mg/kg) (Aydin *et al.*, 1996). The sample of *O. onites*, that originated from the Turkish Antalya region and was found to contain linalool (91 per cent) as a major component, showed no analgesic effects. On the basis of these data and on reports on prostaglandin inhibitory effects of carvacrol (Wagner *et al.*, 1986), Aydin and co-workers consider carvacrol content as related to the analgesic activity of essential oil of *O. onites*. The effects of methanol extracts of *O. majorana* on human platelet anti-aggregant activity, which is related to the well known mechanism of action of NSAID (non-steroid anti-inflammatory drugs) through inhibition of the prostaglandins' metabolic pathway, have been studied by Okazaki *et al.* (1998). They found that *O. majorana* extracts dose-dependently inhibited platelet aggregation induced by collagen (2.0 µg/ml) or ADP (2.0 µg/ml). Successive fractionation of methanol extracts leads to isolation of an active hydroquinon β-D-glucopyranoside, identified as arbutin. This strongly inhibited platelet aggregation was induced by all tested stimulating agents (collagen, ADP, arachidonic acid, thrombin).

Only a few reports on the topical anti-inflammatory effects of *O. vulgare* refer to oregano-herbal mixtures or their decoctions, used in the treatment of inflammation as supporting therapies (Deryabin, 1990; Deryabin, 1991). Podkolzin *et al.* (1986) report on the favourable local effect of insufflation of fine powder mixture of *Hypericum perforatum* and *O. vulgare* (1:1) on the course of rhinitis, that was induced in an animal (rabbit) experiment. In the control animals the rhinitis symptoms were more pronounced and of longer duration, so the powder was proposed as an adjuvant therapy in treatment of rhinitis.

Origanum compactum Benth., a species native to North Africa and locally named 'za'atar', was used traditionally against affections of the respiratory organs as an antispasmodic and anticatarrhal drug and, especially in Morocco, as a spasmolytic drug in the gastrointestinal tract, as antacid, antidiarrhoeal agent, vermifuge and aphrodisiac (van den Brouke and Lemli, 1980; Bellakhdar *et al.*, 1988; Hmamouchi *et al.*, 2000). In order to scientifically validate the traditional medicine data, van den Brouke and Lemli (1980) surveyed extracts of *O. compactum* on antispasmodic effects in different smooth muscle preparations *in vitro*. It was found that water macerates of *O. compactum* significantly inhibited smooth muscle response induced by any of the tested spasmogens (acetylcholine, histamine, serotonine, $BaCl_2$, nicotine ...) in the guinea-pig ileum. The structure–activity relationship revealed that the antispasmodic effect of *O. compactum* was almost completely explained by its essential oil content. Moreover, in the pharmacological inhibition of smooth muscular activity, non-specific and non-competitive mechanism of action was attributed to thymol and carvacrol: they caused both direct musculotropic (muscle relaxant activity) and indirect neurotropic action (inhibition of the nerve action potential) on the smooth muscle (van den Brouke and Lemli, 1980). The same results were obtained by testing the antispasmodic effects of pure active components, i.e. thymol ($ED_{50 \text{ per cent}} = 0.86 \times 10^{-4}$ M) and carvacrol ($ED_{50 \text{ per cent}} = 1.0 \times 10^{-4}$ M). It was concluded that both phenols act as non-competitive Ca^{2+} antagonists, which block nerve fibre conduction and induce musculotropic and neurotropic spasmolyse (van den Brouke and Lemli, 1982).

IMMUNOSTIMULANT, ANTIMUTAGENIC AND ANTICANCER ACTIVITY

Some studies have shown that oregano extracts or herbal mixtures with *Origanum* spp. possess *in vitro* antiviral activity or have immunostimulating effects both *in vitro* and *in vivo*. However, little knowledge has been attained so far on mechanisms of immunomodulating activity or underlying active compounds. It has been shown that ethanol extracts of *O. vulgare* inhibited intracellular propagation of $ECHO_9$ Hill virus and also showed interferon inducing activity *in vitro* (Skwarek et al., 1994). Flavonoid luteoline, a constituent of *Origani herba*, has been considered as responsible for the induction of an interferon-like substance. A mixture of herbal preparation containing rosemary, sage, thyme and oregano (*O. vulgare*) showed radical scavenging activity and inhibition of the human immunodeficiency virus (HIV) infection at very low concentrations (Aruoma et al., 1996). It was suggested that the main active compounds of herbal preparations were carnosol, carnosic acid, carvacrol and thymol. Significant inhibitory effects of *O. vulgare* extracts against HIV-1 induced cytopathogenicity in MT-4 cells were also observed by Yamasaki et al. (1998). According to Krukowski et al. (1998), an increase in immunoglobulin (IgG) levels was observed in reared calves, fed with a conventional concentrate supplemented by a mineral-herbal mixture containing *O. majorana*.

A strong and dose-dependent capacity of inactivating dietary mutagen Trp-P-1 in the *Salmonella typhimurium* TA 98 assay was observed in *O. vulgare* water extracts, that exhibited significant antimutagenic effects *in vitro* (Ueda et al., 1991). *Origanum majorana* aqueous extracts were also able to suppress the mutagenicity of liver-specific carcinogen Trp-P-2 (Natake et al., 1989). When studying the mechanism of suppressing the mutagenicity of Trp-P-2 in *O. vulgare*, it was found that two flavonoids, galangin and quercetin acted as Trp-P-2 specific desmutagens, which neutralised this mutagen during or before mutating the bacteria (*Salmonella typhimurium* TA 98) (Kanazawa et al., 1995). The amounts of galangin and quercetin required for 50 per cent inhibition (IC_{50}) against 20 ng of Trp-P-2 were 0.12 µg and 0.81 µg, respectively. It was also found that quercetin acted as a mutagen at high concentrations (>10 µg/plate), but was a desmutagen when applied at low (>0.1 <10 µg/plate) concentrations. Milic and Milic (1998) have found that isolated phenolic compounds from different spice plants, including *O. vulgare*, strongly inhibited pyrazine cation free radical formation in the Maillard reaction and the formation of mutagenic and carcinogenic amino-imidazoazarene in creatinine containing model systems.

In a literature survey, referring to the anticancer activity of *Origanum* genus, different approaches, testing systems and cell lines have been used by different authors when assessing the carcinogenic potential of plants or their isolated compounds. However, there are no available data on practical/clinical use of oregano in cancer prevention. In 1966 an international project was performed with the aim of screening the native plants of former Yugoslavia for their potential agricultural use in the USA and Yugoslavia (Mayer et al., 1971). In the frame of this project 1466 samples of 754 plant species were analysed for chemical and antitumour activity. According to the results of the Cancer Chemotherapy National Service Center Screening Laboratories (Washington, DC) a high carvacrol (60–85 per cent) containing *O. heracleoticum* (= *O. vulgare* spp. *hirtum* (Link) Ietswaart) was reported to show high antitumour activity. Lam and Zheng (1991) have found that essential oil of *O. vulgare* fed to mice, induced the activity of glutathione S-transferase (GST) in various tissues. The GST enzyme system is involved in detoxification of chemical carcinogens and plays an important role in prevention of carcinogenesis, what would explain the anticancer potential of *O. vulgare* essential oil. This oil exhibited high levels of cytotoxicity (at dilutions of up to 1:10000) against four permanent eukaryotic cell lines including two derived from human cancers (epidermoid larynx carcinoma: Hep-2 and epitheloid cervix carcinoma: HeLa) (Sivropoulou et al., 1996). Other studies, that refer to *in vitro* cytotoxic and/or anti-proliferative effects of *O. vulgare* extracts or isolated compounds (carvacrol, thymol) include those of Bocharova et al. (1999) and He et al. (1997), who observed moderate suppressing activities of *O. vulgare* extracts (CE_{50} = 220 mg/ml) on human ovarian carcinoma cells (CaOv), or of isolated carvacrol and thymol (IC_{50} = 120 µmol/l) on Murine B 16(F10) melanoma cells – a tumour cell line with high metastatic potential.

Antitumour-promoting activity or *in vitro* cytotoxic effects towards different tumour cell lines were attributed also to *O. majorana* extracts or their constituents (Assaf et al., 1987; Okuyama et al., 1995; Hirobe et al., 1998). When studying cytotoxic activity of *O. majorana* water–alcoholic extracts and of isolated compounds (arbutin, methylarbutin and their aglycons – hydroquinone and hydroquinone monomethyl ether) towards cultured rat hepatoma cells (HTC line), a high dose-dependent HTC cytotoxicity of hydroquinone was observed, whilst arbutin was not active (Assaf et al., 1987). At 300 µM hydroquinone caused 40 per cent cellular mortality after 24 h of incubation, but no cells remained viable after 72 h. It has been established that this well known antiseptic of the urinary tract was a more potent cytotoxic compound towards rat hepatoma cells than many classic antitumour agents like azauridin or colchicin, but less than valtrate, a monoterpenic ester of *Valeriana* spp.

INSECT-POLLINATING AND ANTIPARASITIC ACTIVITY

Origanum taxa, especially those that are rich in essential oils, have been extensively studied for their insect-pollinating (Ricciurdelli d'Albore, 1983; Beker et al., 1989) or nectar yielding (Kucherov and Siraeva, 1981; Jovančević et al., 1984; Jablonski, 1986) effects. Although scientifically poorly understood, the traditional knowledge on attracting effects *Origanum* spp. for pollinating insects, especially honeybee (*Apis mellifera*), has been practically exploited since 1877, when the idea of culturing the bee forages with additional but non-marketable values was born (Ayers and Ayers, 1997). Interesting findings, that reveal the very complex mechanism underlying the communication between insects and attracting plants, were reported by Beker et al. (1989). They have observed that honeybees are capable of discriminating between different blends of odours and behave selectively to different parts (leaves, inflorescence) or chemotype (thymol, carvacrol) of *O. syriacum* due to perceiving distinct olfactory stimuli. It was assumed that the aroma blend from the whole plant serves as a long-distance olfactory cue, while the final short-range orientation is dependent on floral odour signals.

Observations from studies on *Origanum* benefit effects in parasite-control in pollinating insects show promising results (Abou Zaid et al., 1987; Mazeed, 1987; Kraus et al., 1994; Gal, 1997; Long et al., 1997), although some authors were sceptical towards the practical significance of essential oils in treatment of parasite infestation (Koeniger, 1991). In laboratory tests *Origanum* oils showed a high acaricidal effect (80–90 per cent mortality) on *Varroa jacobsoni*. Under the subtropical climatic conditions of Israel, a high mortality (85 per cent and 91 per cent) of *Varroa* mites was observed after spring treatment with 20 per cent and 33 per cent oregano oil impregnated in cardboard, respectively. *Origanum* treatment in autumn as well as the use of pure origanum oil during summer was harmful to the bee colonies (Gal et al., 1992; Lensky et al., 1996). The use of essential oil of *O. majorana* in treatment of *V. jacobsoni* has attracted much practical attention from several authors. A combination of formic acid and of essential oil of marjoram has been shown to be very effective in treatment of *Varroa* mites both in laboratory trials and in field experiments (Long et al., 1997). In field experiments, that were carried out under tropical (Vietnam) and temperate (Germany) climatic conditions, formic acid was applied to a tray covered by gauze and placed on the bottom board of the hive while *O. majorana* essential oil was applied to two wood pieces (1.5 ml per piece), that were placed on the top bars of the combs. The combination of essential oil and formic acid, applied at 15 per cent concentration, resulted in 96.24–99.68 per cent mite mortality in tropical climate and in 97.56–99.92 per cent in temperate climatic conditions. Due to the relatively low concentrations of formic acid this combination did not affect bee mortality and was proposed as a promising practical method in the control of *V. jacobsoni*. The highly significant repellent activity and antiparasitic effects of essential oil of marjoram were observed towards *Varroa* mites, which were exposed to test wax tubes with incorporated essential oil at 0.1 per cent and 1 per cent (Kraus et al., 1994). These concentrations of *O. majorana* essential oil were not noxious to honeybees.

Effective antiparasitic activity was observed also when essential oil of *O. majorana* was sprayed onto bees in colonies infested with *V. jacobsoni* at concentration (100 ppm), that was found non toxic to bees (Fathy and Fouly, 1997). *Origanum majorana* essential oil has been shown as a potent acaricidal agent against *Acarapis woodi* (Renie), the acarine disease-causing parasite that invades the tracheal system of the honeybees during winter and early spring. Infestation percentage in the bee colonies, treated with *O. majorana* essential oil (10 drops of oil per piece of cotton wool in a Petri dish, that was put under the combs of infested colonies) was significantly reduced already after 15 days of treatment, and after 30 days of treatment no infestation was found among the tested bees (Abou Zaid et al., 1987; Mazeed, 1987).

In respect of the control of human parasites or parasite-related diseases both *in vitro* and *in vivo* studies were conducted. *O. vulgare* essential oil was studied for its *in vitro* antimalarial activity on *Plasmodium falciparum* (Milhau et al., 1997). It displayed only moderate (IC_{50} = 516 µg/ml after 24 h and 355 µg/ml after 72 h) antiparasitic effects against chloroquine resistant strains of *Plasmodium falciparum* when compared to the more active oils of *Rosmarinus officinalis* or *Myrtus communis* (IC_{50} = 267 µg/ml after 24 h and 149 µg/ml after 72 h). However, the observed efficacy of the tested oils against both chloroquine-resistant and -sensitive strains allowed it to be deduced that the oils could interfere with *P. falciparum* growth by different mechanisms than chloroquine. This preliminary screening of activity together with concomitant analysis of essential oils set the direction toward selection of major components, like carvacrol, that were proposed for future investigations for antimalarial potential.

A clinical study, done by Force *et al.* (2000), showed that emulsified *O. vulgare* diet (600 mg daily) for 6 weeks considerably affected the enteric parasites (*Blastocystis hominis*, *Entamoeba hartmanni*, *Endolimax nana*) and significantly improved the gastrointestinal symptoms in seven of 11 patients, who were positive for *Blastocystis hominis*.

INSECTICIDAL, NEMATICIDAL AND MOLLUSCICIDAL ACTIVITY

Higher plants, especially medicinal and aromatic plants (MAP), are a potential source of new insecticides, and many research groups are trying to prove their activity against noxious pests. Some natural compounds, isolated from these MAP (such as rotenon, pyretrins, and azadirachtin) are already commercially available on the market. A range of active compounds, including terpenoids, flavonoids, tannins, essential oils or their components (like carvacrol), that are present in *O. vulgare* in relative high amounts, were considered as a potential source of natural biocides (Duke, 1992). Investigations on the activity of aromatic plants against stored product- or plant-noxious pests gave diverse results when considering different insect species, their developmental stages (eggs and adults) or the way of application (fumigant, contact) (Regnault-Roger and Hamraoui, 1993a; Shaaya *et al.*, 1993; Regnault-Roger and Hamraoui, 1995; Kalinović *et al.*, 1997; Mateeva *et al.*, 1997; Rakowski and Ignatowicz, 1997; Baricevic *et al.*, 2001). Generally, plants or their essential oils showed more potent activities when applied directly to the insect surface than after fumigant application. Among different aromatic species, the plant essential oils from the Lamiaceae family have the best insecticidal effects against bean weevil *A. canthoscelides obtectus* (Regnault-Roger and Hamraoui, 1993a). Oregano (*O. vulgare* L.) is one of the plants, used traditionally in southern France, to control bean weevil (*A. obtectus* Say) in stored kidney beans (*Phaseolus vulgaris* L.) (Regnault and Hamraoui, 1993b; Bernath and Badulosi, 1997). A high carvacrol containing *O. vulgare* ssp. *hirtum* essential oil showed both fumigant and contact toxicities to bean weevil (*A. obtectus* Say) in laboratory trials (Baricevic *et al.*, 2001). When considering fumigant toxicity, insecticidal effect (mortality rate 82.5 per cent) was observed 6 days after application of high concentrations of oregano essential oil (150 µl per 55 g of beans). When considering contact toxicity, both oregano drug plant and essential oil at all tested concentrations significantly increased the bean weevils' mortality rates in comparison to the controls. Essential oils (5 µl, 15 µl and 30 µl per 55 g of beans) induced 100 per cent mortality of the bean weevil population when applied directly to the surface of the beans (55 g) in Petri dishes. Also, egg laying and hatching was inhibited after treatment of bean weevil with powdered drug plant (0.33 g, 0.66 g, 1.0 g and 2.0 g) or with essential oil at all tested concentrations.

O. vulgare L. susbsp. *hirtum* essential oil with high carvacrol content showed insecticidal activity also against *Drosophila melanogaster* (Karpouhtsis *et al.*, 1998) and a strong ovicidal activity against the eggs of stored product insects *Tribolium confusum* and *Ephestia cautella* (Shaaya *et al.*, 1993) already at low concentrations (2 µl/l or 4 µl/l air), but showed very low fumigant toxicity against adult insects (*Tribolium confusum*, *Tribolium castaneum*, *Ephestia cautella*, *Sitophilus oryzae*) (Shaaya *et al.*, 1993; Shaaya *et al.*, 1997). The exposure to vapours of essential oil from *O. syriacum* var. *bevanii* Ietswaart resulted in 77 per cent and 89 per cent mortality of the eggs of the confused flour beetle (*Tribolium confusum*) and the Mediterranean flour moth (*Ephestia kuehniella*), respectively (Tunç et al., 2000). This oil (1 μl/l air) also showed high fumigant toxicity against females of two greenhouse pests, i.e. carmine spider mite (*Tetranychus cinnabarinus*) and cotton aphid (*Aphis gossypii*) (100 per cent mortality after 48 h and 96 h exposure, respectively) (Tunç and Şahinkaya, 1998). By contrast, only limited insecticidal potential of essential oils of *Origanum creticum*, which showed only moderate contact toxicity ($LD_{>90}$ = 100 μg/larva) against tobacco cutworm (*Spodoptera litura*), was reported by Isman et al. (2001).

Origanum vulgare essential oil (15 per cent) was also tested for its repellent activity against *Culicoides imicola* Kieffer, the vector of the African horse sickness virus. When applied at concentration of 4 mg/m^2 on farm animals (horses), essential oil showed only non-significant repellency for 2 h and was far less effective than synthetic repellent di-ethyl toluamide (DEET), and was not recommended for its use in order to prevent the spread of *Culicoides*-borne pathogens (Braverman and Chizov-Ginzburg, 1997, 1998).

Toxicity and resistance toward nematodes count as important attributes of aromatic plants, that offer new applications in the field of plant health care programmes of sensitive crops, especially when nematicides or resistant cultivars are not available. A high level of resistancy against infestation with root-knot nematode (*Meloidogyne inconita* Chitwood) was observed in *O. vulgare* and *O. majorana* plants, which were free of root galls even after exposure to initial nematode populations of 15 eggs/cm^3 of soil medium in greenhouse experimental conditions (Walker, 1995). However, the root-knot nematodes caused a significant decrease in dry weight of *O. vulgare* but not that of *O. majorana*. Essential oil of *O. majorana*, with terpinen-4-ol (41.6 per cent) as the major compound, affected the soil stages of phytonematodes (*Rotylenchulus reniformis*, *Criconemella* spp., *Hoplolaimus* spp.) and inhibited more than 80 per cent of *Meloidogyne incognita* juvenile hatching compared to about 3.5 per cent at the control (Abd-Elgawad and Omer, 1995). Laboratory trials carried out with *O. vulgare*, *O. majorana* and *O. syriacum* leaf extracts or essential oils showed, that these plants considerably affected the spread of *Meloidogyne* nematodes, either by inhibition of egg hatching (Ramraj et al., 1991; Oka et al., 2000) or by immobilisation and exhibiting toxicity to nematode juveniles (Hashim et al., 1999; Oka et al., 2000). The toxicity increased with increasing concentration and exposure time. *Origanum* extracts or essential oils showed protective effects against root galling also when applied to nematode-sensitive crops. Alagumalai et al. (1997) observed that water extracts of *O. vulgare* dose-dependently diminished the population of *M. incognita* around chickpeas. Oka et al. (2000), who studied the nematicidal effects against *Meloidogyne javanica in vitro* and in pot experiments, found that *O. vulgare* and *O. syriacum* essential oils, when mixed in sandy soil at concentration of 200 mg/kg, reduced the root galling of cucumber seedlings. Similar effects were obtained by carvacrol and thymol at concentration of 150 mg/kg soil.

A strong molluscicidal effect of *O. compactum* ethyl acetate extracts (LC_{90} = 2.00 mg/l) against the schistosomiasis-transmitting snail *Bulinus truncatus* was attributed to the content of flavonoids and terpenoids, that are known to have molluscicidal potential (Hmamouchi et al., 2000). Interesting findings were observed by Vokou et al. (1998), who studied the effects of two subspecies of *O. vulgare* (ssp. *hirtum* and ssp. *vulgare*) on the behaviour of three snail species, native in Greece (*Helix lucorum*, *H. aspersa*, and *Eobania vermiculata*) during the different stages of the foraging cycle. *O. vulgare* ssp. *hirtum*, which contained much higher amounts of essential and was rich in phenolic compounds, considerably affected the snail feeding behaviour, while no significant effects were observed in ssp. *vulgare*. During the encounter stage, a repellent activity of

O. vulgare ssp. hirtum was observed. During the acceptance stage, all snail species tended to reject food types that contained high concentrations of subsp. hirtum essential oil, but at the feeding stage, subsp. hirtum essential oil caused a reduction of daily consumption rates. This is in agreement with Barone and Frank (1999), who found that polar (methanol) extracts of O. vulgare showed only scarce repellent effects on slug (*Arion lusitanicus*) feeding on rape.

TOXICITY

Besides toxicological data, referring to toxicities of single components of essential oils – such as limonene (carcinogenic in male rats) from O. majorana (de Vincenzi and Mancini 1997) – several clinical studies confirmed the allergenic potential of *Origanum* spp. On the basis of clinical history, and of *in vitro* and *in vivo* studies, O. vulgare showed cross-sensitivity with other plants of the Lamiaceae family. The potential allergic response, that could be evoked in sensitive patients after the ingestion of food seasoned with O. vulgare, comprises an increased serum level of specific IgE and induced systemic allergic reactions (Benito et al., 1996). Similarly, perioral dermatitis has been reported to be induced by O. majorana food flavouring (Farkas, 1981). O. vulgare has been shown also to induce allergic contact dermatitis, as clinically evaluated by patch test (Futrell and Rietschel, 1993).

Due to empirically proven emmenagogue and abortifacient effects, excessive use of O. vulgare or O. majorana should be avoided during pregnancy (Brinker, 1998).

REFERENCES

Abd-Elgawad, M.M. and Omer, E.A. (1995) Effect of essential oils of some medicinal plants on phytonematodes. *Anzeiger für Schadlingskunde, Pflanzenschutz, Umweltschutz* 68(4), 82–84.

Abdalla, A.E. and Roozen, J.P. (1999) Effect of plant extracts on the oxidative stability of sunflower oil and emulsion. *Food Chem.* 64(3), 323–329.

Abdalla, A.E., Tirzite, D., Tirzitis, G. and Roozen, J.P. (1999) Antioxidant activity of 1,4-dihydropyridine derivatives in beta-carotene-methyl linoleate, sunflower oil and emulsions. *Food Chem.* 66(2), 189–195.

Abou Zaid, M.I., Mazzed, M.M. and Salem, M.M. (1987) Evaluation of some natural bioactive substances for controlling *Acarapis woodi* (Remie). *Bull. ent. soc. Egypt, Econ ser.* 16,283–287.

Adam, K., Sivropoulou, A., Kokkini, S., Lanaras, T. and Arsenakis, M. (1998) Antifungal activities of *Origanum vulgare* subsp. *hirtum*, *Mentha spicata*, *Lavandula angustifolia*, and *Salvia fruticosa* essential oils against Human Pathogenic Fungi. *J. Agric. Food Chem.* 46(5), 1739–1745.

Adlova, G.P., Denisova, S.V., Ilidzhev, A.K., Smirnova, G.A. and Ratnikova, T.N. (1998) The development of bacterial growth stimulants from plants. *Zhurnal Mikrobiologii, Epidemiologiii Immunobiologii* 1, 13–17.

Aeschbach, R., Loliger, J., Scott, B.C., Murcia, A., Butler, J., Halliwell, B. and Aruoma, O.I. (1794) Antioxidant actions of thymol, carvacrol, 6-gingerol, Zingerone and hydroxytyrosol. *Food chem. Toxicol.* 32(1), 31–36.

Afsharypuor, S., Sajjadi, S.E. and Manesh, M.E. (1997) Volatile constituents of *Origanum vulgare* ssp. *viride* (syn.: *O. heracleoticum*) from Iran. *Planta Medica* 63, 179–180.

Akgül, A. and Bayrak, A. (1987) Constituents of essential oils from *Origanum* species growing wild in Turkey. *Planta medica* 53(1), 114.

Akgül, A. and Kivanç, M. (1988) Inhibitory effects of selected Turkish spices and oregano components on some foodborne fungi. *Int. J. Food Microbiol.* 6, 263–268.

Akgül, A. and Kivanç, M. (1989) Sensitivity of four foodborne moulds to essential oils from Turkish spices, herbs and citrus peel. *J. Sci. Food Agric.* 47, 129–132.

Akgül, A., Kivanç, M. and Sert, S. (1991) Effect of carvacrol on growth and toxin production by *Aspergillus flavus* and *Aspergillus parasiticus*. *Sciences des Aliments* 11, 361–370.

Alagumalai, K., Thiruvalluvan, M., Nagendran, N. and Ramaraj, P. (1997) Effect of leaf extract of *Origanum vulgare* on population build up of *Meloidogyne incognita*. *J. Ecotoxicol Environ. Monit.* 7(2), 151–153.

Anonymus (1997) CO_2 extracts – food and medicine? *Food Marketing & Technology* 11(4), 27–31.

Antoun, N. and Tsimidou, M. (1997) Gourmet olive oils: stability and consumer acceptability studies. *Food Res. Int.* 30(2), 131–136.

Arnold, N., Bellomaria, B., Valentini, G. and Arnold, H.J. (1993) Comparative study of the essential oils from three species of *Origanum* growing wild in the eastern Mediterranean region. *J. Essent. Oil Res.* 5(1), 71–77.

Arras, G. and Picci, V. (1984) Attivitá fungistatica di alcuni olii essenziali nei contronti dei principali agenti di alterazioni post-racolta dei frutti di agrumi. *Rivista della Ortofrutticoltura Italiana* 68, 361–366.

Aruoma, O.I., Spencer, J.P.E., Rossi, R., Aeschbach, R., Khan, A., Mahmood, N., Munoz, A., Murcia, A., Butler, J. and Halliwell, B. (1996) An evaluation of the antioxidant and antiviral action of extracts of rosemary and provencal herbs. *Food Chem. Toxicol.* 34(5), 449–456.

Assaf, M.H., Ali, A.A., Makboul, M.A., Beck, J.P. and Anton, R. (1987) Preliminary study of phenolic glycosides from *Origanum majorana*, quantitative estimation of arbutin, cytotoxic activity of hydroquinone. *Planta-Medica* 53(4), 343–345.

Aureli, P., Costantini, A. and Zolea, S. (1992) Antimicrobial activity of some essential oils against *Listeria monocytogenes*. *J. Food Prot.* 55(5), 344–348.

Aydin, S., Öztürk, Y., Beis, R. and Başer, K.H.C. (1996) Investigation of *Origanum onites, Sideritis congesta* and *Satureja cuneifolia* essential oils for analgesic activity. *Phytother. Res.* 10(4), 342–344.

Ayers, G.S. and Ayers, S. (1997) Bee forages with other uses-Part 1: Plants with nonmarketable value. *Am. Bee J.* 137(7), 526–531.

Azzouz, M.A. and Bullerman, L.B. (1982) Comparative antimycotic effects of selected herbs, spices, plant components and commercial antifungal agents. *J. Food Prot.* 45(14), 1298–1301.

Banias, C., Oreopoulou, V. and Thomopoulos, C.D. (1992) The effect of primary antioxidants and synergists on the activity of plant extracts in lard. *J. Am. Oil Chem. Soc.* 69(6), 520–524.

Baratta, M.T., Dorman, H.J.D., Deans, S.G., Biondi, D.M. and Ruberto, G. (1998a) Chemical composition, antimicrobial and antioxidative activity of laurel, sage, rosemary, oregano and coriander essential oils. *J. Essent. Oil Res.* 10(6), 618–627.

Baratta, M.T., Dorman, H.J.D., Deans, S.G., Figueiredo, A.C., Barroso, J.G. and Ruberto, G. (1998b) Antimicrobial and antioxidant properties of some commercial essential oils. *Flavour Fragrance J.* 13, 235–244.

Baricevic, D., Milevoj, L. and Borstnik, J. (2001) Insecticidal effect of oregano (*Origanum vulgare* L. ssp. *hirtum* Ietswaart) on the dry bean weevil (*Acanthoscelides obtectus* Say). *Int. J. Horticultural Sci.* 7(2), 84–88.

Barone, M. and Frank, T. (1999) Effects of plant extracts on the feeding behaviour of the slug *Arion lusitanicus*. *Ann. Appl. Biol.* 134(3), 341–345.

Başer, K.H.C., Ermin, N., Kurkcuoglu, M. and Tümen, G. (1994) Essential oil of *Origanum hypericifolium* O. Schwarz et P.H. Davis. *J. Essent. Oil Res.* 6(6), 631–633.

Başer, K.H.C., Ozek, T. and Tümen, G. (1995) Essential oil of *Origanum rotundifolium* Boiss. *J. Essent. Oil Res.* 7(1), 95–96.

Başer, K.H.C. and Tümen, G. (1992) Composition of the Essential Oil of *Origanum sipyleum* of Turkish Origin. *J. Essent. Oil Res.* 4, 139–142.

Basilico, M.Z. and Basilico, C. (1999) Inhibitory effects of some spice essential oils on *Aspergillus ochraceus* 3174 growth and ochratoxin A production. *Lett. Appl. Microbiol.* 29(4), 238–241.

Beddows, C.G., Jagait, C. and Kelly, M.J. (2000) Preservation of alpha-tocopherol in sunflower oil by herbs and species. *International J. Food Sci. Nutr.* 51(5), 327–339.

Beker, R., Dafni, A., Eisikowitch, D. and Ravid, U. (1989) Volatiles of two chemotypes of *Majorana syriaca* L. (Labiatae) as olfactory cues for the honeybee. *Oecologia* 79(4), 446–451.

Bellakhdar, J., Passannanti, S., Paternostro, M.P. and Piozzi, F. (1988) Constituents of *Origanum compactum*. *Planta Medica* 54(1), 94.

Benito, M., Jorro, G., Morales, C., Pelaez, A. and Fernandez, A. (1996) Labiatae allergy: systemic reactions due to ingestion of oregano and thyme. *Ann. Allergy, Asthma Immunol.* 76(5), 416–418.

Bernath, J. and Badulosi, S. (1997) Some scientific and practical aspects of production and utilization of oregano in central Europe. In *Proceedings of the IPGRI International Workshop on Oregano*, Oregano, 8–12 May 1996, CIHEAM, Valenzano, Bari, Italy, pp. 76–93.

Beuchat, L.R. (1976) Sensitivity of Vibrio parahaemolyticus to spices and organic acids. *J. Food Sci.* 41, 899–902.

Biondi, D., Cianci, P., Geraci, C., Ruberto, G. and Piattelli, M. (1993) Antimicrobial activity and chemical composition of essential oils from Sicilian aromatic plants. *Flavour and Fragrance J.* 8(6), 331–337.

Bisht, G.S. and Khulbe, R.D. (1995) In vitro efficacy of leaf extracts of certain indigenous medicinal plants against brown leaf spot pathogen of rice. *Indian Phytopathol.* 48(4), 480–482.

Blumenthal, M. (Senior ed.) (1998) The Complete German Commission E Monographs. Therapeutic Guide to Herbal Medicines. W.R. Busse, A. Goldberg, J. Gruenwald (Assoc. Eds.), American Botanical Council, Austin, Texas, pp. 358.

Bocharova, O.A., Karpova, R.V., Kasatkina, N.N., Polunina, L.G., Komarova, T.S. and Lygenkova, M.A. (1999) The antiproliferative activity for tumor cells is important to compose the phytomixture for prophylactic oncology. *Farmacevtski Vestnik* 50, 378–379.

Bohm, A.B. (1988) The minor flavonoids. In J.B. Harborne (ed.), *The flavonoids. Advances in Research since 1980*, Chapman and Hall, New York, p. 352.

Braverman, Y. and Chizov-Ginzburg, A. (1997) Repellency of synthetic and plant-derived preparations for *Culicoides imicola*. *Med. Veterinary Entomol.* 11, 355–360.

Braverman, Y. and Chizov-Ginzburg, A. (1998) Duration of repellency of various synthetic and plant-derived preparations for *Culicoides imicola*, the vector of African horse sickness virus. *Arch. Virol.* 14(Suppl.), 165–174.

Brinker, F. (1998) Herb Contraindications and Drug Interactions. Eclectic Medical Publications, Sandy, Oregon, 263pp.

Bruneton, J. (1999) Pharmacognosie, Phytochimie, Plantes médicinales, 3rd edn, Éditions Tec & Doc, Paris, pp. 540.

Bullerman, B., Lieu, F.Y. and Seier, S.A. (1977) Inhibition of growth and aflatoxin production in cinnamom and clove oils. Cinnamic aldehyde and eugenol. *J. Food Sci.* 42, 1107–1109.

Buznego, M.T., Perez Saad, H. and Carrion, L. (1991) Antiepileptic effect of *Plectranthus amboinicus* (Lour) Spren. (French *Origanum*) in three models of experimental epilepsy. *Epilepsia* 32(Suppl. 1), 35.

Calderone, N.W., Shimanuki, H. and Allen-Wardell, G. (1994) An in vitro evaluation of botanical compounds for the control of the honeybee pathogens *Bacillus larvae* and *Ascosphaera apis*, and the secondary invader *B. alvei*. *J. Essent. Oil Res.* 6(3), 279–287.

Charai, M., Mosaddak, M. and Faid, M. (1996) Chemical composition and antimicrobial activities of two aromatic plants: *Origanum majorana* L. and *O. compactum* Benth. *J. Essent. Oil Res.* 8(6), 657–664.

Charai, M., Faid, M. and Chaouch, A. (1999) Essential oils from aromatic plants (*Thymus broussonetti* Boiss., *Origanum compactum* Benth. and *Citrus limon* (L.) N.L. Burm.) as natural antioxidants for olive oil. *J. Essent. Oil Res.* 11(4), 517–521.

Chatterjee, A., Venkateswaran, N., Basu, D.K. and Chatterjee, H.N. (1958) Some chemical pharmacological and antibacterial properties of *Coleus aromaticus* Benth. *Sci & Cult.* 24, 241–243.

Chipault, J.R., Mizuno, G.R., Hawkins, J.M. and Lundberg, W.O. (1952) The antioxidant properties of natural spices. *Food Res.* 17, 46–55.

Chipault, J.R., Mizuno, G.R. and Lundberg, W.O. (1955) Antioxidant properties of spices in oil-inwater emulsions. *Food Res.* 20, 443–448.

Chipault, J.R., Mizuno, G.R. and Lundberg, W.O. (1956) The antioxidant properties of spices in foods. *Food Technol.* 10, 209–211.

Chung, S.K., Osawa, T. and Kawakishi, S. (1997) Hydroxyl radical-scavenging effects of spices and scavengers from brown mustard (*Brassica nigra*). *Biosci. Biotechnol. Biochem.* 61(1), 118–123.

Colin, M.E., Ducos de Lahitte, J., Larribau, E. and Boué, T. (1989) Activité des huiles essentielles de Labiées sur *Ascosphaera apis* et traitement d'un rucher. *Apidologie* 20, 221–228.

Collier, W.A. and Nitta, Y. (1930) Über die Wirkung ätherischer Öle auf Verschiedene Bakterienarten. *Z. Hyg. Infektionskranth.* 111, 301–304.

Conner, D.E., Beuchat, L.R., Worthington, R.E. and Hitchcock, H.L. (1984) Effects of essential oils and oleoresins of plants on ethanol production, respiration and sporulation of yeasts. *Int. J. Food Microbiol.* 1(2), 63–74.

Conner, D.E. and Beuchat, L.R. (1984a) Effects of essential oils from plants on growth of food spoilage yeasts. *J. Food Sci.* 49, 429–434.

Conner, D.E. and Beuchat, L.R. (1984b) Sensitivity of heat-stressed yeasts to essential oils of plants. *Appl. Environ. Microbiol.* 47(2), 229–233.

Conner, D.E. and Beuchat, L.R. (1985) Recovery of heat-stressed yeasts in media containing plant oleoresins. *J. Appl. Bacteriol.* 59, 49–55.

Craig, W.J. (1999) Health-promoting properties of common herbs. *Am. J. Clin. Nutr.* 70(suppl.), 491S–499S.

Curtis, O.F., Shetty, K., Cassagnol, G. and Peleg, M. (1996) Comparison of the inhibitory and lethal effects of synthetic versions of plant metabolites (anethole, carvacrol, eugenol and thymol) on food spoilage yeast (*Debaromyces hansenii*). *Food Biotechnol.* 10(1), 55–73.

Daferera, D.J., Ziogas, B.N. and Polissiou, M.G. (2000) GC-MS analysis of essential oils from some Greek aromatic plants and their fungitoxicity on *Penicillium digitatum*. *J. Agric. Food. Chem.* 48(6), 2576–2581.

Daouk, R.K., Dagher, S.M. and Sattout, E.J. (1995) Antifungal activity of the essential oil of *Origanum syriacum* L. *J. Food Prot.* 58(10), 1147–1149.

Dapkevicius, A., Venskutonis, R., van-Beek, T.A. and Linssen, J.P.H. (1998) Antioxidant activity of extracts obtained by different isolation procedures from some aromatic herbs grown in Lithuania. *J. Sci. Food Agric.* 77(1), 140–146.

Daw, Z.Y., El-Baroty, G.E. and Ebtesam, A.M. (1994) Inhibition of *Aspergillus parasiticus* growth and aflatoxin production by some essential oils. *Chem. Mikrobiol. Technol. Lebensm.* 16(5/6), 129–135.

Deans, S.G. and Ritchie, G. (1987) Antibacterial properties of plant essential oils. *Int. J. Food Microbiol.* 5, 165–180.

Deans, S.G. and Svoboda, K.P. (1990) The antimicrobial properties of marjoram (*Origanum majorana* L.) volatile oil. *Flavour and Fragrance J.* 5(3), 187–190.

Deans, S.G., Svoboda, K.P., Gundidza, M. and Brechany, E.Y. (1992) Essential oil profiles of several temperate and tropical aromatic plants: their antimicrobial and antioxidant activities. *Acta-Horticulturae* 306, 229–232.

de Felice, M., de Leonardis, T. and Comes, S. (1993) Flavouring of edible oils. Effects on autoxidation. *Industrie Alimentari* 32(313), 249–253.

de Vincenzi, M. and Mancini, E. (1997) Monographs on botanical flavouring substances used in foods. Part VI. *Fitoterapia* 68(1), 49–61.

Deighton, N., Glidewell, S.M., Deans, S.G. and Goodman, B.A. (1993) Identification by EPR spectroscopy of Carvacrol and Thymol as the major sources of free radicals in the oxidation of plant essential oils. *J. sci. Food Agric.* 63, 221–225.

Deryabin, A.M. (1990) Pharmaceutical preparation for treatment of mastitis in animals and humans. PCT International Patent Application, WO 90/13305 A1, 16 pp.

Deryabin, A.M. (1991) Medicinal agent and method for treatment of mastitis in animals and humans. *United States Patent*, US 5 061 491, 6pp.

Didry, N., Dubreuil, L. and Pinkas, M. (1993) Antimicrobial activity of thymol, carvacrol and cinnamaldehyde alone or in combination. *Pharmazie* 48, 301–304.

Dorman, H.J.D. and Deans, S.G. (2000) Antimicrobial agents from plants: Antibacterial activity of plant volatile oils. *J. Appl. Microbiol.* 88(2), 308–316.

Dorofeev, A.N., Khort, T.P., Rusina, I.F. and Khmel'nitskii, Y.V. (1989) Search for antioxidants of plant origin and prospects of their use. *Sbornik Nauchnykh Trudov Gosudarstvennyi Nikitskii Botanicheskii Sad* 109, 42–53.

Duke, J.A. (1992) Biting the biocide bullet. In L.F. James, R.F. Keeler, E.M. Bailey, P.R. Cheeke, M.P. Hegarty (eds), *Proceedings of the Third International Symposium*, Poisonous plants, pp. 474–478.

Economou, K.D., Oreopoulou, V. and Thomopoulos, C.D. (1991) Antioxidant activity of some plant extracts of the family Labiatae. *J. Am. Oil Chem. Soc.* 68(2), 109–113.

El-Alim, S.S.L.A., Lugasi, A., Hovari, J. and Dworschak, E. (1999) Culinary herbs inhibit lipid oxidation in raw and cooked minced meat patties during storage. *J. Sci. Food Agric.* 79(2), 277–285.

El-Maksoud, A., Aboul-Fotouh, G.E., Allam, S.M. and Zied, R.M.A. (1999) Effect of marjoram leaves (*Majorana hortensis* L. [*Origanum majorana*]) as a feed additive on the performance of Nile tilapia (*Oreochromis niloticus*) fingerlings. *Egyptian Journal of Nutrition and Feeds* 2(1), 39–47.

Farag, R.S., Daw, Z.Y. and Abo-Raya, S.H. (1989) Influence of some spice essential oils on *Aspergillus parasiticus* growth and production of aflatoxins in a synthetic medium. *J. Food Sci.* 54, 74–76.

Farkas, J. (1981) Perioral dermatitis from marjoram, bay leaf and cinnamon. *Contact Dermatitis.* 7(2), 121.

Fathy, H.M. and Fouly, A.H. (1997) The effect of natural volatile oil plants on *Apis mellifera* honey bee and on the *Varroa jacobsoni* in the bee colonies. *Apiacta* 32(1), 5–12.

Fleisher, A. and Fleisher, Z. (1988) Identification of Biblical Hyssop and origin of the traditional use of oregano-group herbs in the mediterranean region. *Econ.-Bot.* 42(2), 232–241.

Force, M., Sparks, W.S. and Ronzio, R.A. (2000) Inhibition of enteric parasites by emulsified oil of oregano *in vivo. Phytother. Res.* 14(3), 213–214.

Frankel, E.N., Huang, S.W., Kanner, J. and German, J.B. (1994) Interfacial phenomena in the evaluation of antioxidants: bulk oils vs. emulsions. *J. Agric. Food Chem.* 42, 1054–1059.

Futrell, J.M. and Rietschel, R.L. (1993) Spice allergy evaluated by results of patch tests. *Cutis.* 52(5), 288–290.

Gal, H., Slabezki, Y. and Lensky, Y. (1992) A preliminary report on the effect of Origanum oil and thymol applications in honey bee (*Apis mellifera* L.) colonies in a subtropical climate on population levels of *Varroa jacobsoni. Bee Sci.* 2(4), 175–180.

Gerhardt, U. and Schröter, A. (1983) Antioxidative Wirkung von Gewürzen. *Gordian* 83(9), 171–176.

Gergis, V., Spiliotis, V. and Poulos, C. (1990) Antibacterial activity of essential oils from Greek *Sideritis* species. *Pharmazie* 45, 70.

Guérin, J.C. and Réveillère, H.P. (1985) Activité antifongique d'extraits végétaux à usage thérapeutique. II. Étude de 40 extraits sur 9 souches fongiques. *Ann. Pharmaceutiques françaises* 43(1), 77–81.

Gunther, K.D. (1991) Spices in animal feeds. *Int. Pig Top.* 6, 25–27.

Hao, Y.Y., Brackett, R.E. and Doyle, M.P. (1998) Efficacy of plant extracts in inhibiting *Aeromonas hydrophila* and *Listeria monocytogenes* in refrigerated cooked poultry. *Food Microbiology* 15(4), 367–378.

Hammer, K.A., Carson, C.F. and Riley, T.V. (1998) In-vitro activity of essential oils, in particular *Melaleuca alternifolia* (tea tree) oil and tea tree oil products, against *Candida* spp. *J. Antimicrob. Chemother.* 42, 591–595.

Hammer, K.A., Carson, C.F. and Riley, T.V. (1999) Antimicrobial activity of essential oils and other plant extracts. *J. Appl. Microbiol.* 86(6), 985–990.

Harvala, C. and Skaltsa, H. (1986) Contribution à l'étude chimique d'*Origanum dictamnus* L. 1re communication. *Plantes médicinales et phytothérapie* Tome XX(4), 300–304.

Hashim, E.F., Seham, K.A.A. and Kheir, A.A. (1999) Nematicidal activity of some labiataceous plant extracts on *Meloidogyne incognita*. *Ann. Agric. Sci. Cairo* 44(1), 447–457.

He, L., Mo, H., Hadisusilo, S., Qureshi, A.A. and Elson, C.E. (1997) Isoprenoids suppress the growth of Murine B16 Melanomas in vitro and in vivo. Biochemical and molecular roles of nutrients. *Am. Soc. Nutrit. Sci.* 668–674.

Herrmann, K., Schütte, M., Müller, H. and Dismer, R. (1981) Über die antioxidative Wirkung von Gewürzen. *Dtsch. Lebensm. – Rdsch.* 77, 134–138.

Herrmann, K. (1994) Antioxidant effects of important substances – plant phenols and carotenoids in spices. *Gordian* 94(7–8), 113–117.

Hirobe, C., Qiao, Z.S., Takeya, K. and Ibokawa, H. (1998) Cytotoxic principles from *Majorana syriaca*. *Nat. Med.* 52(1), 74–77.

Hitokoto, H., Morozumi, S., Wauke, T., Sakai, S. and Kurata, H. (1980) Inhibitory effects of spices on growth and toxin production of toxigenic fungi. *Appl. Environ. Microbiol.* 39, 818–822.

Hmamouchi, M., Lahlou, M. and Agoumi, A. (2000) Molluscicidal activity of some Moroccan medicinal plants. *Fitoterapia* 71(3), 308–314.

Hook, E.B. (1978) Dietary cravings and aversions during pregnancy. *Am. J. Clin. Nutr.* 31, 1355–1362.

Houghton, P.J., Zarka, R., Delasheras, B. and Hoult, J.R.S. (1995) Fixed oil of *Nigella sativa* and derived thymoquinone inhibit eicosanoid generation in leukocytes and membrane peroxidation. *Planta Medica* 61, 33–36.

Hussein, A.S.M. (1990) Antibacterial and antifungal activities of some Libyan aromatic plants. *Planta Medica* 56, 644–645.

Ismaiel, A.A. and Pierson, M.D. (1990a) Inhibition of germination, outgrowth, and vegetative growth of *Clostridium botulinum* 67B by spice oils. *J. food prot.* 53(9), 755–758.

Ismaiel, A.A. and Pierson, M.D. (1990b) Effect of sodium nitrite and origanum oil on growth and toxin production of *Clostridium botulinum* in TYG broth and ground pork. *J. food prot.* 53(11), 958–960.

Isman, M.B., Wan, A.J. and Passreiter, C.M. (2001) Insecticidal activity of essential oils to the tobacco cutworm, *Spodoptera litura*. *Fitoterapia* 72, 65–68.

Izzo, A.A., Carlo, G., Biscardi, D., Fusco, R., Mascolo, N., Borrelli, F., Capasso, F., Fasulo, M.P. and Autore, G. (1995) Biological screening of Italian medicinal plants for antibacterial activity. *Phytother. Res.* 9(4), 281–286.

Jablonski, B. (1986) Nectar secretion and honey productivity of important honey plants in polish conditions. *Pszczelnicze Zeszyty Naukowe* 30, 195–205.

Jerkovic, I., Mastelic, J. and Milos, M. (2001) The effect of air-drying on glycosidically bound volatiles from seasonally collected oregano (*Origanum vulgare* L. ssp. *hirtum*) from Croatia. *Nahrung* 45(1), 47–49.

Jovančević, R., Popovic, B. and Ivezic, D. (1984) Melliferous plants from the Lamiaceae family distributed in the Lim river basin (Yugoslavia). *Poljoprivreda i sumarstvo* (Yugoslavia) 30(2–3), 65–84.

Kalinović, I., Martinčić, J., Rozman, V. and Guberac, V. (1997) Insecticidal activity of substances of plant origin against stored product insects. *Ochrana Rostlin* 33(2), p. 135–142.

Kanazawa, K., Kawasaki, H., Samejima, K., Ashida, H. and Danno, G. (1995) Specific desmutagens (antimutagens) in Oregano against dietary carcinogen, Trp-P-2, are galangin and quercetin. *J. Agric. Food Chem.* 43(2), 404–409.

Karanika, M.S., Komaitis, M. and Aggelis, G. (2001) Effect of aqueous extracts of some plants of Lamiaceae family on the growth of *Yarrowia lipolytica*. *International J. Food Microbiol.* 64(1–2), 175–181.

Karpouhtsis, I., Pardali, E., Feggou, E., Kokkini, S., Scouras, Z.G. and Mavragani-Tsipidou, P. (1998) Insecticidal and genotoxic activities of oregano essential oils. *J. Agric. Food Chem.* 46(3), 1111–1115.

Kellner, W. and Kober, W. (1954) Möglichkeiten der Verwendung ätherischer Öle zur Raumdesinfektion. 1. Die Wirkung gebräuchlicher ätherischer Öle auf Testkeime. *Arzneim. Forsch.* 4, 319–323.

Kikuzaki, H. and Nakatani, N. (1989) Structure of a new antioxidative phenolic acid from oregano *Origanum vulgare* L. *Agric. Biol. Chem.* 53(2), 519–524.

Kivanç, M. and Akgül, A. (1986) Antibacterial activities of essential oils from Turkish spices and citrus. *Flavour and Fragrance J.* 1, 175–179.

Kivanç, M., Akgül, A. and Dogan, A. (1991) Inhibitory and stimulatory effects of cumin, oregano and their essential, oils on growth and acid production of *Lactobacillus plantarum* and *Leuconostoc mesenteroides*. *International J. Food Microbiol.* 13(1), 81–85.

Koeniger, N. (1991) Auf der Suche nach Lösungen des Varroaproblems (2. Teil). *Schweizerische Bienen Zeitung* 114(10), 578–581.

Kokkini, S. (1997) Taxonomy, diversity and distribution of *Origanum* species. In *Proceedings of the IPGRI International Workshop on Oregano*, oregano, 8–12 May 1996, CIHEAM, Valenzano, Bari, Italy, pp. 2–12.

Korczak, J., Flaczyk, E. and Pazola, Z. (1988) Effects of spices on stability of meat products kept in cold storage. *Fleischwirtschaft* 68(1), 64–66.

Kraus, B., Koeniger, N. and Fuchs, S. (1994) Screening of substances for their effect on *Varroa jacobsoni*: attractiveness, repellency, toxicity and masking effects of ethereal oils. *J. Apicultural Research* 33(1), 34–43.

Krukowski, H., Nowakowicz-Debek, B., Saba, L. and Stenzel, R. (1998) Effect of mineral-herbal mixtures on IgG blood serum level in growing calves. *Roczniki Naukowe Zootechniki* 25(4), 97–103.

Kucherov, E.V. and Siraeva, S.M. (1981) Nectar plants of late summer. *Pchelovodstvo* 8, 18.

Kuebel, K.R. and Tucker, A.O. (1988) Vietnamese Culinary herbs in the essential oil in the United States. *Econ. Bot.* 42, 413–419.

Kunz, B. (1994) Gewürze zur Haltbarkeitsverlängerung von Brot. *Gordian* 94(4), 53.

Kunz, K., Weidenbörner, M. and Kunz, B. (1995) Die Nutzting von Gewürzen in weizenbrot zur kontrolle der lebensmittelrelevanten schimmelpilze *Cladosporium herbarum*, *Eurotium repens*, *Penicillium expansum* und *Rhizopus stolonifer*. *Chem. Mikrobiol. Technol. Lebensm.* 17(1–2), 1–5.

Kurita, N., Miyaji, M., Kurane, R. and Takahara, Y. (1981) Antifungal activity of components of essential oils. *Agric. Biol. Chem.* 45, 945–952.

Lagouri, V., Blekas, G., Tsimidou, M., Kokkini, S. and Boskou, D. (1993) Composition and antioxidant activity of essential oils from Oregano plants grown wild in Greece. *Zeitschrift für Lebensmittel Untersuchung und Forschung* 197(1), 20–23.

Lam, L.K.T. and Zheng, B. (1991) Effect of essential oils on glutathione S-transferase activity in mice. *J. Agric. Food Chem.* 39, 660–662.

Lamaison, J.L., Petitjean-Freytet, C. and Carnat, A. (1990) Teneurs an acide rosmarinique, en dérivés hydroxycinnamiques totaux et activité antioxydante chez les Apiacées, les Borraginacée et les Lamiacées médicinales. *Ann. Pharm. Fr.* 48(2), 103–108.

Lamaison, J.L., Petitjean-Freytet, C. and Carnat, A. (1991) Lamiacées médicinales à propriété antioxydantes, sources potentielles d'acide rosmarinique. *Pharm. Acta Helv.* 66(7), 185–188.

Lamaison, J.L., Petitjean-Freytet, C., Duke, J.A. and Walker, J. (1993) Hydroxycinnamic derivative levels and antioxidant activity in North American Lamiaceae. *Plantes médicinales etphytothérapie*, Tome XXVI(2), 143–148.

Laughton, M.J., Halliwell, B., Evans, P.J. and Hoult, J.R.S. (1989) Antioxidant and pro-oxidant actions of the plant phenolics quercetin, gossypol, and myricetin. *Biochem. Pharmacol.* 38, 2859–2865.

Launchbaugh, K.L. and Provenza, F.D. (1994) The effect of flavor concentration and toxin dose on the formation and generalization of flavor aversions in lambs. *J. Anim. Sci.* 72(1), 10–13.

Lensky, Y., Slabezki, Y., Gal, H., Gerson, U., Dechmani, U. and Jirasavetakul, U. (1996) Integrated control of varroa mite (*Varroa jacobsoni* Oud.) in *Apis mellifera* colonies in Thailand. German Israel Agricultural Research Agreement for Developing Countries, Rehovot, Israel: Final report, 96pp.

Lis-Balchin, M., Deans, S. and Hart, S. (1996) Bioactivity of New Zealand medicinal plant essential oils. *Acta-Horticulturae* 426, 13–30.

Llewellyn, G.C., Burkett, M.L. and Eadie, T. (1981) Potential mold growth, aflatoxin production and antimycotic activity of selected natural spices and herbs. *J. Assoc. Official Anal. Chem.* 64(4), 955–960.

Lolos, M., Oreopoulou, V. and Tzia, C. (1999) Oxidative stability of potato chips: effect of frying oil type, temperature and antioxidants. *J. Sci. Food and Agric.* 79(11), 1524–1528.

Long, L.T., Koeniger, N. and Fuchs, S. (1997) Varroa treatment with combination of formic acid and oil of marjoram: laboratory tests and field experiments. *Apidologie.* 28(3–4), 179–181.

Madsen, H.L. and Bertelsen, G. (1995) Spices as antioxidants. *Trends Food Sci. Technol.* 6(8), 271–277.

Madsen, H.L., Nielsen, B.R., Bertelsen, G. and Skibsted, L.H. (1996) Screening of antioxidative activity of spices. A comparison between assays based on ESR spin trapping and electrochemical measurement of oxygen consumption. *Food Chem.* 57(2), 331–337.

Madsen, H.L., Bertelsen, G. and Skibsted, L.H. (1997) Antioxidative activity of spices and spice extracts. In S.J. Risch and C.T. Ho (eds), *Flavor chemistry and antioxidant properties*, spices, Washington DC: American Chemical Society, pp. 176–187.

Marinova, E.M. and Yanishlieva, N.V. (1997) Antioxidative activity of extracts from selected species of the family Lamiaceae in sunflower oil. *Food Chem.* 58(3), 245–248.

Maruzzella, J.C. and Liguori, L. (1958) The in vitro antifungal activity of essential oils. *J. Am. Pharm. Assoc.* 47, 250.

Maruzzella, J.C. and Sicurella, N.A. (1960) Antibacterial activity of essential oil vapours. *J. Am. Pharm. Assoc.* 49, 692.

Mateeva, A., Stratieva, S. and Andonov, D. (1997) The effect of some plant extracts on *Acanthoscelides obtectus* Say. In Proceedings of the 49th International symposium on crop protection, Gent, Belgium, 16 May1997, University Gent, 62/2b, p. 513–515.

Mayer, E., Sadar, V. and Spanring, J. (1971) New Crops Screening of Native Plants of Yugoslavia of Potential Use in the Agricultures of the USA and SFRJ. University of Ljubljana, Biotechical Faculty, Final Technical Report. Printed by Partizanska knjiga Ljubljana, 210 pp.

Mazeed, M.M. (1987) Controlling acarine mites with natural materials. *Gleanings in Bee Culture* 115(9), 517, 520.

Melegari, M., Severi, F., Bertoldi, M., Benvenuti, S., Circetta, G., Morone Fortunato, I., Bianchi A., Leto, C. and Carruba, A. (1995) Chemical characterization of essential oils of some *Origanum vulgare* L. sub-species of various origin. *Rivista Italiana EPPOS* 16, 21–28.

Milhau, G., Valentin, A., Benoit, F., Mallié, M., Bastide, J.M., Pélissier, Y. and Bessière, J.M. (1997) In vitro antimalarial activity of eight essential oils. *J. Essent. Oil Res.* 9(3), 329–333.

Milic, B.L. and Milic, N.B. (1998) Protective effects of spice plants on mutagenesis. *Phytother. Res.* 12(Suppl. 1), S3–S6.

Milos, M., Mastelic, J. and Jerkovic, I. (2000) Chemical composition and antioxidant effect of glycosidically bound volatile compound from oregano (*Origanum vulgare* L. ssp. *hirtum*). *Food Chem.* 71, 79–83.

Mirovich, V.M., Peshkova, V.A., Shatokhina, R.K. and Fedoseev, A.P. (1989) Phenolcarboxylic acids of *Origanum vulgare*. *Khimiya Prirodnykh Soedinenii* 25(6), 850–851.

Møller, J.K.S., Madsen, H.L., Aaltonen, T. and Skibsted, L.H. (1999) Dittany (*Origanum dictamnus*) as a source of water-extractable antioxidants. *Food Chem.* 64(2), 215–219.

Montes-Belmont, R. and Carvajal, M. (1998) Control of *Aspergillus flavus* in maize with plant essential oils and their components. *J. Food Prot.* 61(5), 616–619.

Moure, A., Cruz, J.M., Franco, D., Domínguez, J.M., Sineiro, J., Domínguez, H., Núnez, M.J. and Parajó, J.C. (2001) Natural antioxidants from residual sources. *Food Chem.* 72(2), 145–171.

Muller-Riebau, F., Berger, B. and Yegen, O. (1995) Chemical composition and fungitoxic properties to phytopathogenic fungi of essential oils of selected aromatic plants growing wild in Turkey. *J. Agric. Food Chem.* 43(8), 2262–2266.

Nakatani, N. and Kikuzaki, H. (1987) A new antioxidative glucoside isolated from oregano (*Origanum vulgare* L.). *Agric. Biol. Chem.* 51(10), 2727–2732.

Nakatani, N. (1992) Natural antioxidants from spices. *ACS Symposium Series* 507, 72–86.

Nakatani, N. (1997) Antioxidants from spices and herbs. In Natural antioxidants: chemistry, health effects and applications (F. Shahidi, Ed.), AOCS Press, Champaign USA, 64–75.

Natake, M., Kanazawa, K., Mizuno, M., Ueno, N., Kobayashi, T., Danno, G. and Minamoto, S. (1989) Herb-water extracts markedly suppress the mutagenicity of Trp-P-2. *Agric. Biol. Chem.* 53(5), 1423–1425.

Nguyen, U., Frakman, G. and Evans, D.A. (1991) Process for extracting antioxidants from Labiatae herbs. United States Patent, US5017397.

Nielsen, P.V. and Rios, R. (2000) Inhibition of fungal growth on bread by volatile components from spices and herbs, and the possible application in active packaging, with special emphasis on mustard essential oil. *Int. J. Food Microbiol.* 60, 219–229.

Oka, Y., Nacar, S., Putievsky, E., Ravid, U., Yaniv, Z. and Spiegel, Y. (2000) Nematicidal activity of essential oils and their components against the root-knot nematode. *Phytopathology* 90(7), 710–715.

Okazaki, K., Nakayama, S., Kawazoe, K. and Takaishi, Y. (1998) Antiaggregant effects on human platelets of culinary herbs. *Phytother. Res.* 12(8), 603–605.

Okuyama, T., Matsuda, M., Masuda, Y., Baba, M., Masubuchi, H., Adachi, M., Okada, Y., Hashimoto, T., Zou, L.B. and Nishino, H. (1995) Studies on cancer bio-chemoprevention of natural resources. X. Inhibitory effect of species on TPA-enhanced 3H-choline incorporation in phosphdipids of C3H10T1/2 cells and on TPA-induced mouse ear edema. *Clin. Pharm. J.* 47(5), 421–430.

Özcan, M. (1998) Inhibitory effects of spice extracts on the growth of *Aspergillus parasiticus* NRRL 2999 strain. *Zeitschrift fur Lebensmittel Untersuchung und Forschung* 207(3), 253–255.

Özcan, M. and Akgül, A. (1995) Antioxidant activity of extracts and essential oils from Turkish spices on sunflower oil. *Acta Alimentaria* 24(1), 81–90.

Palic, A., Krizanec, D. and Dikanovic-Lucan, Z. (1993) Antioxidative Eigenschaften von Gewürzen in Rohwürsten. *Fleischwirtschaft* 73(6), 684–687.

Panizzi, L., Flamini, G., Cioni, P.L. and Morelli, I. (1993) Composition and antimicrobial properties of essential oils of four Mediterranean Lamiaceae. *J. Ethnopharmacol.* 39, 167–170.

Paster, N., Juven, B.J., Shaaya, E., Menasherov, M., Nitzan, R., Weisslowicz, H. and Ravid, U. (1990) Inhibitory effect of oregano and thyme essential oils on moulds and foodborne bacteria. *Lett. Appl. Microbiol.* 11, 33–37.

Paster, N., Menasherov, M., Shaaya, E., Juven, B. and Ravid, U. (1993) The use of essential oils applied as fumigants to control mycotoxigenic fungi attacking stored grain. *Hassadeh* 74(1), 25–27.

Paster, N., Menasherov, M., Ravid, U. and Juven, B. (1995) Antifungal activity of oregano and thyme essential oils applied as fumigants against fungi attacking stored grain. *J. Food Prot.* 58(1), 81–85.

Pearson, D.A., Frankel, E.N., Aeschbach, R. and German, J.B. (1997) Inhibition of endothelial cell-mediated oxidation of low-density lipoprotein by rosemary and plant phenolics. *J. Agric. Food Chem.* 45(3), 578–582.

Pellecuer, J., Jacob, M., Buochberg, M.S. and Allegrini, J. (1980) Therapeutic value of the cultivated mountain savory (*Satureia montana* L.: Labiatae). *Acta Horticulturae* 96(187), 35–39.

Podkolzin, A.A., Petrov, A.P., Didenko, V.I., Ovchinnikov, Yu, M. and Antonov, A.V. (1986) Effect of St. John's wort and wild marjoram on the course of acute rhinitis in experiment. *Vestnik Otorinolaringologii* 5, 40–44.

Pol, I.E. and Smid, E.J. (1999) Combined action of nisin and carvacrol on *Bacillus cereus* and *Listeria monocytogenes*. *Lett. Appl. Microbiol.* 29(3), 166–170.

Prudent, D., Perineau, F., Bessiere, J.M., Michel, G.M. and Baccou, J.C. (1995) Analysis of the essential oil of wild oregano from Martinique (*Coleus aromaticus* Benth.) – evaluation of its bacteriostatic and fungistatic properties. *J. Essent. Oil Res.* 7(2), 165–173.

Rakowski, G. and Ignatowicz, S. (1997) Effects of some plant extracts on fecundity and longevity of the dry bean weevil, *Acanthoscelides obtectus* Say (Coleoptera: Bruchidae). *Polskie Pismo Entomologiczne* 66(1–2), 161–167.

Ramraj, P., Alagumalai, K. and Hepziba, C.S.S. (1991) Effect of leaf extract of *Origanum vulgare* (Fam. *Lamiaceae*) on the hatching of eggs of *Meloidogyne incognita*. *Indian J. Nematology* 21(2), 156–157.

Regnault-Roger, C. and Hamraoui, A. (1993a) Influence d'huiles essentielles aromatiques sur *Acanthoscelides obtectus* Says, bruche du haricot (*Phaseolus vulgaris* L.). *Acta botanica Gallica* 140(2), 217–222.

Regnault-Roger, C. and Hamraoui, A. (1993b) Efficency of plants from south of France used as traditional protectant of *Phaseolus vulgaris* L. against its bruchid *Acanthoscelides obtectus* Say. *J. stored Prod. Res.* 29(3), 259–264.

Regnault-Roger, C. and Hamraoui, A. (1995) Comparison of the insecticidal effects of water extracted and intact aromatic plants on *Acanthoscelides obtectus*, a bruchid beetle pest of kidney beans. *Chemoecology* 5–6(1), 1–5.

Remmal, A., Bouchikhi, T., Rhayour, K., Ettayebi, M. and Tantaoui-Elaraki, A. (1993) Improved method for the determination of antimicrobial activity of essential oils in agar medium. *J. Essent. Oil Res.* 5(2), 179–184.

Ricciardelli-d'Albore, G. (1983) Wild insects and honeybees as pollinators of some Labiatae of herbal interest (*Origanum majorana, Origanum vulgare, Rosmarinus officinalis, Salvia officinalis, Salvia sclarea*) in a specialized area. *Redia* 66, 283–293.

Saito, Y., Kimura, Y. and Sakomoto, T. (1976) The antioxidant effect of some spices. *J. Jap. Soc. Food Nutr.* 29, 404–408.

Sawabe, A. and Okamoto, T. (1994) Natural Phenolics as Antioxidants and Hypotensive Materials. *Bulletin of the Institute for Comprehensive Agricultural Sciences, Kinki University* No. 2, 1–11.

Schmitz, S., Weidenbörner, M. and Kunz, B. (1993) Herbs and spices as selective inhibitors of mould growth. *Chem. Mikrobiol. Technol. Lebensm.* 15(6), 175–177.

Scortichini, M. and Rossi, M.P. (1989) In vitro activity of some essential oils toward *Erwinia amylovora* (Burrill) Winslow et al. *Acta Phytopathologica et Entomologica Hungarica* 24(3–4), 423–431.

Scortichini, M. and Rossi, M.P. (1993) In vitro behaviour of *Erwinia amylovora* towards some natural products showing bactericidal activity. *Acta Horticulturae* 338, 191–198.

Shaaya, E.U., Ravid, N., Paster, B., Juven, U., Zisman, U. and Pissarev, V. (1991) Fumigant toxicity of essential oils against four major stored-product insects. *J. Chem. Ecol.* 17(3), 499–504.

Shaaya, E., Ravid, U., Paster, N., Kostjukovsky, M., Menasherov, M. and Plotkin, S. (1993) Essential oils and their components as active fumigants against several species of stored product insects and fungi. *Acta Horticulturae* 344, 131–137.

Shaaya, E., Kostjukovski, M., Eilberg, J. and Sukprakarn, C. (1997) Plant oils as fumigants and contact insecticides for the control of stored-product insects. *J. Stored Prod. Res.* 33(1), 7–15.

Shelef, L.A., Jyothi, E.K. and Bulgarelli, M.A. (1984) Growth of enteropathogenic and spoilage bacteria in sage-containing broth and foods. *J. Food Sci.* 49, 737–740.

Shimoni, M., Putievsky, E., Ravid, U. and Reuveni, R. (1993) Antifungal activity of volatile fractions of essential oils from four aromatic wild plants in Israel. *J. Chem. Ecol.* 19(6), 1129–1133.

Sirnik, M. and Gorišek, M. (1983) Effect of some Yugoslav and imported spices on growth of selected microorganisms in foods. *Tehnologija Mesa* 24(4), 120–122.

Sivropoulou, A., Papanikolaou, E., Nikolaou, C., Kokkini, S., Lanaras, T. and Arsenakis, M. (1996) Antimicrobial and cytotoxic activities of *Origanum* essential oils. *J. Agric. Food Chem.* 44(5), 1202–1205.

Skaltsa, H. and Harvala, C. (1987) Contribution à l'étude chimique d'*Origanum dictamnus* L. 2e communication (Glucosides des feuilles). *Plantes médicinales et phytothérapie* Tome XXI(1), 56–62.

Skandamis, P.N. and Nychas, G.J.E. (2000) Development and evaluation of a model predicting the survival of *Escherichia coli* O157:H7 NCTC 12900 in homemade eggplant salad at various temperatures, pHs, and Oregano essential oil concentrations. *Appl. Environ. Microbiol.* 66(4), 1646–1653.

Skandamis, P., Tsigarida, E. and Nychas, J.E. (2000) Ecophysiological attributes of *Salmonella typhimurium* in liquid culture and within a gelatin gel with or without the addition of oregano essential oil. *World J. Microbiol. Biotechnol.* 16, 31–35.

Skwarek, T., Tynecka, Z., Glowniak, K. and Lutostanska, E. (1994) Plant inducers of interferons. *Herba Polonica* 40(1–2), 42–49.

Smid, E.J., Dewitte, Y. and Gorris, L.G.M. (1995) Secondary plant metabolites as control agents of postharvest *Penicillium* rot on tulip bulbs. *Postharvest Biol. Technol.* 6, 303–312.

Soulèlès, C. (1990) Sur les flavonoides d'Origanum dubium. *Plantes médicinales et phytothérapie*, Tome XXIV, 3, 175–178.

Stehmann, J.R. and Brandão, M.G.L. (1995) Medicinal plants of Lavras Novas (Minas Gerais, Brazil). *Fitoterapia* 66(6), 515–520.

Stenzel, R., Widenski, K. and Saba, L. (1998) Growth and development of calves receiving mash with addition of herb for 3 months. *Ann. Univ. Mariae Curie Sklodowska, Sectio EE Zootechnica* 16, 101–106.

Stiles, J.C., Sparks, W. and Ronzio, R.A. (1995) The inhibition of *Candida albicans* by oregano. *J. Appl. Nutr.* 47(4), 96–102.

Takácsová, M., Príbela, A. and Faktorová, M. (1995) Study of the antioxidative effects of thyme, sage, juniper and oregano. *Nahrung* 39(3), 241–243.

Tantaoui-Elaraki, A., Ferhout, H. and Errifi, A. (1993) Inhibition of the fungal asexual reproduction stages by three Moroccan essential oils. *J. Essent. Oil Res.* 5(5), 535–545.

Thanassoulopoulos, C.C. and Yanna, L. (1997) On the biological control of *Botrytis cinerea* on kiwifruit cv. "Hayward" during storage. In E. Sfakiotakis and J. Porlingis (eds), *Proceedings of the third International Symposium on kiwifruit*, Thessaloniki, Greece *Acta Horticulturae*, No. 444, 757–762.

Thompson, D.P. (1989) Fungitoxic activity of essential oil components on food storage fungi. *Mycologia* 81(1), 151–153.

Thompson, D.P. (1990) Influence of pH on the fungitoxic activity of naturally occuring compounds. *J. Food Prot.* 53, 428–429.

Tsigarida, E., Skandamis, P. and Nychas, G.J.E. (2000) Behaviour of *Listeria monocytogenes* and autochthonous flora on meat stored under aerobic, vacuum and modified atmosphere packaging conditions with and without the presence of oregano essential oil at 5 °C. *J. Appl. Microbiol.* 89(6), 901–909.

Tsimidou, M., Papavergou, E. and Boskou, D. (1995) Evaluation of oregano antioxidant activity in mackerel oil. *Food Res. Int.* 28(4), 431–433.

Tucker, A.O. (1986) Botanical nomenclature of Culinary Herbs and Potherbs. In L.E. Craker and J.E. Simon (eds), *Herbs, Spices and Medicinal Plants: Recent Advances in Botany, Hoticulture and Pharmacology, Vol. 1*, The Oryx Press Arizona, USA, pp. 33–80.

Tümen, G. and Başer, K.H.C. (1993) The essential oil of *Origanum syriacum* L. var. *bevanii* (Holmes) Ietswaart. *J. Essent. Oil Res.* 5, 315–316.

Tunç, I. and Şahinkaya, Ş. (1998) Sensitivity of two greenhouse pests to vapours of essential oils. *Entomol. Exp. et Appl.* 86(2), 183–187.

Tunç, I., Berger, B.M., Erler, F. and Dağli, F. (2000) Ovicidal activity of essential oils from five plants against two stored-product insects. *J. Stored Prod. Res.* 36, 161–168.

Ueda, S., Kuwabara, Y., Hirai, N., Sasaki, H. and Sugahara, T. (1991) Antimutagenic capacities of different kinds of vegetables and mushrooms. *J. Jpn. Soc. Food Sci. Technol.* 38(6), 507–514.

Ultee, A., Gorris, L.G.M. and Smid, E.J. (1998) Bactericidal activity of carvacrol towards the food-borne pathogen *Bacillus cereus*. *J. Appl. Microbiol.* 85(2), 211–218.

Ultee, A., Kets, E.P.W. and Smid, E.J. (1999) Mechanisms of action of carvacrol on the food-borne pathogen *Bacillus cereus*. *Appl. Environ. Microbiol.* 65(10), 4606–4610.

Ultee, A., Kets, E.P.W., Alberda, M., Hoekstra, F.A. and Smid, E.J. (2000) Adaptation of the food-borne pathogen *Bacillus cereus* to carvacrol. *Arch. Microbiol.* 174(4), 233–238.

van Den Broucke, C.O. and Lemli, J.A. (1980) Antispasmodic activity of *Origanum compactum*. *Planta Medica* 38, 317–331.

van Den Broucke, C.O. and Lemli, J.A. (1982) Antispasmodic activity of *Origanum compactum*. Part 2. Antagonistic effect of thymol and carvacrol. *Planta Medica* 45(3), 188–190.

Vekiari, S.A., Tzia, C., Oreopoulou, V. and Thomopoulos, C.D. (1993a) Isolation of natural antioxidants from Oregano. *Rivista Italiana Delle Sostanze Grasse* 70(1), 25–28.

Vekiari, S.A., Oreopoulou, V., Tzia, C. and Thomopoulos, C.D. (1993b) Oregano flavonoids as lipid antioxidants. *J. Am. Oil Chem. Soc.* 70(5), 483–487.

Vichi, S., Zitterl-Eglseer, K., Jugl, M. and Franz, C. (2001) Determination of the presence of antioxidants deriving from sage and oregano extracts added to animal fat by means of assessment of the radical scavenging capacity by photochemiluminescence analysis. *Nahrung* 45(2), 101–104.

Villalba, J.J. and Provenza, F.D. (1996) Preference for flavored wheat straw by lambs conditioned with intraruminal administrations of sodium propionate. *J. Anim Sci.* 74(10), 2362–2368.

Vokou, D., Vareltzidou, S. and Katinakis, P. (1993) Effects of aromatic plants on potato storage: sprout suppression and antimicrobial activity. *Agric. Ecosystems Environ.* 47(3), 223–235.

Vokou, D., Tziolas, M. and Bailey, S.E.R. (1998) Essential oil mediated interactions between oregano plants and Helicidae grazers. *J. Chem. Ecol.* 24(7), 1187–1202.

Vokou, D. and Liotiri, S. (1999) Stimulation of soil microbial activity by essential oils. *Chemoecology* 9(1), 41–45.

Wagner, H., Wierer, M. and Bauer, R. (1986) In vitro-hemmung der prostaglandin-biosynthese durch etherische öle und phenolische verbindungen. *Planta Medica* 53, 184–187.

Walker, J.T. (1995) Garden herbs as hosts for southern rootknot nematode [*Meloidogyne incognita* (Kofoid & White) Chitwood, race 3]. *HortScience* 30(2), 292–293.

Yadava, R.N. and Khare, M.K. (1995) A triterpenoid from *Majorana Hortensis*. *Fitoterapia* 66(2), 185.

Yamaguchi, F., Yoshimura, Y., Nakazawa, H. and Ariga, T. (1999) Free radical scavenging activity of grape seed extract and antioxidants by electron spin resonance spectrometry in an H_2O_2/NaOH/DMSO system. *J. Agric. Food Chem.* 47, 2544–2548.

Yamasaki, K., Nakano, M., Kawahata, T., Mori, H., Otake, T., Ueba, N., Oishi, I., Inami, R., Yamane, M., Nakamura, M., Murata, H. and Nakanishi, T. (1998) Anti-HIV-1 activity of herbs in Labiatae. *Biol. Pharm. Bull.* 21(8), 829–833.

Yanishlieva, N.V. and Marinova, E.M. (1995) Antioxidant activity of selected species of the family Lamiaceae grown in Bulgaria. *Nahrung* 39(5–6), 458–463.

Yanishlieva, N.V. and Marinova, E.M. (1998) Activity and mechanism of action of natural antioxidants in lipids. *Recent Res. Dev. Oil Chem.* 2(1), 1–14.

Yaniv, Z., Dafni, A., Friedman, J. and Palevitch, D. (1987) Plants used for the treatment of diabetes in Israel. *J. Ethnopharmacol.* 19(2), 145–151.

Yeomans, M.R. (1996) Palatability and the micro-structure of feeding in humans: the appetizer effect. *Appetite* 27(2), 119–133.

Zaika, L.L. and Kissinger, J.C. (1981) Inhibitory and stimulatory effects of oregano on *Lactobacillus plantarum* and *Pediococcus cerevisiae*. *J. Food Sci.* 46(4), 1205–1210.

Zaika, L.L., Kissinger, J.C. and Wasserman, A.E. (1983) Inhibition of lactic acid bacteria by herbs. *J. Food Sci.* 48(5), 1455–1459.

Zava, D.T., Dollbaum, C.M. and Blen, M. (1998) Estrogen and progestin bioactivity of foods, herbs, and spices. *Proc. Soc. Exp. Biol. Med.* 217(3), 369–378.

Part 6

Uses of Oregano in the food industry

9 Processing, effects and use of Oregano and marjoram in foodstuffs and in food preparation

Seija Marjatta Mäkinen and Kirsti Kaarina Pääkkönen

INTRODUCTION

Origanum (wild marjoram, oregano, Mexican sage, *Origanum vulgare*, *Labiateae*, *Lamiaceae*), is a herbal plant, which according to mythology, was planted in a crevice of a mountain by the goddess Aphrodite. The name *Origanum* is derived from two Greek words; *oros* means mountain and *ganos* means joy; it is a herb of joy. It grows wild on hillsides of Mediterranean countries. Oregano has been a popular herb especially in Mexican, Italian and in Greek cuisine. Oregano was cultivated in southern Finland in the 1600s, and nowadays it is one of those rare herbal plants which grows wild in Nordic countries, it is also a popular decorative plant in flower gardens and in the yards of Finnish summer cottages. It grows wild on the islands of archipelagos of the Baltic Sea and fishermen have always used it to season a special delicacy made from Baltic herring.

Oregano and marjoram (sweet marjoram, *Majorana hortensis M.*, *Origanum majorana L. Labiateae*, *Lamiaceae*) have a similar kind of aroma. The ancient Romans and Greeks made head wreaths of marjoram as well as of oregano to decorate bridal couples. It symbolized love, honor and happiness. Hippocrates recommended marjoram as an aphrodisiac herb. Aristoteles told that after eating a snake, a tortoise will eat oregano as an immediate dessert. That is why oregano is said to be an effective antidote to snake-bite.

In the Middle Ages monks brought marjoram to Europe, and Europeans started to use marjoram both as a medicinal herb and as a seasoning in porridges, puddings and cakes, and at the beginning of the New Ages marjoram was used for strewing chambers and in sweat bags and powders for perfuming linen.

William Shakespeare (1564–1616) writes in All's Well That Ends Well: "Indeed, sir, she was the sweet marjoram of the sallet, or rather, the herb of grace," and marjoram was eaten, indeed, already in salads during those times.

Marjoram has been one of the most popular herbs in Finland and it is used just in ordinary everyday Finnish traditional cuisine. It is a must in such traditional dishes as pea soup and in kale soup and in many other dishes made from kale. When animals were still slaughtered only at farm houses in Finland, marjoram was cultivated in home gardens and used in autumn when most of animals were slaughtered, and it seasoned home-made sausages and all kind of dishes made from blood and inner organs.

Although the first pizzeria was founded in 1895 in New York (Tucker and Maciarello, 1994), oregano was quite an unknown herb to the Americans until the Second World War, when soldiers returning home from Italy had got the taste of it in Italian pizza. Thereafter, it spread very fast all over the world within pizza cuisine.

According to Tucker and Maciarello (1994) per capita importation of oregano into the United States grew 3800 per cent from 1940 to 1985! These authors have written a review article on botany, chemistry and cultivation of different taxa of oregano (Tucker and Maciarello, 1994).

IS OREGANO A NUTRITIONALLY VALUABLE HERB?

We do not eat herbs in such amounts that they would be a significant source of vitamins or minerals in our daily diet. According to McCance and Widdowson's food composition tables (Holland et al., 1991) oregano contains significant amounts of the following vitamins: vitamin E, B_6, riboflavin, niacin, folate, panthotenate and biotin, although there is no reliable information on the concentrations, the same comments concern with such minerals as iron, copper, sulphur, chlorine, iodine, and selenium. Lagouri and Boskou (1996) have found that non-polar fraction of oregano extracts contain α-, β-, γ- and δ-tocopherol, which are involved with the antioxidant capacity of oregano. γ-tocopherol content was significantly higher than other tocopherol homologs. According to Brune et al.'s (1989) experiments in Sweden, oregano inhibits iron absorption and the effect is caused by its galloyl substances and the inhibition is in proportion to its content of galloyl groups. There are analyzed values (1 mg/100 g fresh leaves) for C-vitamin (45 mg), thiamin (0.07 mg), and for carotene (0.81 mg). Its potassium content is 33 times that of sodium. One hundred grams of fresh oregano leaves contain 310 mg of calcium, 53 mg of magnesium, 39 mg of phosphorus, 0.9 mg of zinc and 0.3 mg of manganese (Holland et al., 1991). Compared to sweet basil, for instance, oregano contains 65 per cent more energy (oregano 66 kcal/100 g; basil 40 kcal/100 g) and more than two-fold that much of fat (oregano 2 g fat/100 g; basil 0.8 g/100 g), although the fat content is small as in the herbs in general.

OREGANO HAS EFFECTS ON PALATABILITY, FOOD INTAKE, FOOD AVERSIONS AND ALLERGY

Oregano has an ability to increase appetite in humans. According to Gray et al. (1997) tomato dressing in a pasta dish seasoned with 0.27 per cent oregano increased the palatability and the intake of food compared with an unseasoned control food. However, the concentration of oregano is of importance. In Gray's (Gray et al., 1997) experiments, already doubling the amount of oregano (0.56 per cent) reduced the food intake and eating rate as well.

Avoid oregano during pregnancy

So called conditioned flavor aversions caused by foodstuffs are rather common with humans and with animals, if a gastrointestinal malaise is experienced after eating some novel food. Oregano is a herb which has been used in animal experiments to study food aversions (Launchbaugh and Provenza, 1994). However, pregnant women can develop aversions even without any nausea or ill-feeling. Oregano is one of the most common foodstuffs, which have the ability to cause aversions during pregnancy. Although pica and cravings are also common phenomena during pregnancy, oregano causes rather an aversion than pica or craving symptoms (Hook, 1978, 1980). Therefore we would recommend not to use oregano as a seasoning in any food during pregnancy.

Oregano can cause systemic allergic reactions

According to Benito *et al.* (1996) there were no cases described in medical literature of systemic allergic reactions due to oregano before their report on three cases. Benito *et al.* (1996) concluded that plants belonging to the *Labiateae* family seem to show cross-sensitivity on the basis of clinical history and *in vitro* and *in vivo* test results. However, already in 1993 Futrell and Rietschel have published results of their patch tests of spice allergies, and 4 of their 55 patients with suspected contact dermatitis showed positive results with oregano.

OREGANO HAS EFFECTIVE ANTIOXIDATIVE PROPERTIES

Current worldwide research is focused in finding new and safe antioxidants to prevent oxidative deterioration of fatty foodstuffs and to minimize damage of biological membranes of living cells caused by oxidative reactions. Many studies have concentrated upon spice and herb constituents.

Already in the 1950s Chipault *et al.* (1952) measured antioxidative activities of different spices and found that oregano was among those spices, which have an ability to retard oxidation of lard. Later on Chipault *et al.* (1955) found that oregano was effective also in oil-in-water emulsion and when it was added as ground into different types of foods, its antioxidant capacity was best in mayonnaise and in french dressing (Chipault *et al.*, 1956). Nakatani and Kikuzaki (1987) and Kikuzaki and Nakatani (1989) reported that antioxidative activity of water soluble fraction of methanol extraction of oregano leaves was due to two new compounds and it was comparable to BHA (butylated hydroxy anisol). Tsimidou and Boskou (1994) have written a whole chapter by name "Antioxidant Activity of Essential Oils from the Plants of the *Lamiaceae* Family" in the book Spices, Herbs and Edible Fungi edited by Charalambous (1994). Carvacrol is one of the compounds responsible for antioxidant capacity of various herbs, and according to Tsimidou and Boskou (1994) "oregano plants" are rich in carvacrol according to which the various taxa could be differentiated.

Recent research has confirmed that oregano indeed is an effective antioxidant. In studies done by Baratta *et al.* (1998) oregano oil showed highest antioxidative activity compared with laurel, sage, rosemary and coriander oils. The antioxidant activity was measured by the modified thiobarbituric acid reactive species (TBARS) assay, using egg yolk and rat liver as oxidable substrates. Dorman *et al.* (1995) made measurements of antioxidant capacity of oregano oil among other herb oils finding it one of the most antioxidative ones, and suggested that feeding trials *in vivo* upon polyunsaturated fatty acid metabolism during vital periods of life span, in particular fetal/neonate and aging periods, ought to be done. Chung *et al.* (1997) have detected that oregano as well as brown mustard, thyme, clove and allspice exhibit strong hydroxy-radical scavenging effects. In chromatographic and spectorophotometric analyzes by Vekiari *et al.* (1993), oregano contains flavonoids (apigenin, eriodictyol, dihydrokaempferol and dihydroquercetin), which were effective lipid antioxidants in lard and in vegetable oils under storage and during frying.

Madsen *et al.* (1996) report that total phenol content of oregano extract correlates linearly with the antioxidant activity as measured by oxygen depletion, but not with free radical scavenging effect. They conclude that extracts of the investigated spices (*Labiateae* family spices) contain components with at least two different antioxidative mechanisms. In their experiments Turkish oregano and Chilean oregano showed the highest activity, while marjoram as well as basil had the lowest one.

Lagouri and Boskou (1996) did, indeed, publish a study, in which they showed that also a non-polar fraction of oregano is responsible for its antioxidant capacity due to its contents of tocopherol homologs. They also reported that the non-polar fraction was able to suppress the mutagenity of Trp-P-2, a dietary carcinogen. In this work four different species of oregano, *O. vulgare* subsp. *hirtum*, *Satureja thymbram*, *O. dictamnus* and *Origanum onites* were investigated.

OREGANO CONTAINS ANTIMICROBIAL SUBSTANCES

In addition to antioxidative properties of various spice substances, food scientists are currently searching for new natural substances from spices and herbs, which would have uses as natural food preservatives. For example investigations on prolongation of storage life of vegetables and root vegetables like potato have involved herbs and spices and their essential oils. Prevention of deterioration due to microbial growth in warm countries for example in Mediterranean area, where spices and herbs belong to every-day cuisine would be easily applied in food science.

Already in 1948 it was shown by Dold and Knapp (1948) that spices belonging to Lamiaceae family had antimicrobial activity. Inhibitory ability of ethanol extracts of oregano against *Clostridium botulinum* was reported by Huhtanen in 1980 and by Ueda *et al.* in 1982, and reduction of aflatoxin production in culture medium of *Aspergillus paraciticus* by ground oregano was shown by Salzer (1982) in the same year. Salmeron *et al.* (1990) have reported that although oregano stimulates the growth of both *A. paraciticus* and *Aspergillus flavus*, at the same time it acts as antiaflatogenics.

Antimicrobial activity of oregano has been found at least against such organisms as *Bacillus cereus*, *Acinetobacter baumanii*, *Aeromonas veronii*, bg *sobria*, *Candida albicans*, *Candida lipolytica*, *Enterococcus faecalis*, *Escherichia coli*, *Klebsiella pneumoniae*, *Pseudomonas aeruginosa*, *Salmonella enterica* subsp., *Enterica typhimurium*, *Salmonella enteritidis*, *Serratia marcescens*, *Staphylococcus aureus*, *Aspergillus ochraceus*, *A. flavus*, *A. paraciticus*, *Clostridium perfringens*, *Erwinia carotovora*, *Listeria monocytogenes*, *Lactobacillus plantarum*, *Leuconostoc mesenteroides*, *Debaryomyces hansenii*, *Hansenula anomala*, *Kloeckera apiculata*, *Lodderomyces elongiporus*, *Rhodotorula rubra*, *Saccharomyces cerevisiae*, *Torulopsis glabrata*.

The main results of the studies made mainly in the 1980s and 1990s and which were involved with antimicrobial effects of oregano are listed in Table 9.1.

THE FINNISH HERB STUDY

Oregano (*O. vulgare* L.) and marjoram (*O. majorana* L.) were among the herbal plants included in a large Finnish Herb study in 1983–1985 (Mäkinen *et al.*, 1986). The research institutes participating in the study were: Department of Horticulture, Department of Food Chemistry and Technology, Department of Agricultural

Table 9.1 Experiments of antimicrobial effects of oregano

Microbe	Study design	Effects	Publication
Bacillus cereus	1–3 mM carvacrol 30 min incubation	Reduction of viable cell numbers exponentially, changed permeability of cell membranes to K^+, H^+ leading to decrease of ATP-pool	Ultee et al., 1999
Acinetobacter baumanii Aeromonas veronii Candida albicans Enterococcus faecalis Escherichia coli Klebsiella pneumoniae Pseudomonas aeruginosa Salmonella enterica subsp Enterica serotype typhimurium Serratia marcescens Staphylococcus aureus	Fifty two essential plant oils at 0.03–2% (v/v) were tested for activity against microbes	Oregano inhibited all organisms, conclusion: "plant essential oils and extracts may have role as pharmaceuticals and preservatives"	Hammer et al., 1999
Salmonella enteritidis	Essential oil of oregano 0.5; 1.0, 2.0 v/w, 5, 10, 15, 20 °C pH 4.3 and pH 5.3 in traditional Greek salad TARMA, storage temperatures: 5, 10, 15, 20 °C	Death of microbe, death rate depended on pH, storage temperature and oil concentrations	Koutsoumanis et al., 1999
Aspergillus ochraceus	Essential oil of mint, basil, sage, coriander and oregano 0, 500, 750, 1000 ppm in culture media; incubation for 7, 14 or 21 days at 25 °C	Complete inhibition of fungal growth and ochratoxin production at 750 ppm for 14 d incubation and at 1000 ppm for 21 d incubation. Conclusion: "some herbs essential oils could be used instead of synthetic antifungal products"	Basilico and Basilico, 1999
Bacillus cereus	carvacrol	Dose related inhibition of growth, total inhibition of growth ≥0.75 mmol/L	Ultee et al., 1998
Staphylococcus aureus Salmonella enteritidis	Inoculation of fresh fish fillets dressed with olive oil, lemon juice and oregano	Oregano both bacteriostatic and bactericidal. Highest death rate was at 30 °C. Spores were 2–3-fold less sensitive than vegetative cells. Susceptibility was least at pH 7	Tassou et al., 1996

Table 9.1 (Continued)

Microbe	Study design	Effects	Publication
Aspergillus parasiticus	Thirteen essential oils of herbs, 0.1% essential oil in growth medium	Oregano oil inhibited mycelial growth and of production of aflatoxin. Aflatoxin synthesis was inhibited at higher extent than mycelial growth	Tantaoui-Elaraki and Beraoud, 1994
Erwinia carotovora	Essential oil of lavender, mint, spearmint, rosemary, sage and oregano, sprout suppression, inhibition of microbial growth	Turkish oregano (O. onites) and especially Greek oregano (O. vulgare ssp. hirtum) possessed antimicrobial activities against E. carotovora but no inhibition of potato sprouting	Vokou et al., 1993
Listeria monocytogenes	Thirteen spices screened for growth inhibition; at 24 °C and at 4 °C using concentration gradient plate method	0.5–1.0% w/v oregano was bacteriostatic, but at 1% conc. (w/v) in sterile meat slurry little effect on growth	Ting and Deibel, 1992
Lactobacillus plantarum Leuconostoc mesenteroides	Growth inhibition of cumin and oregano and stimulation of acid production, essential oil at 150, 300 and 600 ppm	Oregano and its essential oil at all concentrations inhibited growth of both microbes, but stimulated acid production of Lactobacillus plantarum	Kivanc et al., 1991
Nine foodborne fungi	Growth inhibition of ten Turkish spices, essential oil of oregano and carvacrol in culture media of pH 3.5 and 5.5	Oregano at 1.0, 1.5, and 2.0% w/v levels affected all fungi; essential oil and carvacrol at 0.025% and at 0.05% (w/v) completely inhibited growth of all fungi; oregano and sodium chloride (8% w/v) were synergistic	Akgul and Kivanc, 1988
Eight strains of yeasts	Changes of sensitivity to eight essential oils of spices	Oregano oil was inhibitory only for Rhodotorula rubra, it also inhibited its pigment production	Conner and Beuchat, 1984
Seven mycotoxin producing molds	Antifungal effects of 16 ground herbs and spices and other substances at 2% potato dextrose agar	Oregano completely inhibited growth of all mycotoxigenic molds	Azzouz and Bullerman, 1982
Aspergillus flavus Aspergillus parasiticus	Ground spices and herbs as substrates for mycelial growth, sporulation and aflatoxin production	Oregano was antiaflatoxigenic	Llewellyn et al., 1981

Economics and Department of Nutrition of Helsinki University, Department of Chemistry and Biochemistry of Turku University, Technical Research Centre of Finland, and Research Laboratories of the State Alcohol Company, Finland.

Connected with this study Nykänen (1986a) showed in flavor studies, that marjoram grown outdoors in the middle of Finland (Sahalahti 61°29'N, 24°20'E) and in the northern part (Kaamanen 69°04'N, 27°07'E) of Finland had the same amount of flavor, but plants cultivated under cover had *c.* 50 per cent higher amount of volatiles. A total of 56 compounds were identified, of which 18 were reported for the first time. Nykänen (1986b) analyzed also the composition of oregano grown in four different districts of Finland (Helsinki 60°15'N, 25°00'E; Sahalahti 61°29'N, 24°20'E; Haapavesi 64°10'N, 25°20'E; Kaamanen 69°04'N, 27°07'E) and identified a total of 82 compounds, of which 37 were components not reported before. The oregano samples were divided into three types on the basis of the oil. Germacrene-D was the main component in the hydrocarbon type, and carvacrol in the phenol type. The intermediate type contained equal amounts of both germacrene-D and carvacrol. Oregano in flower gave higher yield of oil than did the herb in budding (Nykänen, 1986b).

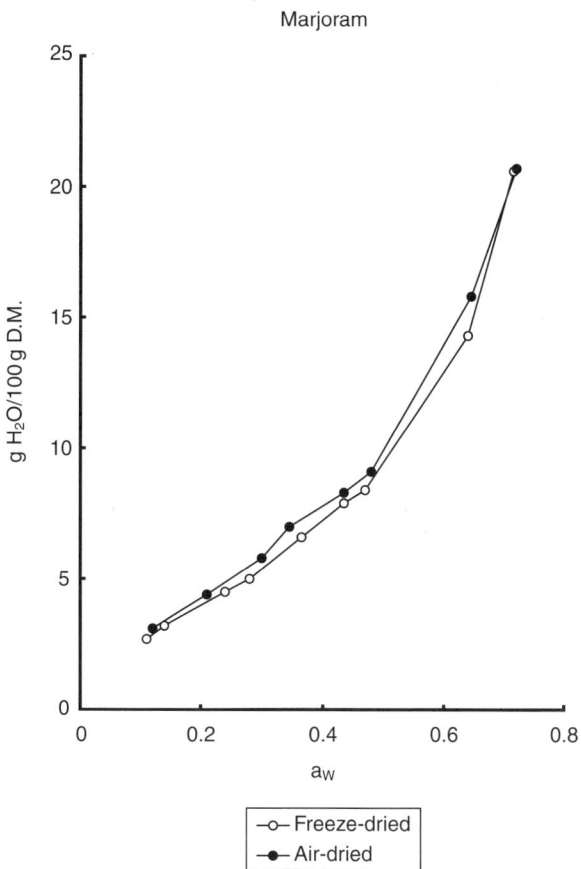

Figure 9.1 Adsorption isotherms of freeze-dried and air-dried marjoram at 23 °C.

For drying, packaging and for sensory studies the herbs were grown in southern Finland (Helsinki 60°15'N, 25°00'E); marjoram was harvested at time of bud formation and oregano when in bloom. Herbs were either dried immediately after harvesting or were frozen and stored at −20 °C and freeze-dried within 2 weeks. For convection drying temperature was 35–37 °C and for freeze-drying 30 °C. The corresponding drying times were 24 h and 12 h (Pääkkönen et al., 1990). The moisture content of fresh herbs was 85 per cent for marjoram and 75 per cent for oregano. After drying with heated air, the moisture content for marjoram was 9 per cent, and 7 per cent for oregano, but only 5 per cent after freeze-drying.

The adsorption isotherms of two marjoram powders dried with different methods showed that drying method does not affect the water-holding capacity of the dried product (Figure 9.1) (Pääkkönen et al., 1990).

IS IT MICROBIOLOGICALLY SAFE TO USE OREGANO?

In connection with the Finnish study of herbs (Mäkinen et al., 1986), Malmsten et al. (1991) tested the microbiological quality of marjoram (O. majorana L.) and of oregano (O. vulgare L.). Molds and aerobic sporeformers, especially B. cereus, were detected in most samples; fecal indicator organisms were found infrequently. Incidence of microbial groups in dried herbs after storage of one and two years is shown in Table 9.2. Aerobic sporeformers were presented from 350 to 4.2×10^3/g dried marjoram and from 50 to 1.0×10^3/g dried oregano. B. cereus was present in all samples, counts ranging from 100 to 3.8×10^3/g dried marjoram and from 1.0 to 3×10^3/g dried oregano. These levels are not high enough to cause food poisoning. Anaerobic spore-formers were randomly distributed, and counts were low; from 5 to 30 organisms/g dried herbs. Coliforms and fecal streptococci were found in both freeze-dried and air-dried samples, but only sporadically and at very low counts. Molds and yeasts were found in almost all samples, and they were approximately 10-fold higher in air-dried than in freeze-dried marjoram. Increasing the storage time from 1 year to 2 increased 10-fold the number of aerobic sporeformers in freeze-dried and in air-dried oregano.

The presence of the aerobic sporeformers did not depend on the types of packages. There was a decline in sensory quality when stored at 35 °C, but it was rather due to chemical changes than to microbiological deterioration (Pääkkönen et al., 1990). The original fresh material was the essential factor determining the quality of the dried herbs. Nevertheless, the drying method, type of packaging, and storage conditions also had clear effects on the microbiological quality of the herbs (Malmsten et al., 1991). According to Silva et al. (1999) oregano maintained similar microbiological quality in the dry chamber (1660 UFC/g) and in the shut room (1700 UFC/g), in both glass jars and polyethylene bags. After 1 year oregano maintained desirable microbiological characteristics when packaged in polyethylene or glass and stored in the dry chamber or in shut room. When stored in paper bags contamination could not be prevented.

Przezdziecka and Baldwin (1971) have made sensory evaluation experiments with herbs and spices and found out that the threshold intensity of taste and aroma of marjoram is "low" when compared with those of garlic ("high") and basil ("intermediate"). The results of odor and taste evaluation for marjoram dried with different methods and stored at 23 °C and at 35 °C in the Finnish herb study (Pääkkönen et al., 1990) are shown in Figure 9.2. Both odor and taste of freeze-dried marjoram as well as of air-dried

Table 9.2 Microbial counts of freeze-dried and air-dried marjoram and oregano after storage at 23 °C (Results of counts on five replicates)

Microbial group	Storage time process								
	1 Year					2 Years			
	Freeze-dried*			Air-dried (Glass jar)	All samples (Range)	Freeze-dried*			Air-dried (Glass jar)	All samples (Range)
	Glass jar	Vacuum	Nitrogen			Glass jar	Vacuum	Nitrogen		

Microbial group	Glass jar	Vacuum	Nitrogen	Air-dried (Glass jar)	All samples (Range)	Glass jar	Vacuum	Nitrogen	Air-dried (Glass jar)	All samples (Range)
Marjoram										
APC	6×10^5	2×10^6	7×10^5	8×10^5	$1.4 \times 10^5 - 2.9 \times 10^7$	3×10^5	3×10^6	4×10^6	2×10^6	$9.1 \times 10^4 - 5.9 \times 10^6$
Coliforms	<3	<3	<3	7	<3–35	<3	<3	<3	<3	<3
E. coli	<3	<3	<3	<3	<3	<3	<3	<3	<3	<3
Fecal streptococci	<10	<10	<10	14	<10–90	<10	<10	<10	<10	<10
Molds and yeasts	220	190	790	7×10^3	$50 - 2.3 \times 10^4$	64	510	29	2×10^3	$<10 - 8.1 \times 10^3$
Aerob. sporeform.	1×10^3	2×10^3	2×10^3	3×10^3	$350 - 4.2 \times 10^3$	1×10^3	1×10^3	1×10^3	4×10^3	$650 - 1.8 \times 10^4$
Bacillus cereus	510	1×10^3	990	2×10^3	$100 - 3.8 \times 10^3$	1×10^3	770	720	3×10^3	$500 - 9.5 \times 10^3$
Clostridia	<10	<10	<10	<10	<10–20	<10	<10	<10	<10	<10
Oregano										
APC	2×10^5	6×10^5	7×10^5	1×10^5	$1.0 \times 10^4 - 2.6 \times 10^6$	7×10^4	1×10^6	8×10^4	2×10^5	$4.3 \times 10^4 - 1.7 \times 10^6$
Coliforms	<3	<3	<3	<3	<3–4	<3	<3	<3	<3	<3–43
E. coli	<3	<3	<3	<3	<3	<3	<3	<3	<3	<3
Fecal streptococci	<10	<10	<10	<10	<10	<10	<10	<10	<10	<10
Molds and yeasts	91	4×10^3	510	180	$30 - 4.8 \times 10^4$	15	510	150	2×10^3	$<10 - 8.0 \times 10^3$
Aerob. sporeform.	170	360	160	150	$50 - 1.0 \times 10^3$	2×10^3	2×10^3	3×10^3	3×10^3	$1.1 \times 10^3 - 3.9 \times 10^3$
Bacillus cereus	170	200	150	100	50–600	2×10^3	2×10^3	2×10^3	2×10^3	$1.1 \times 10^3 - 1.3 \times 10^4$
Clostridia	<10	17	<10	<10	<10–650	<10	<10	<10	<10	<10

Note
* Freeze-dried values are given in Geometric mean.

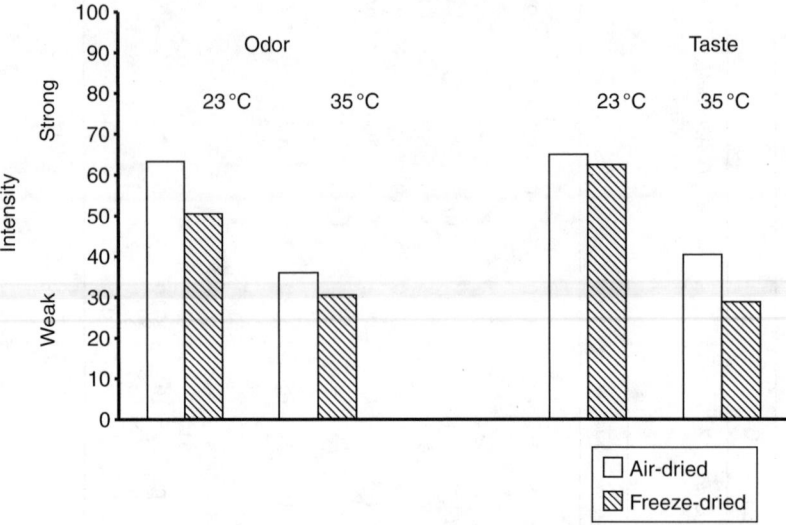

Figure 9.2 Odor and taste of air-dried and freeze-dried marjoram after 12 months in glass jars at 23 °C and 35 °C. Means with unlike superscripts are significantly different at the 99% level using Tukey's test. 0 = weak; 100 = strong intensity.

marjoram were sensitive to storage temperature. Neither odor nor taste of freeze-dried marjoram were affected by 2 years' storage in glass jars at 23 °C, but the intensity of odor and taste of marjoram dried with heated air was greater ($p < .01$) after one year's storage and even after two years' storage than immediately after drying (Figure 9.3). The erratic changes in intensity scoring of both the air-dried and freeze-dried marjoram indicated alteration in taste of marjoram during the storage (Pääkkönen *et al.*, 1990).

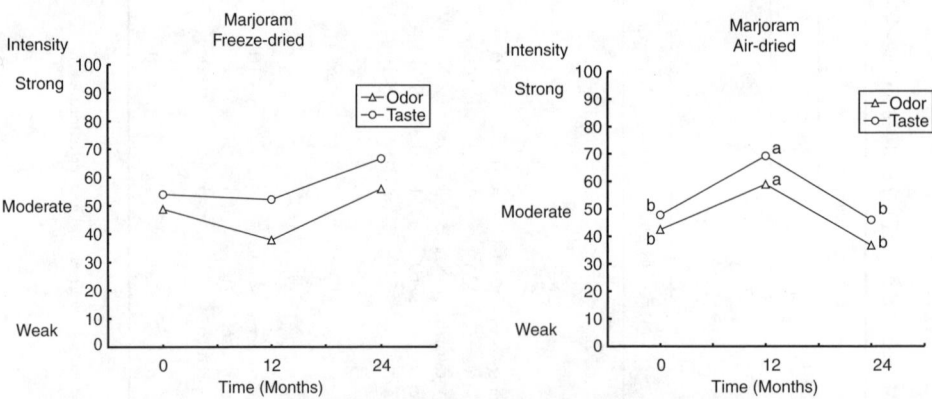

Figure 9.3 Odor and taste of freeze-dried and air-dried marjoram immediately after processing and after 1 and 2 years in glass jars at 23 °C. Means with unlike superscripts are significantly different at the 99% level using Tukey's test. 0 = weak; 100 = strong intensity.

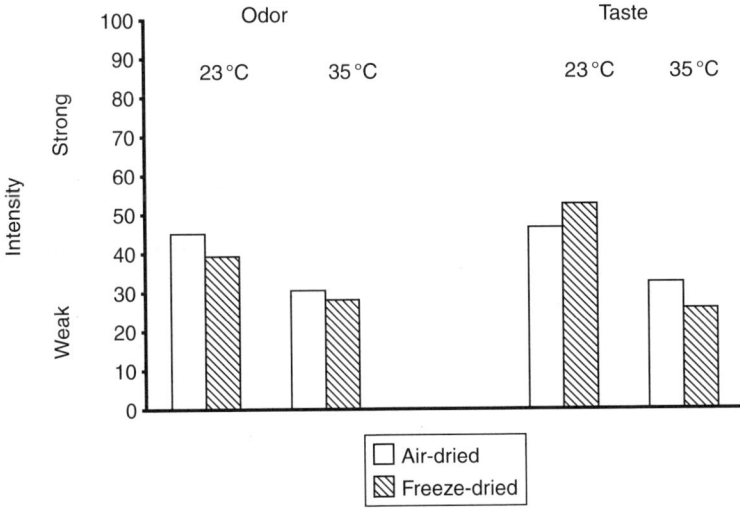

Figure 9.4 Odor and taste of air-dried and freeze-dried oregano after 12 months in glass jars at 23 °C and 35 °C. Means with unlike superscripts are significantly different at the 99% level using Tukey's test. 0 = weak; 100 = strong intensity.

The results of the sensory evaluation of oregano dried with different methods and stored in glass jars at 23 °C and 35 °C are shown in Figure 9.4. The drying method or packaging did not have any effect on either odor or taste of dried oregano, but detrimental effect of the elevated storage temperature was obvious (Pääkkönen *et al.*, 1990).

Freeze-dried marjoram exhibited more intense green color than air-dried marjoram. After 9 month's storage in light or at raised temperature, the color tone of the freeze-dried marjoram had changed only slightly. It was the drying process that affected color substantially and more than any of the other storage condition investigated (Pääkkönen *et al.*, 1990).

DRYING CONDITIONS DETERMINE THE SENSORY QUALITY OF DRIED PRODUCT

To enhance preservation, herbal plants are subjected to various treatments including hot air drying, microwave drying, infrared drying, and freeze-drying. In hot air-drying temperature, air velocity and humidity have to be specific for each plant. Undesirable effects may occur, e.g. loss of aromatic substances and changes in color due to improper conditions. In connection of Finnish Herb Study Pääkkönen *et al.* (1990) studied the effect of drying and packaging on the quality of herb. The intensity scoring of odor and of taste was higher for dried than for the fresh herb. Deans *et al.* (1991) used domestic microwave oven (650 W, 2450 MHz) for drying sweet marjoram (*O. majorana* L.).

Although there was a noticeable reduction of microbial counts, the effects on volatile profile was not acceptable; major constituents were reduced and some even destroyed and two new compounds were formed. According to Hälvä (1987), the concentration of volatile oil in oregano decreased from 2.55 per cent to 1.94 per cent in drying process. According to Nykänen (1986a,b) monoterpene hydrocarbon and monoterpene alcohol contents of marjoram and of oregano are very much the same, but sesquiterpene hydrocarbons are far more numerous in oregano.

Drying at home

If you anyhow want to dry herbs at home, it is possible to use microwave oven. Drying in microwave oven stores the color better than drying in air or in electric oven. If you harvest herbs such as oregano from your own garden, do it in the morning, when the content of aromatic oil in the leaves is at highest. Wash herbs with cold water, let them dry in air. If you put them in microwave oven when they still are wet, they start to cook instead of dehydrating. Therefore place the sprigs between paper towels, put in a microwave oven and heat with full power until the leaves are completely dry, e.g. brittle. The paper will trap the water released from the sprigs and prevent the herb leaves cooking. Usually a batch of herbs does not need more than 3 min to get dry. Take them off from the oven when they still have their green color. Let them cool, strip the sprigs and pack the leaves in airtight dark-colored jars. The appearance of water droplets inside a jar reveals that drying has been incomplete. The jars are stored in cool, dark place.

Drying herbs reduces the growth of microorganisms and prevents some damaging biochemical reactions. Drying, however, gives rise to a number of unavoidable changes in appearance, taste and smell.

MARJORAM AND OREGANO ARE USED IN MANY FOODS

More than 60 species of oregano are used as spices, and there are great differences on the contents of aromatic compounds according to growing place for example between the taste of European oregano and Mexican oregano. Mexican type has a pungent flavor and it is a popular herb in Mexican cuisine in pizzas and in barbecue dressings. European type is used in meat, sausages, salads, stewings, dressings and in soups. Already in ancient times, before the cultivation of hop started, oregano was used for seasoning of beer.

Foodstuff industry use oregano oil and oregano resin both in foods and in beverages and also in cosmetics. Oregano oil is used in alcoholic beverages, baked goods, meats and meat products, condiments and relishes, milk products, processed vegetables, snack foods, and fats and oils. Marjoram, too, is used in many foods and beverages in food industry; meat sauces, canned foods, vinegar, vermouths and bitters are often seasoned with marjoram.

Oregano and marjoram are popular herbs among vegetarian dishes

Oregano goes well with cabbage, kale, chard, tomatoes, mushrooms, zucchini (courgettes), broccoli, beans, tomato, pepper, onions, aubergine, and also with potatoes.

Pies and casserols made from vegetables get more flavor when seasoned with oregano or marjoram, and they are popular ingredients in low-calorie vegetarian dishes. Marjoram gives more aroma to such vegetable dishes like pea soup and other pea dishes, squash and stews made from mixed vegetables, mushrooms and asparagus. Oregano is a good substitute for table salt, and those who want to decrease their intake of salt, are advised to season their food with oregano.

Meat dishes

From all the different meat varieties, *lamb* meat dishes are the ones which are most often seasoned with oregano. For example, the following recipes found in Healthy Home Cooking series of Time-Life Books for Fresh Ways with Lamb (1990) contain oregano: lamb meat salads, marinated lamb meat cutlets, lamb meat noisettes with julienned vegetables, lamb sausages on skewers, lamb roasts, lamb steaks, lamb shanks, lamb lasagna, lamb terrain, lamb and leek parcels, lamb timbales and lamb meat loaf. Marjoram is included in the following dishes: lamb cutlets, lamb steak, lamb and mushroom burgers, lamb moussaka, baked stuffed onions with lamb meat, lamb timbales. The other spices and herbs used with oregano were black pepper, parsley, marjoram, capers, thyme, sage, rosemary, basil, garlic, coriander and cayenne pepper. Those spices and herbs used within marjoram were garlic, parsley, thyme, black pepper, chives, bay leaf, nutmeg, and horseradish.

Although oregano is not a very common herb used to season *pork* dishes, it can be used within pork, too. It goes well with pork roast and with Mediterranean type dishes made from pork fillet and stuffed wine leaves and it is very well suited with dishes made from inner organs like kidneys and lungs as well as from ham. Marjoram is often used in homemade pork meat sausages, and in pork fillet stuffings.

Fish and shellfish dishes

Oregano is a herb used in seasoning also fish and shellfish. Oregano goes well with white fishes like grayling, cod, pearch and pike and with other low-fat fish.

In Time-Life Books, Healthy Home Cooking series for Fresh ways with Fish and Shellfish (1989), oregano is found in recipes for baked fish and for fish gratins, fish salads, and stuffings for lobsters. Fish timbales and spicy dressings for fish dishes are also often seasoned with oregano.

Marjoram is found in recipes like salmon steaks with pepper and in turbot salad.

Pizzas and pasta dishes

Pizza cuisine spread all over the world after the second world war. Oregano is the most common spice for pizza. Time-Life Books for Healthy Home Cooking series for Fresh Ways with Pasta (1990) dishes contain altogether 108 recipes and oregano is included in ten of them.

Parsley, garlic, thyme, black pepper, basil, pepper flakes, capers, coriander, ginger root and cayenne pepper are the other spices used in these recipes for pasta and macaroni dishes along with oregano. The main foodstuff in these pasta and macaroni dishes seasoned with oregano can be cheese, tomatoes, clams, broccoli, pork, chicken, mushrooms, beef and fish like sole. Oregano is put in pasta dishes as such or is used to season the dressing. There are actually not any foodstuff one could not season with oregano.

Salad dressings

Oregano is a herb giving flavor to many salad dressings. It goes well with garlic as well as with onions, lemon juice or vinegar, oil and with other herbs like thyme and parsley. Also mustard dressings like french dressing can be seasoned with oregano. Along with oregano black pepper is a spice very often included in salad dressings. The main ingredient in the salad can be a vegetable like beans, broccoli, pasta, pollenta, aubergine, tomato, peppers, courgettes or chicken, fish, ham, clams etc.

Terrines and pates

Terrines and pates are nowadays popular dishes served in gourmet restaurants, but not at all that popular in home cooking, since they are tedious and time consuming to prepare.

In Time-Life Books Healthy Home Cooking there is a special cooking book for Fresh Ways with Terrines and Pâtés (1991). Such dishes like Herbed Pork and Veal Terrine, Tomato Sirloin Loaf, Salmon Coulibiaca, Pea and Tomato Timbales, contain oregano and Red Pepper, Spinach and Mushroom Loaf marjoram as well.

PRINCIPLES OF SEASONING WITH HERBS

Although herbs are not the main ingredients of the dishes, they have important uses both as seasonings and enhancing the aroma and smell of many dishes. In many cookbooks the authors write that the herb is added "according to taste," and actually it is often quite impossible to tell, what is the exact amount of the herb for a certain dish. As it is written in this article, the taste and odor of oregano and marjoram is affected by such as plant variety, growing and climatic conditions, drying and packaging etc. Anyhow, there are some general rules when seasoning with herbs.

A mild-tasting foodstuff; cooked fish for example, is not seasoned as strongly as meat or salmon, which has a natural strong flavor. A cold dish needs always more herbs than a warm one, and a frozen one the most. Soups, which are eaten as such, need always less herbs than stuffings and dressings, in which herbs are used to add the flavor of the basic foodstuff. It is good to remember that oregano and marjoram have stronger aroma when dried compared with the fresh ones. Two teaspoons of fresh herb correspond 1/4 teaspoon of the dried one. Sugar and salt may enhance the flavor of the herb. The fresh herb is best to add just before serving, since the aromatic substances do not stand high temperatures; they evaporate easily or get destroyed or change in flavor.

REFERENCES

Akgul, A. and Kivanc, M. (1988) Inhibitory effects of selected Turkish spices and oregano components on some foodborne fungi. *Int. J. Food Microbiol.* 6(3), 263–268.

Azzouz, M.A. and Bullerman, L.B. (1982) Comparative antimycotic effects of selected herbs, spices, plant components and commercial antifungal agents. *J. Food Prot.* 45(14), 1298–1301.

Baratta, M.T., Dorman, H.J.D., Deans, S.G., Biondi, D.M. and Ruberto, G. (1998) Chemical composition, antimicrobial and antioxidative activity of laurel, sage, rosemary, oregano and coriander essential oils. *J. Ess. Oil Res.* 10(6), 618–627.

Basilico, M.Z. and Basilico, J.C. (1999) Inhibitory effects of some spice essential oils on *Aspergillus ochraceus* NRRL 3174 growth and ochratoxin A production. *Appl. Microbiol.* 29(4), 238–241.

Benito, M., Jorro, G., Morales, C., Pelaez, A. and Fernandez, A. (1996) Labiatae allergy: systemic reactions due to ingestion of oregano and thyme. *Ann. Allergy, Asthma Immunol.* 76(5), 416–418.

Brune, M., Rossander, L. and Hallberg, L. (1989) Iron absorption and phenolic compounds: importance of different phenolic structures. *Eur. J. Clin. Nutr.* 43(8), 547–557.

Charalambous, G. (ed.) (1994) *Spices, Herbs and Edible Fungi.* Elsevier Science B.V., The Netherlands, 764p.

Chipault, J.R., Mizuno, G.R., Hawkins, J.M. and Lundberg, W.O. (1952) *Food Res.,* 17, 46–55. Ref. Nakatani, N. Antioxidant and antimicrobial constituents of herbs and spices. In G. Charalambous (ed.), *Spices, Herbs and Edible Fungi.* Elsevier Science B.V., The Netherlands, pp. 251–271, 1994.

Chipault, J.R., Mizuno, G.R., Hawkins, J.M. and Lundberg, W.O. (1955) *Food Res.,* 20, 443–447. Ref. Nakatani, N. Antioxidant and antimicrobial constituents of herbs and spices. In G. Charalambous (ed.), *Spices, Herbs and Edible Fungi.* Elsevier Science B.V., The Netherlands, pp. 251–271, 1994.

Chipault, J.R., Mizuno, G.R. and Lundberg, W.O. (1956) *Food Technol.,* 10, 209–211. Ref. Nakatani, N. Antioxidant and antimicrobial constituents of herbs and spices. In G. Charalambous (ed.), *Spices, Herbs and Edible Fungi.* Elsevier Science B.V., The Netherlands, pp. 251–271, 1994.

Chung, S.-K., Osawa, T. and Kawakishi, S. (1997) Hydroxyl radical-scavenging effects of spices and scavengers from brown mustard (*Brassica nigra*). *Biosci. Biotechnol. Biochem.* 61(1), 118–123.

Conner, D.E. and Beuchat, L.R. (1984) Sensitivity of heat-stressed yeasts to essential oils of plants. *Appl. Environm. Microbiol.* 47(2), 229–233.

Deans, S.G., Svoboda, K.P. and Bartlett, M.C. (1991) Effects of microwave oven and warm-air drying on the microflora and volatile oil profile of culinary herbs. *J. Ess. Oil Res.* 3, 341–347.

Dold, H. and Knapp, Z. (1948) *Hyg. Infektionskr.,* 127, 51. Ref. Nakatani, N. Antioxidant and antimicrobial constituents of herbs and spices. In G. Charalambous (ed.), *Spices, Herbs and Edible Fungi.* Elsevier Science B.V., The Netherlands, pp. 251–271, 1994.

Dorman, H.J.D., Deans, S.G., Noble, R.C. and Surai, P. (1995) Evaluation *in vitro* of plant essential oils as natural antioxidants. *J. Ess. Oil Res.* 7(6), 645–651.

Futrell, J.M. and Rietschel, R.L. (1993) Spice allergy evaluated by results of patch tests. *Cutis* 52(5), 288–290.

Gray, R.W., Mitchell, C.J., True, S. and Yeomans, M.R. (1997) Independet effects of palatability and within-meal pauses on intake and appetite ratings in human volunteers. *Appetite* 29, 61–76.

Hälvä, S. (1987) Studies on production techniques of some herb plants. I. Effect on Agryl P 17 mulching on herb yield and volatile oils of basil (*Ocimum basilicum* L.) and marjoram (*Origanum majorana* L.). *J. Agric. Sci. Finland* 59(1), 31–36.

Hammer, K.A., Carson, C.F. and Riley, T.V. (1999) Antimicrobial activity of essential oils and other plant extracts. *J. Appl. Microbiol.* 86(6), 985–990.

Holland, B., Unwin, I. and Buss, D. (eds) (1991) In McCance and Widdowson's *The Composition of Foods. Vegetables, Herbs and Spices.* Bath, U.K.

Hook, E.B. (1978) Dietary cravings and aversions during pregnancy. *Am. J. Clin. Nutr.* 31(8), 1355–1362.

Hook, E.B. (1980) Influence of pregnancy on dietary selection. *Intern. J. Obesity* 4(4), 338–340.

Huhtanen, C.N. (1980) Inhibition of *Clostridium botulinum* by spice extracts and aliphatic alcohols. *J. Food Prot.* 43, 195–197.

Kikuzaki, H. and Nakatani, N. (1989) *Agric. Biol. Chem.* 53, 519–524. Ref. Nakatani, N. Antioxidant and antimicrobial constituents of herbs and spices. In G. Charalambous (ed.), *Spices, Herbs and Edible Fungi.* Elsevier Science B.V., The Netherlands, pp. 251–271, 1994.

Kivanc, M., Akgul, A. and Dogan, A. (1991) Inhibitory and stimulatory effects of cumin, oregano and their essential oils on growth and acid production of *Lactobacillus plantarum* and *Leuconostoc mesenteroides*. *Int. J. Food Microbiol.* 13(1), 81–85.

Koutsoumanis, K., Lambropoulou, K. and Nychas, G.J. (1999) A predictive model for the non-thermal inactivation of *Salmonella enteritidis* in a food model system supplemented with a natural antimicrobial. *Int. J. Food Microbiol.* 49(1–2), 63–74.

Lagouri, V. and Boskou, D. (1996) Nutrient antioxidants in oregano. *Int. J. Food Sci. Nutr.* 47(6), 493–497.

Launchbaugh, K.L. and Provenza, F.D. (1994) The effect of flavor concentration and toxin dose on the formation and generalization of flavor aversions in lambs. *J. Animal Sci.* 72(1), 10–13.

Llewellyn, G.C., Burkett, M.L. and Eadie, T. (1981) Potential mold growth, aflatoxin production, and antimycotic activity of selected natural spices and herbs. *J. Assoc. Offic. Anal. Chem.* 64(4), 955–960.

Madsen, H.L., Nielsen, B.R., Bertelsen, G. and Skibsted, L.H. (1996) Screening of antioxidative activity of spices. A comparison between assays based on ESR spin trapping and electrochemical measurement of oxygen consumption. *Food Chem.* 57(2), 331–337.

Mäkinen, S., Hälvä, S., Pääkkönen, K., Huopalahti, R., Hirvi, T., Ollila, P., Nykänen, I. and Nykänen, L. (1986) *Maustekasvitutkimus*. The Academy of Finland, Report SA 01/813, Helsinki, Finland.

Malmsten, T., Pääkkönen, K. and Hyvönen, L. (1991) Packaging and storage effects on microbiological quality of dried herbs. *J Food Sci.* 56(3), 873–875.

Nakatani, N. and Kikuzaki, H. (1987) *Agric. Biol. Chem.* 51, 2727–2732. Ref. Nakatani, N. Antioxidant and antimicrobial constituents of herbs and spices. In G. Charalambous (ed.), *Spices, Herbs and Edible Fungi*. Elsevier Science B.V., The Netherlands, pp. 251–271, 1994.

Nykänen, I. (1986a) High resolution gas chromatographic – mass spectrometric determination of the flavour composition of marjoram (*Origanum majorana* L.) cultivated in Finland. *Z. Lebensm. Unters. Forsch.* 183, 172–176.

Nykänen, I. (1986b) High resolution gas chromatographic – mass spectrometric determination of the flavour composition of wild marjoram (*Origanum vulgare* L.) cultivated in Finland. *Z. Lebensm. Unters. Forsch.* 183, 267–272.

Pääkkönen, K., Malmsten, T. and Hyvönen, L. (1990) Drying, packaging, and storage effects on quality of basil, marjoram and wild marjoram. *J Food Sci.* 55(5), 1373–1377, 1382.

Przezdziecka, T. and Baldwin, Z. (1971) Sensory characterization of spices. *Roczniki Instytuty Przemyslu Miesnego* 8, 45–46.

Salmeron, J., Jordano, R. and Pozo, R. (1990) Antimycotic and antiaflatoxigenic activity of oregano (*Origanum vulgare*, L.) and thyme (*Thymus vulgaris* L.). *J. Food Prot.* 53, 697–700. Ref. Nakatani, N. Antioxidant and antimicrobial constituents of herbs and spices. In G. Charalambous (ed.), *Spices, Herbs and Edible Fungi*. Elsevier Science B.V., The Netherlands, pp. 251–271, 1994.

Salzer, U.-J. (1982) *Fleischwirtsch.* 62, 885–887. Ref. Nakatani, N. Antioxidant and antimicrobial constituents of herbs and spices. In G. Charalambous (ed.), *Spices, Herbs and Edible Fungi*, Elsevier Science B.V., The Netherlands, pp. 251–271, 1994.

Silva, F., Casali, V.W.D., Lima, R.R. and Andrade, N.J. (1990) Postharvest quality of *Achillea millefolium* L., *Origanum vulgare* L., and *Petroselinum crispum* (Miller) A.W. Hill in three types of packaging. *Revista Brasileira de Plantas Medicinais* 2(1), 37–41.

Tantaoui-Elaraki, A. and Beraoud, L. (1994) Inhibition of growth and aflatoxin production in *Aspergillus paraciticus* by essential oils of selected plant materials. *J. Environ. Pathol., Toxicol. & Oncol.* 13(1), 67–72.

Tassou, C.C., Drosinos, E.H. and Nychas, G.J.E. (1996) Inhibition of resident microbial flora and pathogen inocula on cold fresh fish fillets in olive oil, oregano, and lemon juice under modified athmosphere or air. *J. Food Prot.* 59(1), 31–34.

Time-Life Books (1989) Healthy Home Cooking. *Fresh Ways with Fish and Shellfish*. Amsterdam, 144p.
Time-Life Books (1990) Healthy Home Cooking. *Fresh Ways with Lamb*. Amsterdam, 143p.
Time-Life Books (1990) Healthy Home Cooking. *Fresh Ways with Pasta*. Amsterdam, 144p.
Time-Life Books (1991) Healthy Home Cooking. *Fresh Ways with Terrines and Pâtés*. Amsterdam, 144p.
Ting, E.W.T. and Deibel, K.E. (1992) Sensitivity of Listeria monocytogenes to species at two temperatures. *J. Food Safety* 2(2), 2129–2137.
Tsimidou, M. and Boskou, D. (1994) Antioxidant activity of essential oils from the plants of the *Lamiaceae* family. In G. Charalambous (ed.), *Spices, Herbs and Edible Fungi*. Elsevier Science B.V., The Netherlands, pp. 273–284, 1994.
Tucker, A.O. and Maciarello, M.J. (1994) Oregano: botany, chemistry, and cultivation. *Devel. Food Sci.* 34, 439–456.
Ueda, S., Yamashita, H., Nakajima, M. and Kuwabara, Y. (1982) Nippon Shokuhin Kogyo Gakkaishi, 29, 111–116. Ref. Nakatani, N. Antioxidant and antimicrobial constituents of herbs and spices. In G. Charalambous (ed.), *Spices, Herbs and Edible Fungi*. Elsevier Science B.V., The Netherlands, pp. 251–271, 1994.
Ultee, A., Gorris, L.G. and Smid, E.J. (1998) Bactericidal activity of carvacrol towards the food-borne pathogen *Bacillus cereus. J. Appl. Microbiol.* 85(2), 211–218.
Ultee, A., Kets, E.P. and Smid, E.J. (1999) Mechanisms of action of carvacrol on the food-borne pathogen *Bacillus cereus. Appl. Environm. Microbiol.* 65(10), 4606–4610.
Vekiari, S.A, Oreopoulou, V., Tzia, C. and Thomopoulos, C.D. (1993) Oregano flavonoids as lipid antioxidants. *J. Am. Oil Chem. Soc.* 70(5), 483–487.
Vokou, D., Vareltzidou, S. and Katinakis, P. (1993) Effects of aromatic plants on potato storage: sprout suppression and antimicrobial activity. *Agric. Ecosyst. Environm.* 47(3), 223–235.

Part 7
Biotechnology

10 The biotechnology of Oregano (*Origanum* sp. and *Lippia* sp.)

Spiridon E. Kintzios

INTRODUCTION

Although probably the world's commercially most valued spice, oregano is yet a novel target for biotechnology. The current focus of relevant applications is the establishment of clonal populations, which could offer a means to overcome the problems associated with germplasm variability. However, additional perspectives include the use of cell cultures for the scaled-up production of useful pharmaceuticals and the enhancement of breeding activities (e.g. *in vitro* selection for disease resistance, exploitation of somaclonal variation).

So far, a number of researchers have reported on the establishment of tissue cultures from oregano plants and, in some cases, the regeneration of plantlets. The present review focuses on the tissue culture of the *Origanum* sp. which is the commercially most important oregano species worldwide. The biotechnology of *Lippia* sp., the Mexican oregano is less extensively reviewed, due to the limited number of relative reports in the literature.

Origanum sp.

Explant source

Explants used for the establishment of oregano callus cultures include hypocotyls and cotyledons (Matsubara *et al.*, 1996), shoot apexes (Kurtis and Shetty, 1995; Shetty *et al.*, 1996), nodal segments (Baricevic *et al.*, 1997), roots (Kumari and Pardha Saradhi, 1992) and leaves (Alves-Pereira and Fernandes-Ferreira, 1998). Nevertheless, most researchers prefer to use explants derived from *in vitro* grown seedlings, such as established clonal lines (Yang and Shetty, 1998).

Explant disinfection

Leaf and stem explants have been regularly immersed in 70 per cent ethanol (0.5–2 min) usually followed by surface-sterilization in a 0.5–1 per cent (v/v) solution of sodium-hypochlorite or calcium hypochlorite (10–15 min) immersion (Baricevic, 1997; Alves-Pereira and Fernades-Ferreira, 1998). Seeds can be surface-disinfected in 0.05 per cent (w/v) mercuric chloride solution for 5 min (Kumari and Pardha Saradhi, 1992). For micropropagation purposes, several investigators prefer to use explants derived from aseptically grown seedlings (e.g. Ueno and Shetty, 1998). In my experience (unpublished data), leaf explants derived from field-grown plants can be optimally surface-sterilized containing 1–2 per cent Tween-80, while disinfection for 12 min in 0.1 per cent (w/v) mercuric chloride solution led to total explant necrosis.

Culture media

The basal medium of Murashige and Skoog (MS) (1962) has been used in an almost exclusive manner for the initiation of oregano tissue cultures. Media were solidified with various gelling agents, such as agar (0.8 per cent) or gellan-gum (0.3 per cent) and were supplemented with 2–3 per cent sucrose. Alves-Pereira and Fernades-Ferreira (1998) reported a higher induction of callus from the leaves of *Oregano vulgare* ssp. *virens* on MS medium than Gamborg's B_5 medium (Gamborg *et al.*, 1968), which has also been used in some experiments (Kumari and Pardha Saradhi, 1992). Elevated concentrations of the gelling agents have been associated with a lower incidence of vitrification (see next section) (Curtis and Shetty, 1996). Baricevic (1997) reported on the beneficial effect of sodium hydrogen phosphate ($NaH_2PO_4 \cdot 12\ H_2O$; 0.174 g l^{-1}) on the micropropagation of oregano (*O. vulgare* subsp. *heracleoticum*) from axillary buds.

Culture conditions

For culture initiation, oregano explants have been incubated at 23–27 °C, over a 16 h photoperiod and under a photosynthetic photon flux density (PPFD) of 100 μmol $m^{-2}\ s^{-1}$ provided by cool white florescent tubes. For the multiplication of shoot cultures through axillary bud proliferation, some investigators (e.g. Baricevic, 1997; Ueno and Shetty, 1998) suggested an incubation temperature of 20–24 °C under a PPFD of 31–40 μmol $m^{-2}\ s^{-1}$.

Morphogenesis in vitro

Callus induction and morphology

Auxins have been commonly used for callus induction from various oregano explant sources (Becker, 1970), usually in combination with a cytokinin, such as benzyladenine (BA) and at a 0.25–1 mg l^{-1} concentration. Kumari and Paradha Saradhi (1992) managed to standardize a protocol for rapid multiplication of oregano plants from cotyledonary callus cultures. According to their research, callus and root induction were both favoured by increased α-naphaleneacetic acid (NAA) concentrations (10^{-5} M) in the medium, although the best callus induction medium was supplemented with 10^{-7} M 2,4-dichlorophenoxyacetic acid (2,4-D). Calli obtained in media supplemented with 2,4-D varied from compact, nodular, friable, to gelatinous depending on the concentration of 2,4-D. At higher concentrations of 2,4-D calli were friable and gelatinous, albeit induced at a lower rate. Increasing 2,4-D concentrations also affected negatively root induction. When calli were transferred onto media with BA, alone or in combination with NAA, formation of localized green patches occurred on the calli which formed shoots within a few days. Best shoot induction was obtained on medium with 10^{-6} M NAA + 10^{-6} M BA. The shoot inducing ability of the calli was retained after 20 subcultures (approx. 18 months). Rooting of the 15–20 mm long shoots was achieved in half-strength B_5 liquid medium supplemented with 1 per cent sucrose and either 10^{-6} M indolebutyric acid (IBA) or NAA, two auxins which are widely employed to induce rooting. Root induction was suppressed with elevated BA concentrations. Rooted shoots were initially hardened for 18 days in half-strength B_5 medium, then for 28 days in the same medium deprived of sucrose and other organics and finally re-established for 18 days in a peat:vermiculite:soil (1:1:2) mixture at $28 + 2$ °C under continuous illumination.

Alves-Pereira and Fernades-Ferreira (1998) also reported the occurrence of two types of callus tissues induced from leaves of *O. vulgare* ssp. *virens*. Green, friable calli (G-calli) were induced on MS medium + 0.25 mg l^{-1} 2,4-D + 0.5 mg l^{-1} BA. Increasing the 2,4-D concentration to 1 mg l^{-1} led to the formation of dark calli with abnormal root primordia (R-calli).

Organogenesis and micropropagation

Plantlets have been regenerated from oregano callus on MS medium supplemented with NAA and BA, usually at low concentrations (0.1 mg l^{-1}) (Matsubara et al., 1996). Most frequently, however, micropropagation of oregano clones involved axillary bud induction from shoot apex explants on MS + 1 mg l^{-1} BA (Ueno and Shetty, 1998). Baricevic (1997) pointed out that oregano tissue culture has to be set up with axillary buds, because apical cuttings frequently enter the fructification stage *in vitro*, thus weakening the vigour of vegetative shoots. Rooting of micropropagated shoots can be achieved on growth regulator-free MS medium at a relative high rate (>95 per cent) within 10–14 days (Figure 10.1). Micropropagated plants tend to acclimatize rapidly (3 weeks) in the greenhouse at a high rate (>95 per cent) and usually are more vigorous than seed-propagated plants (Baricevic, 1997).

Somatic embryogenesis

The occurrence of embryogenic oregano tissue cultures has never been reported in the literature. Recently, my research group has observed the formation of globular stage somatic embryo on callus cultures induced from oregano leaf explants on MS supplemented with a partially characterized oligosaccharide from *Hibiscus cannabinus* (kenaf) (Figure 10.2).

Use of bacteria for the prevention of hyperhydricity

Hyperhydricity or vitrification is a physiological malformation affecting plants propagated via tissue culture conditions. Due to a combination of chlorophyll deficiency, poor lignification and excessive hydration of tissues, hyperhydrated plants are usually enlarged, thick, translucent and brittle. Kalidas Shetty (Laboratory of Food Biotechnology, University of Massachusetts) and his collaborators prevented hyperhydricity in oregano shoot cultures by inoculating them with non-specific, polysaccharide-producing soil bacteria, such as *Pseudomonas* sp., *Pseudomonas* sp. F, *P. mucidolens* and *Beijerinkia indica* (Shetty et al., 1996; Bela et al., 1998; Perry et al., 1999). Hyperhydricity was prevented after 10–15 days of inoculation and lasted over eight subculture cycles without any re-inoculation after the initial one. The bacterial treatments

Figure 10.1 Micropropagated oregano plant (derived from an axillary bud) with aerial roots on growth regulator-free MS medium.

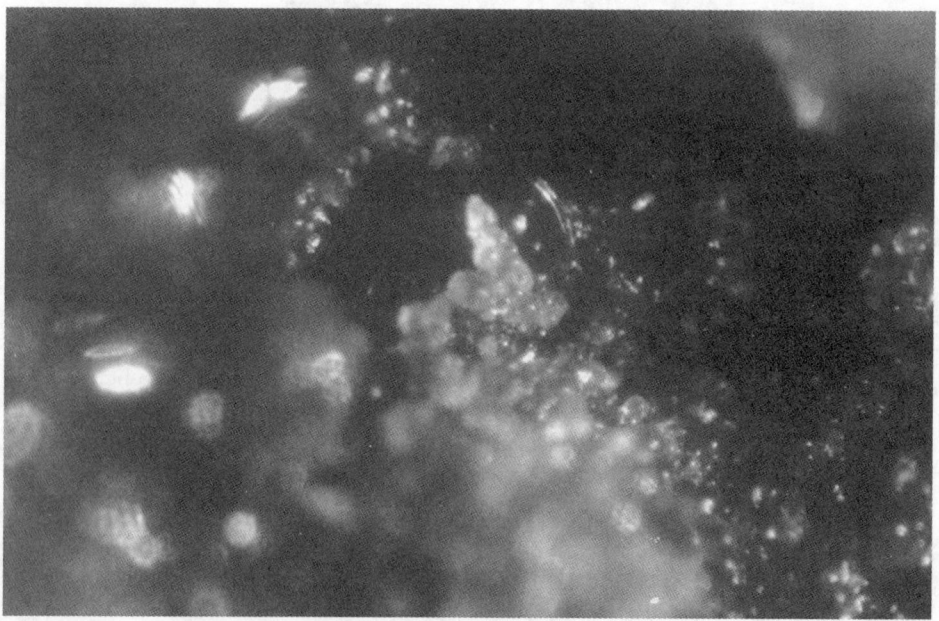

Figure 10.2 Formation of globular stage somatic embryo on callus cultures (induced from oregano leaf explants) on MS supplemented with a partially characterized oligosaccharide from *Hibiscus cannabinus* (kenaf).

stimulated chlorophyll and total phenolics accumulation, while reducing plant growth (Ueno and Shetty, 1997, 1998).

In vitro *secondary metabolite accumulation*

Studies on production of secondary metabolites by *in vitro* cultures of *Origanum* species are scarce. Yang and Shetty (1998) used tissue culture-generated shoot-based clonal lines in order to investigate the role of proline-linked pentose phosphate pathway in stimulating rosmarinic acid (RA) production. RA biosynthesis in *O. vulgare* clonal line O-1 was stimulated in response to proline, proline precursors (ornithine and arginine) and a proline analogue (azetidine-2-carboxylate, A2C). Exogenous treatment of cells with proline and proline precursors in the presence or absence of A2C significantly enhanced RA content; higher levels of endogenous proline were observed compared with the control. A2C treatment alone stimulated highest levels of RA without any increase in endogenous proline. The stimulation of RA synthesis suggested that deregulation and enhancement of proline synthesis or proline oxidation may be important for RA biosynthesis. RA-stimulating compounds also enhanced total phenolics and hardened stem tissues, indicating possible lignification due to polymerization of phenolic metabolites.

Alves-Pereira and Fernades-Ferreira (1998) compared the essential oil and hydrocarbon content of hydrodistillates from leaves and calli from *O. vulgare* ssp. *virens*. They found that callus tissues did not contain either mono- or sesquiterpenoids; *n*-alkanes were the main compounds found, but their respective concentration depended on callus morphology (G-calli and R-calli, see above). It has been hypothesized that the differentiation of abnormal root primordia in R-calli, together with a higher concentration of 2,4-D in the culture medium was correlated with an increased napthalene (0.5 µg g^{-1} dry wt) and eicosane (0.9 µg g^{-1} dry wt) accumulation and a decreased squalene content. Finally, Socorro *et al*. (1998) compared the essential oil content of *O. bastetanum* plants collected from the wild in Spain or grown *in vitro*. Micropropagated plants produced smaller essential oil yields which had a slightly different composition to that of wild plants.

Lippia sp.

Juliani *et al*., (1999) reported on a method for the micropropagation of *Lippia junelliana* (Mold.) Tronc. (a native plant of Argentina) from shoot tips or nodal segments. Proliferating microshoot cultures were obtained by placing shoot tips or nodal segments on full strength MS supplemented with 4.4 mM BA or 0.04 mM IBA plus 4.4 mM BA. The rooting of shoots was better on full-strength MS medium without growth regulators. Rooted plantlets were successfully acclimatized to soil. The shoot cultures showed a lower essential oil accumulation in comparison with parent plants. Essential oil accumulation was closely related with growth and showed a negative correlation with shoot proliferation.

CONCLUSION

In vitro culture of cells and organs offers the opportunity to clonally micropropagate oregano lines with improved traits. It can further contribute, in a very significant degree, to sustaining elite oregano germplasm, a task of increasing importance in view of the rapid genetic erosion of this species. The use of tissue culture for the production of essential oil components is not yet feasible from a commercial point of view. Nevertheless, further achievements in this field could potentially contribute toward this goal and open new perspectives in high grade flavour preparation, perfumery, cosmetics and liquor industries.

REFERENCES

Alves-Pereira, I.M.S. and Fernades-Ferreira, M. (1998) Essential oils and hydrocarbons from leaves and calli of *Origanum vulgare* ssp. *virens*. *Phytochemistry* 48, 795–799.

Baricevic, D., Zupancic, A., Erzen-Vodenik, M. and Seliskar, A. (1997) *In situ* and *ex situ* conservation of natural resources of medicinal and aromatic plants in Slovenia. In situ in ex situ ohranjanje naravnih izvorov zdravilnih in aromaticnih rastlin v Sloveniji. *Sjemenarstvo* 14, 23–29.

Baricevic, D. (1997) Experiences with oregano (Origanum spp.) in Slovenia. In S. Padulosi (ed.), Promoting the conservation and use of unterutilized and neglected crops – Oregano, 14, 111–121.

Becker, H. (1970) *Biochem. Physiol. Pflanzen* 161, 425–441.

Bela, J., Ueno, K. I. and Shetty, K. (1998) Control of hyperhydricity in anise (*Pimpinella anisum*) tissue culture by *Pseudomonas* spp. *J. Herbs, Spices Med. Plants* 6, 57–67.

Curtis, O.F. and Shetty, K (1996) Growth medium effects of vitrification, total phenolics, chlorophyll, and water content of in vitro propagated oregano clones. *Acta Horticulturae* 426, 489–497.

Gamborg, O.L., Miller, R.A. and Ojima, K. (1968) Nutrient requirements of suspension cultures of soybean root cells. *Exp. Cell Res* 50, 151–158.

Juliani, H.R. Jr., Koroch, A.R., Juliani, H.R. and Trippi, V.S. (1999) Micropropagation of *Lippia junelliana* (Mold.) Tronc. *Plant Cell, Tiss. Org. Cult.* 59(3), 175–179.

Kumari, N. and Pardha Saradhi, P. (1992) Regeneration of plants from callus cultures of *Origanum vulgare* L. *Plant Cell Rep.* 11, 476–479.

Matsubara, S., Ino, M., Murakami, K., Kamada, M. and Ishihara, I. (1996) Callus formation and plant regeneration of herbs in Perilla family. *Scientific Reports of the Faculty of Agriculture, Okayama University*, 85, 23–30.

Murashige, T. and Skoog, F. (1962) A revised method for rapid growth and bioassays with tobacco tissue cultures. *Physiol. Plant.* 15, 472–497.

Perry, P.L., Ueno, K. and Shetty, K. (1999) Reversion to hyperhydration by addition of antibiotics to remove *Pseudomonas* in unhyperhydrated oregano tissue cultures. *Process Biochem.* 34, 717–723.

Shetty, K., Carpenter, T.L., Curtis, O.F. and Potter, T.L. (1996) Reduction of hyperhydricity in tissue cultures of oregano (*Origanum vulgare*) by extracellular polysaccharide isolated from *Pseudomonas* spp. *Plant Science* (Limerick) 120(2), 175–183.

Socorro, O., Tarrega, I. and Rivas, F. (1998) Essential oils from wild and micropropagated plants of *Origanum bastetanum*. *Phytochemistry* 48, 1347–1349.

Ueno, K.I. and Shetty, K. (1997) Effect of selected polysaccharide-producing soil bacteria on hyperhydricity control in oregano tissue cultures. *Appl. Environ. Microbiol.* 63, 767–770.

Ueno, K. and Shetty, K. (1998) Prevention of hyperhydricity in oregano shoot cultures is sustained through multiple subcultures by selected polysaccharide-producing soil bacteria without re-inoculation. *Appl. Microbiol. Biotechnol.* 50, 119–124.

Yang, R.H. and Shetty, K. (1998) Stimulation of rosmarinic acid in shoot cultures of oregano (*Origanum vulgare*) clonal line in response to proline, proline analogue, and proline precursors. *J. Agricult. Food Chemy.* 46, 2888–2893.

Part 8

Miscellaneous

11 Bibliometric analysis of agricultural and biomedical bibliographic databases with regard to medicinal plants genera *Origanum* and *Lippia* in the period 1981–1998

Tomaž Bartol and Dea Baričevič

INTRODUCTION

The topic of medicinal or spice plants has been scattered amid a huge number of resources. There is even no common name for these groups of plants and products. Information can be found under such different captions as condiments, culinary herbs, drug plants, essential oil plants, ethnobotany, flavouring crops, herbaceous agents, herbal drugs, herbal medicine, herbs, medicinal plants, spice plants, spices, traditional medicine, to mention only some of them. Genera *Lippia* and especially *Origanum* may in a broader sense also be covered by all of the above broader terms.

Papers on the subject are dispersed among dozens of international or national journals. Books and reports are published in probably all countries of the world. National and international meetings are organized on regular basis. New standards and patents are submitted yearly. Previously a great part of the above information became sooner or later available via some of the bibliographic services. Such information was initially published in specialized printed bibliographies. More than a couple of decades ago these started to be replaced by electronic databases available online via special providers. Such electronic services were quite efficient but, albeit they represented huge improvement over paper bibliographies, rather expensive to be used for routine and superficial or unfocused searches. More than a decade ago many of those databases became widely available on CD-ROMs. Now the same bibliographic data is becoming increasingly available via the World Wide Web. There is a rapid development of possibilities of accessing the full texts. The WWW, moreover, offers an instant access to data far surpassing traditional, professional and scientific information.

All of the above data collection and dissemination media, with the exception of printed bibliographies, which cannot match efficiency of electronic data storage, still continue on fairly equal basis. Some bibliographic resources are nevertheless available via licensed online providers only. Others are accessible also through CD-ROMs and may additionally be accessed for a fee or free of charge on the WWW.

Because of the heterogeneity of non-bibliographic-WWW resources, we did not consider those in our investigation. We checked on several instances the enormous quantity of natural-products-related pages. As it is well known such information is presented in many different forms ranging from traditional articles to well organized professional or scientific databases, and all the way to cook-books and numerous pages that offer medicinal plants and products as commodities on commercial basis.

Such data can bibliometrically not be estimated in as accurate a way as bibliographic data. The bibliographic data have been, in the instance of our investigation, compiled on some consistent basis for more than two decades and still offer a fairly high level of retrieval uniformity.

We thus present the data as available via some standard bibliographic databases that are well known to most scientists who are active in the field of medicinal plants. We focus on agricultural (Agricola, Agris, CAB Abstracts), food (Food Science and Technology – FSTA) and biomedical (Embase, Medline) databases. The comparison of these groups has only rarely been presented in scientometric literature. A great amount of bibliometric or scientometric research focused on Institute of Scientific Information's Science Citation Index (SCI). Chemical Abstracts and Biological Abstracts have also been covered by similar examinations, and so have Medline and to some extent Embase. Agricultural coverage, however, has not been so comprehensive.

REVIEW OF LITERATURE

There are many articles that analyze bibliographic databases so we mention only a few selected early and recent references that focus mostly on agricultural and biomedical databases and their comparison. We will also include aspects of food and human nutrition that can be placed in the agricultural field in a broader sense but is also covered by a specialized database.

Large quantity of scientometric analyzes take as a subject (co) citations data with one of the pioneers being Garfield (1964, op. cit. Small and Griffith, 1974). There are other possibilities such as co-word and co-classification analyzes, bibliographic coupling, mapping of science etc. Some later analyzes may include many different topics, such as co-author-based scientific co-operation (Glanzel et al., 1999), structure of international scientific co-operation networks (Gomez et al., 1999), analysis of certain area in a certain country (Garcia-Lopez, 1999) that combines mapping and citation analysis, ranking of citation by countries (Bonitz et al., 1999), and many more.

No end-user can be expected to conduct searches in all potentially relevant databases. There are too many such sources. Some of them are rather expensive and each online hit (record) must be paid for. Also, many users, especially those involved with interdisciplinary sciences, already experience retrieval anxiety or frustration (Kuhlthau, 1997). This may lead to a poor *recall* (number of retrieved relevant documents as a fraction of number of all relevant documents that exist in a database) as a better *precision* (number of retrieved relevant documents as a fraction of all documents that were retrieved in a search) is sought. Precision and recall as classic bibliometric indexes were introduces by Cleverdon (1962).

Much research was also carried out with regard to identifying scatter or ranking of information among possible information resources. Bradford determined as early as 1934 that on a given subject a certain amount of articles will be found in a few core journals, with the same amount of articles scattering in a greater number journals less dedicated to this subject, and again the same amount of articles scattering among many other journals dedicated to other subjects. Bradford scatter patterns were studied also in other items such as bibliographic references (Martyn and Slater, 1964) in abstract journals or databases (Tenopir, 1982). Multiple database searches, however, will invariably yield multiple occurrences (occurrences of same records in several different databases). A simple Jaccard coefficient or index (Jaccard, 1912) may serve as a good indicator of such co-appearance.

As there has been much investigation of differences among different databases we present only some research relevant to medicinal plants and agricultural and biomedical databases. Ogawa and Kangohri (1989) compared herbal medicine coverage by Medline, Embase and some other databases. Gupta *et al.* (1990) focused on Indian journal papers on medicinal and aromatic plants covered by an Indian database medicinal and aromatic plants abstracts (MAPA) and studied some aspects of medicinal plants bibliographic data. Zhang (1994) investigated the presence of Chinese traditional medicine in Medline. Bartol (1999) tested Bradford bibliograph with regard to international scatter of articles on medicinal plant genus *Salvia*, and *Salvia*-related journals available in the libraries in Slovenia. Some documentation literature on medicinal plants is also dedicated to computer applications and database construction what was not a subject of our analysis.

More bibliometric research, dedicated to other subjects, involved agricultural and medical databases. Some comparison between Agricola, Agris and CAB was presented already in 1980 (Juliano-Longo and Dantas-Machado). Agris was compared with FSTA in order to assess its coverage of food industry (Lofflerova and Pessrova, 1982). Deselaers (1986) observed that the three principal agricultural databases cover approximately the same scope with differences in terms of special subjects with suggestion that a more intensive co-operation would reduce unnecessary efforts and costs needed in maintaining such information services. Chen (1989) analyzed overlapping between Agricola, Agris and CAB for tropical agriculture and found that overlapping was very low. Thomas (1990) also described overlap between these databases and conceded that there is no single source that allows locating all references to a particular topic in agriculture. Comparison between the three databases was also presented by Patil and Kumar (1998). Principal food database FSTA was used as a reference comparison for a newer food-related product Frosti database on food and drink industry (Clarke, 1998). Growth of food science and technology literature in terms of a comparison between an Indian institute and the world was investigated by Seetharam and Rao (1999).

Databases can sometimes be evaluated by user surveys with respect to different quality criteria, such as coverage, harmonization or output formats what was the case in the comparison of CAB with an 'ideal' theoretical database (Wilson, 1998). There are several articles that deal with indexing and classification characteristics of agricultural databases. Weintraub (1992) presented some difficulties, which arise when using same descriptors in all three agricultural databases, and emphasized a need for a unified thesaurus and standardized terminology. A few other articles were dedicated to an agricultural meta-thesaurus such as Andre's (1992). This project, however, seems to remain dormant. Some other search tools are also being developed with regard to harmonizing searches, such as the meta-data vocabulary developed by the U.S. Department of Agriculture to locate certain food, agriculture and natural resources material in different databases (Cortez, 1999).

In the area of biomedical databases, more scientometric data can be found. Analyses of biomedical databases frequently include other life sciences databases such as biological abstracts (BIOSIS), Chemical Abstracts, and SCI. Search quality of Medline and Embase was compared by Soremark (1990). Medline and Embase were compared with regard to drug literature (van Putte, 1991). Multi-database searches on pharmaceutical products in four databases, including Medline and Embase, were performed by Sodha and Amelsvoort (1994). They found that most of the unique citations did not contain the search term in the title or abstract. Brown (1998a) suggested the use of at least two complementary biomedical databases in order to conduct comprehensive search in biomedicine. Embase and Medline along with some other databases were investigated by Matthews et al. (1999) with regard to communicating risk to patients in primary care. In study on retrieval of pharmaceutical information Brown (1998b) found Embase superior to Medline, with the former being sometimes avoided because of higher search costs. He gives possibilities of some cost saving measures in Embase searches such as free title displays.

There are some articles that analyze both medical and agricultural databases. Medline, CAB, and FSTA were compared with regard to veterinary literature (Matsubuchi, 1987). Coverage of literature on mycotoxins in several databases, including Agricola, Agris, CAB and Medline was examined by Datta (1988). Agricola, CAB, Embase and Medline were evaluated with respect to human nutrition (Nixon, 1989). Medline, Embase and CAB abstracts were compared with regard to a specific subject related to obesity in humans (Parsons et al., 1999).

DATABASES AND SERVICES

The objective of our research was examination of scatter of references on genera *Origanum* and *Lippia* in all-major international agricultural and biomedical bibliographic databases in the period 1981–1998. Year 1999 was excluded because of possible indexing delay in some databases. To maximize our results (bibliometric recall) some other databases such as chemical abstracts and biological abstracts (BIOSIS) could also be used. Such searches covering almost 20 years and yielding up to thousand references would nevertheless be too costly so we decided to systematically compare only agricultural and biomedical databases. Here we do not wish to forget the natural-products oriented data-base Napralert (NAtural PRoducts ALERT) which is maintained by the Programme for Collaborative Research in the Pharmacological Sciences (PCRPS) at the College of Pharmacy, University of Illinois at Chicago. This database was not consulted because we tried to establish the scatter of a specific multidisciplinary topic as is evidenced in non-specialized databases.

Bibliographic databases described bellow offer similarly structured bibliographic information (author, document title, source, publication data, subject headings, abstract) on journal articles, books, conference papers, patents, standards, theses, etc. These databases are accessible in different formats. All are available for a fee both online and on CD-ROMs, or on tape for in-house installation, through several online providers or commercial vendors. Some are in addition accessible free of charge on the WWW.

The three general agricultural databases Agris, Agricola and CAB Abstracts cover all aspects of agriculture and allied disciplines, including production and protection aspects of animal and plant sciences, food and feed, and animal and human nutrition, forestry, wood technology, aquaculture and fisheries, farming and farming systems, agricultural economics, rural sociology, extension, education, information, legislation, earth, environmental sciences, etc. FSTA is the world's leading food and human nutrition database. Medline and Embase (Biobase) both cover immense field of biomedicine.

AGRICOLA

Agricola is a product of the National Agricultural Library (NAL). NAL, the largest agricultural library in the world, is part of the Agricultural Research Service of the U.S. Department of Agriculture, and is one of four National Libraries in the U.S. AGRICOLA contains over 3 million bibliographic records produced from 1970 to the present. It is also accessible free of charge on the WWW.
NAL: http://www.nalusda.gov
Agricola: http://www.nalusda.gov/ag98

AGRIS

Agris is part of the World Agricultural Information Centre (WAICENT) information system of the Food and Agriculture Organization (FAO) of the United Nations. Data are co-operatively compiled by 140 national and international centres and sent to the FAO for central processing. Agris contains over 2 million bibliographic records produced from 1975 to the present. Agris is accessible free of charge on the WWW.
FAO: http://www.fao.org
Agris: http://www.fao.org/agris/default32.htm

CAB ABSTRACTS

CAB Abstracts are compiled by CAB International (CABI) which is an international organization based in United Kingdom. CABI evolved from the Commonwealth Agricultural Bureau (CAB) and its predecessor the Imperial Agricultural Bureau (IAB). The CAB Abstracts database contains over 3 million bibliographic records produced from 1973 to the present. The database is available for a fee.
CABI: http://www.cabi.org

FOOD SCIENCE AND TECHNOLOGY

Food Science and Technology Abstracts (FSTA) are compiled by International Food Information Service (IFIS). IFIS, which is based in United Kingdom and Germany, is the leading international service for food science, food technology and human nutrition. FSTA database contains over half a million bibliographic records on food and nutrition related subjects produced from 1969 to the present. The database is available for a fee.
IFIS: http://www.ifis.co.uk

EMBASE, BIOBASE

Both Embase and Biobase are compiled by an international publisher Elsevier Science. EMBASE, the Excerpta Medica database, contains over 8 million records on biomedicine and drug-related literature from 1974 to the present. The BIOBASE provides international coverage of the basic biological sciences. It contains over 1 million records from 1994 to the present. Bibliographic records are available for a fee.
Elsevier: http://www.elsevier.nl

MEDLINE

Medline (Medlars Online) is compiled by the National Library of Medicine (NLM). NLM, which is one of the four National Libraries in the United States, is the world's largest medical library. The Medline database contains over 9 million records on biomedicine from 1966 to the present. Free of charge access is provided by two NLM Web-based products Pub Med and Internet Grateful Med.
NLM: http://www.nlm.nih.gov
PubMed: http://www.ncbi.nlm.nih.gov/entrez/query.fcgi
Grateful Med: http://igm.nlm.nih.gov

We obtained the data from CD-ROMs and commercial online services and we harmonized and then compiled the data into two experimental databases. In our analyses we sometimes coupled the three general agricultural databases (Agris, Agricola and CAB Abstracts) into the agricultural group, and the biomedical databases (Medline and Embase + Biobase) into the biomedical group. In further text we refer to the database CAB Abstracts as CAB. Embase and Biobase are both produced by Elsevier. We received references from both Embase and Biobase in such a form where there were no duplications so we coupled the two databases into one group, presented by an acronym *eb* (Embase + Biobase). In tables and figures we usually refer to the above databases and groups in terms of acronyms, which we formulated for this purpose (Table 11.1). There is distinction between upper and lower case for clarity purpose. The upper case is used for a database group and lower case for a single database. We used the symbols + and * in the Boolean syntax when we investigated overlap of databases or co-occurrence of

Table 11.1 Abbreviations and symbols used in tables in figures with regard to databases and database groups

Abb./Sym.	Database or group of databases
ac	Agricola
AG	Agricultural database group (Agricola, Agris, CAB Abstracts)
as	Agris
BM	Biomedical database group (Embase, Biobase, Medline)
cb	CAB Abstracts
eb	Embase + Biobase
fs	FSTA
me	Medline
+	Boolean OR
*	Boolean AND

terms or records in different databases. Boolean AND (*) implies that the record is present in all databases or groups, while OR (+) indicates the presence of the record in at least one of the specified databases or groups. We use the terms *reference* and *record* synonymously. The term reference, however, indicates a primary bibliographic source (e.g. a book or articles), whereas the term record indicates a reference to such primary source in a particular database.

Our main aim was to assess the following database features for both genera *Origanum* and *Lippia* separately:

- number of all yearly new references in the observed period regardless of the database (exclusion of duplicates and multiple occurrences), and possible growth patterns;
- number of references in each particular database and group of databases;
- number of references with regard to the type of the primary source;
- number of single and multiple occurrences of the same reference (the number of databases indexing the same document);
- overlap (co-occurrences) of same references with regard to databases and groups of databases;
- scatter of journal articles in journals and identification of core journals;
- some indexing characteristics and differences among databases.

DATA COMPILATION AND EXPERIMENTAL DATABASE CONSTRUCTION

Search syntax

To achieve optimal search precision we identified the names most commonly used with regard to both genera. With *Lippia* the Latin term was sufficient (a). There were a few instances when the term Mexican oregano was used and was not preceded by *Lippia*. Such records were later identified and transferred from the *Origanum* database. With *Origanum* we had to use both scientific names Origanum and Majorana. Accordingly we also used the English terms oregano and marjoram due to the terminological inconsistency with regard to wild marjoram (oregano, *Origanum vulgare*) and sweet marjoram (marjoram, *Origanum majorana*, *Majorana hortensis*). The terms were coupled with the Boolean OR to achieve the presence in a record of at least one of the specified terms (b).

a   Lippia;
b   Origanum or Majorana or oregano or marjoram.

We searched with free text terms to maximize recall. Hundred per cent precision had to be achieved by subsequent manual deletion of noise records. Some noise was due to an editor's name Marjoram, and a serial title containing Majorana.

Downloading

We downloaded bibliographic data separately for *Lippia* and *Origanum* and created tables with a word processor. We selected fields such as Author(s), Title, Source,

Document Type, Descriptors (Key-words), Abstract, Classification Codes, and Language. Author and Title field were structured generally in the same way even though there existed some differences which will be explained. Source and Document Type fields had different structures so they had to be harmonized. There was of course no point in harmonizing Abstract, Descriptor and Classification fields which are specific in each database.

Source (SO): For the document source there existed for the same type of record as many as three different fields in different databases. There may have been a separate field for the data on conference and a separate field for the bibliographic description of the proceedings. If the proceedings were published in a journal there was a separate field for the series data. We had to manually merge the fields for each particular record.

Document type (DT) and Publication year (PY): Document types had very different presentations in databases. In some cases there was no DT type, so we had to insert a blank field or column for addition of later data. Also there was sometimes no separate PY field. The publication year was included in the Source field. In such a case sorting was impossible so we had to insert another field and manually add data for PY.

We set up an experimental database table using program MS-excess where we defined the above fields and also added some more fields such as a field for the database from which a particular record was extracted. We uploaded all worktables into the experimental database. Then more systemic harmonization was needed.

With presentation of some authors there existed differences among databases that impaired direct sorting.

Example

>   AU: Broucke-CO-van-den; Lemli-JA
>   AU: Van-Den-Broucke,-C.O.; Lemli,-J.A.
>   AU: Van den Broucke CO. Lemli JA.

One would expect at least (English) titles to be uniform. However, it turned out that especially CAB Abstracts documentation services frequently modified author titles what again impaired uniform sorting.

Examples

>   CAB: Wild insects and honeybees as pollinators of some Lamiaceae of herbal interest...
>   Agricola Insects as pollinators of some Lamiaceae of herbal interest...
>   Agris   About flavonoids from *Origanum dubium*.
>   CAB   The flavonoids of *Origanum dubium*.
>   CAB   Oregano production in Cuba: an alternative to importation.
>   FSTA   Production of oregano in Cuba: an alternative to importation.

Unified sorting according to the source was not possible with journals, let alone with conferences and books of proceedings. Also, there sometimes existed original language journal title in one database and English translation of journal title in another database.

Examples

> Journal,-Association-of-Official-Analytical-Chemists
> Journal-Association of Official-Analytical-Chemists
> Journal-of-the-Association-of-Official-Analytical-Chemists
> J-Assoc-Off-Anal-Chem
> Journal-of-AOAC-International
> Shokuhin-Eiseigaku-Zasshi = Journal-of-the-Food-Hygienic-Society-of-Japan.

For this reason we had to devise special journal codes that we then attributed to each particular journal article. All different representations of a journal title had to be identified. A new field had to be introduced in our experimental database, and the unified journal codes inputted manually.

We also wished to analyze the references according to the document type so we had to harmonize the types to enable sorting. We chose the following document types:

am	monograph article (contribution, part of a monograph)
ap	proceedings article (paper)
aj	journal aricle
m	monograph
mp	book of proceedings
ma	book of abstracts
t	thesis
s	standard
p	patent

When all the data were downloaded and uploaded we performed many cross-reference sortings and corrections to achieve as much uniformity as possible. We thus obtained a database of some 1600 *Origanum* records and a database of some 420 *Lippia* records. The two databases contained all the multiple occurrences so they had not as yet reflected the actual number of different references.

For further analyzes another database for both *Origanum* and *Lippia* had to be created for filtering of duplicates or multiple occurrences. To retain the data on the frequency of occurrences, however, we set up in the two databases a matrix for inclusion of data on each database record occurrence. For each reference we thus obtained a record supplied with all of the above bibliographic data plus the data on databases in which the particular record was indexed. These two experimental databases permitted further analyses with results described in the next chapter. The terms duplication (multiple occurrence), overlap, co-occurrence or co-appearance are usually used synonymously. From Origanum we also transferred Lippia-only references which were retrieved because they sometimes employed the term of oregano (e.g. Mexican oregano as synonym for *Lippia graveolens*).

Duplication among databases and database groups was assessed with Jaccard similarity coefficient J_s.

$$J_s = a/(a + b + c), \tag{11.1}$$

where
a = number of items common to (shared by) resources;
b = number of items unique to the first resource; and
c = number of items unique to the second resource.

RESULTS OF ANALYSIS

After the filtering of multiple occurrences 935 different *Origanum* and 240 different *Lippia* references, indexed by at least one of the databases, were established. As an introduction to results we first present yearly growth of new references on genera *Lippia* and *Origanum*. All duplications have been filtered so the numbers in Table 11.2 and Figure 11.1 represent actual yearly output by each genus. We may observe with both genera constant yearly increase that is, however, much more evident with *Origanum* than with *Lippia*.

The references were synthesized from different publications. 213 *Lippia* (89 per cent) and 761 *Origanum* (81 per cent) references were extracted from journal articles, followed by monographs (Table 11.3). Many of those monographs could additionally be classified as unconventional (gray) literature and were issued mostly as research reports by research institutions. There were also several theses, standards and patents with regard to *Origanum*.

Table 11.2 Yearly growth of new *Lippia* and *Origanum* references

Year	Origanum	Lippia
1981	27	12
1982	32	8
1983	33	9
1984	43	12
1985	33	14
1986	38	6
1987	41	17
1988	33	9
1989	46	8
1990	55	11
1991	49	14
1992	50	9
1993	82	25
1994	57	10
1995	74	13
1996	71	21
1997	85	19
1998	86	23
Total	935	240

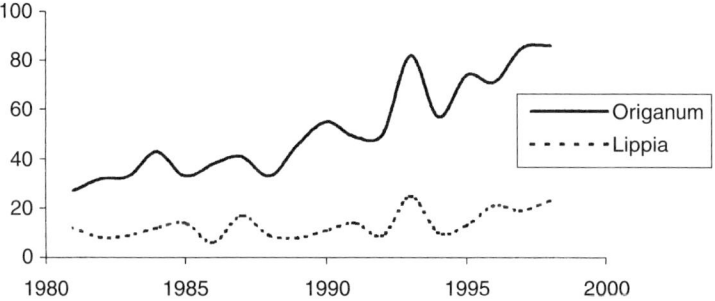

Figure 11.1 Yearly growth of new Lippia and Origanum references.

Table 11.3 Lippia and Origanum references by document type (DT)

DT	Lippia	Origanum
aj	213	761
am	3	15
ap	8	79
m	14	45
ma	–	4
mp	1	1
p	–	15
s	–	10
t	1	5
Total	240	935

We further present *Lippia* and *Origanum* references as indexed by each particular database (Figure 11.2). It can be observed that CAB abstracts (CAB), followed on both instances by Agris, serve as the database with the highest number of records on both genera, accounting for some 62 per cent *Lippia* and 54 per cent of all *Origanum* records. The agricultural databases (Agricola, Agris, CAB) explain some 90 per cent, and 77 per cent, and biomedical databases (Embase + Biobase, Medline) some 23 per cent and 16 per cent respective *Lippia* and *Origanum* references, with Embase/Biobase containing twice as many records as Medline. FSTA contains, in proportion, a much higher number of references on Origanum (315) than Lippia (16) what logically ensues from the culinary/spice aspects of Origanum/oregano.

Table 11.4 shows the number of instances (multiple occurrences – duplicates or higher-order repeats) when the same record was indexed by different databases, and by one database only (single occurrence). There were five Origanum references (presented in the last part of this chapter) that had been indexed by all six databases, and as many as 581 references that were indexed by only one database. That implies that among several potential bibliographic resources as much as 62 per cent references were retrievable in one database only.

Table 11.5 shows which databases, and in what number, accounted for single occurrences. Among agricultural databases CAB accounted for some 39 per cent of single and 24 per cent of all occurrences what implies that if CAB were not consulted some three

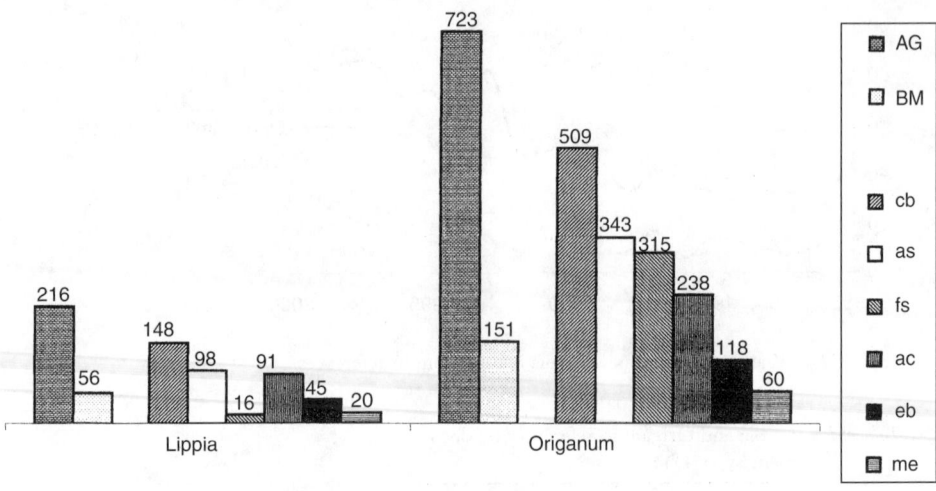

Figure 11.2 *Lippia* and *Origanum* references in respective databases and groups of databases.

Table 11.4 Number of occurrences of the same record in databases

Occurrence	Lippia	Origanum
6	–	5
5	2	19
4	16	56
3	35	104
2	52	170
1	135	581
Total	240	935

quarters of possibly relevant references would be retrieved. Among the two biomedical databases Medline and Embase/Biobase the later contains twice as many single occurrences.

In the following text we present overlap of records among particular databases and groups of databases. We display some results in a form of Jaccard coefficient (form. /1/) which implies potential relevance of all sources. As FSTA seems to be a less relevant source of *Lippia*-related references we excluded FSTA from *Lippia* analyses. We thus present results separately for *Origanum* and separately for *Lippia*.

ORIGANUM

We present double and triple overlap among all agricultural databases and between the two biomedical databases (Table 11.6), between the agricultural and biomedical group, and between the two groups and FSTA (Table 11.7). We can see a fairly low overlap between agricultural databases, averaging some 30 per cent (Jaccard coefficient). Duplication between FSTA and each agricultural database alone was even lower (15–17

Table 11.5 Single occurrence of a record in databases

Database	Origanum	Lippia
ac	28	13
as	118	31
cb	230	69
fs	165	5
eb	28	3
me	12	14
Total	581	135

Table 11.6 Occurrence of *Origanum* references in either database (+: Boolean OR), co-occurrence in databases (*: Boolean AND), and Jaccard coefficient (*/+)

	database		*	+	*/+
	ac	as	152	429	0.35
	ac	cb	157	590	0.26
	as	cb	175	677	0.26
ac	as	cb	113	723	0.16
	eb	me	28	150	0.19
	fs	ac	82	470	0.17
	fs	as	84	573	0.15
	fs	cb	110	713	0.15

Table 11.7 Occurrence of *Origanum* references in either database group (+: Boolean OR), cooccurrence in database group (*: Boolean AND), and Jaccard coefficient (*/+)

	database/group		*	+	*/+
	BM	AG	99	770	0.13
	AG	fs	144	889	0.16
	BM	fs	40	424	0.09
AG	BM	fs	35	935	0.04

per cent). The co-occurrence of a reference in all three agricultural databases was mere 16 per cent. Between the two biomedical databases co-occurrence (19 per cent) was ever weaker than between particular agricultural bases.

Duplications between the groups were also very low (Table 11.7). The coefficients were assessed along principles where each reference need have appeared in at least one database in both groups, so e.g. the possible BM/AG combinations (Boolean AND) were *ac/eb, ac/me, as/eb, as/me, cb/eb, cb/me, ac/as/eb, ac/as/me, ac/cb/eb, ac/cb/me, as/cb/eb, as/cb/me, ac/as/cb/eb, ac/as/cb/em, ac/as/cb/eb/em*. There were only 4 per cent of records that were retrievable in both an agricultural and biomedical database along with FSTA, what shows high level of interdisciplinarity.

We further assessed primary resources where *Origanum*-related documents were issued. Among 935 different references there were altogether 761 journal articles (Table 11.3). These were published in 367 different journals. Twenty six core journals published five

Table 11.8 List of 26 different journals with five or more published articles on Origanum

Journal title	No. of articles
Journal of Essential Oil Research	41
Acta Horticulturae	28
Flavour and Fragrance Journal	15
Perfumer and Flavoris	15
Rastitel'nye Resursy	15
Planta Medica	15
Journal of Food Protection	13
Journal of Food Science	12
Journal of Agricultural and Food Chemistry	11
Z. fuer Lebensmittel Untersuchung und Forschung	11
Phytochemistry	9
Fleischwirtschaft	9
Journal of AOAC Internation	9
Nahrung	9
Alimentaria	8
Chemie Mikrobiologie Technologie der Lebensmittel	7
Fitoterapia	7
Gemuese Munchen	7
Journal of Chemical Ecology	7
Food Chemistry	6
Herba Polonica	6
Journal of Home and Consumer Horticulture	6
Journal of Spices and Aromatic Crops	6
International Journal of Food Microbiology	5
Journal of the Science of Food and Agriculture	5
Plantes Medicinales et Phytotherapie	5
Total	289

articles each or more what comprises 289 articles (Table 11.8). Journal of Essential Oil Research alone accounted for 41 articles. There were also 11, 29, 42, and 258 journals that published four, three, two, and one article respectively in the entire 1981–1998 period.

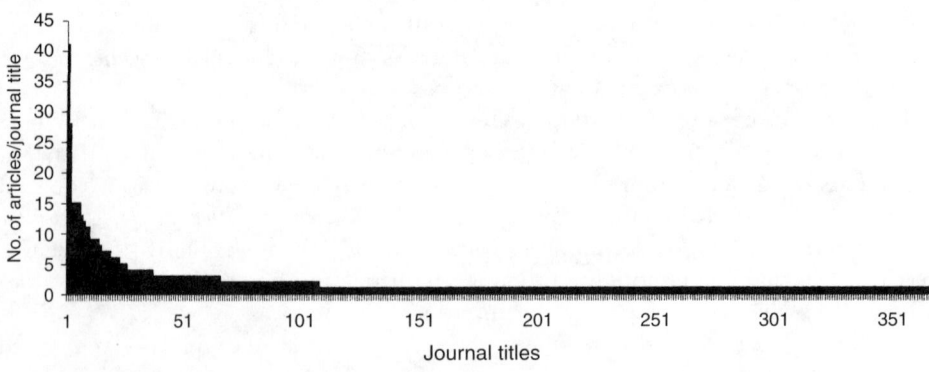

Figure 11.3 Origanum-articles by number of articles per journal title.

We present, in the form of the Bradford bibliograph of scatter, all the journals by number of published articles (Figure 11.3). On the left side of the column chart the journals with the highest number of articles are presented in a decreasing order of articles per journal beginning with the Journal of Essential Oil Research with 41 articles. In the middle and on the right sight of the x-axis there are 258 bars that represent those journals with an only article in the entire period.

LIPPIA

In *Lippia* similar pattern can be observed as in *Origanum*.

Due to the low number of *Lippia*-related FSTA references (16) co-occurrence was estimated only for biomedical agricultural databases and groups (Table 11.9). The highest overlap was between Agris and Agricola, and again the lowest between the biomedical databases. The overlap between all three agricultural databases was just as in Origanum – 16 per cent, what shows interestingly similar patterns.

Among the 240 different *Lippia*-related references there were 213 journal articles. These were published in 109 journals. Among the 213 articles there were 80 single-journal occurrences, 15 double, 3 triple occurrences, etc. The ten core journals (five or more articles per journal) accounted for 90 articles each, with the Journal of Essential Oil Research alone accounting for as many as 20. This was the same journal as in

Table 11.9 Occurrence of *Lippia* references in either database or database group (+: Boolean OR), co-occurrence in databases or database groups (*: Boolean AND), and Jaccard coefficient (*/+)

	database		*	+	*/+
	ac	as	55	134	0.41
	ac	cb	56	183	0.31
	as	cb	45	201	0.22
ac	as	cb	35	216	0.16
	eb	me	9	56	0.16
	AG	BM	37	235	0.16

Table 11.10 List of 10 different journals with five or more published articles on *Lippia*

Journal title	No. of articles
Journal of Essential Oil Research	20
Phytologia	16
Journal of Natural Products	10
Phytochemistry	9
Planta Medica	8
Journal of Ethnopharmacology	6
Revista Brasileira de Farmacia	6
Essenze Deriv. Agrum.	5
Flavour and Fragrance Journal	5
Insect Science and its Application	5
Total	90

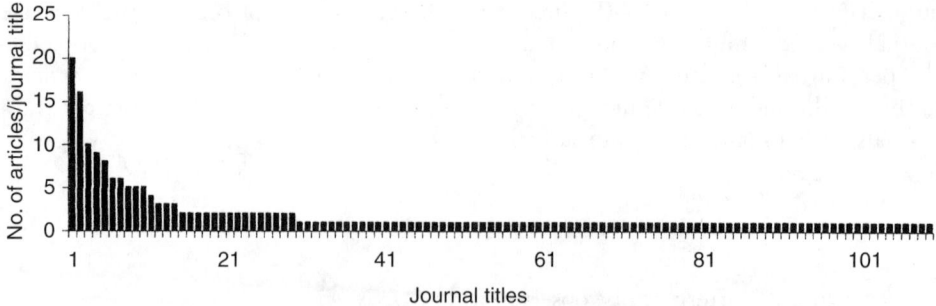

Figure 11.4 *Lippia*-references by number of articles per journal title.

Origanum. The other high frequency *Lippia* journal articles (Table 11.10), however, were mostly obtained from different journals than those in *Origanum* (Table 11.8).

In *Origanum*, there were 26 journals with five or more articles. In *Lippia*, there were ten. The lower number is to be ascribed not only to the overall lower number of *Lippia* references but probably also to the less interdisciplinary characteristics of this genus. This could also be observed in the significantly lower number of *Lippia* references in the food-oriented database FSTA. The chart on scatter of articles (Figure 11.4), however, shows similar patterns as in *Origanum* (Figure 11.3).

CHARACTERISTICS OF INDEXING

We present the titles of five *Origanum*-related documents indexed by all six databases, and two *Lippia*-related documents indexed by five databases. There was no *Lippia* document indexed by all six databases (see Table 11.4).

Origanum documents

Control of *Aspergillus flavus* in maize with plant essential oils and their; components.
  Journal of Food Protection. 1998
Bactericidal activity of carvacrol towards the food-borne pathogen *Bacillus cereus*.
  Journal of Applied Microbiology. 1998
Nutrient antioxidants in oregano.
  International Journal of Food Sciences and Nutrition. 1996
Detection of enterotoxigenic *Clostridium perfringens* in spices used in Mexico by dot blotting using a DNA probe.
  Journal of Food Protection. 1998
Independent effects of palatability and within-meal pauses on intake and appetite ratings in human volunteers.
  Appetite. 1997

Lippia documents:

Hernandulcin: an intensely sweet compound discovered by review of ancient literature.
  Science. 1985

(+)-4 beta-hydroxyhernandulcin, a new sweet sesquiterpene from the leaves and flowers of *Lippia dulcis*.
Journal of Natural Products. 1992.

We further present, on the example of *Origanum*, a few selected examples of multiple indexing of the same document and thus some differences that exist among the above databases. Organism description, if any, is presented in bold typescript.
Explanation of field names as presented in each particular database:

DE	descriptor (a controlled term)
ID	identifier (free-language term)
ENDE	indexer-assigned descriptor
ENC	computer-assigned (broader) descriptor
OD	organism descriptor
BT	broader term

Document Title 1: *Nutrient antioxidants in Oregano*

AGRICOLA – DE: *Origanum-vulgare*. satureja-. *Origanum-dictamnus*. *Origanum-onites*. nutrient-content. food-composition. antioxidants-. alpha-tocopherol-. beta-tocopherol-. peroxide-value. protection-. dried-foods. leaves-. species-differences. ID: satureja-thymbra. gamma-tocopherol-. delta-tocopherol-

AGRIS – ENDE: *antioxidants-; *tocopherols-; *vitamin-e; *vitamins-; *Origanum-vulgare*; *satureja-; *Origanum-*; *culinary-herbs; *terpenoids- ENC: crops-; flavouring-crops; labiatae-; origanum-; terpenoids-; triterpenoids-; vitamin-e; vitamins- ID: *Origanum-dictamnus; Origanum-onites*; satureja-thymbra

CAB Abstracts – DE: antioxidants-; tocopherols-; vitamin-E; vitamins-; culinary-herbs; estimation-; plant-; terpenoids-; composition- OD: *Origanum-vulgare*; Satureja; *Origanum-dictamnus*; *Origanum-onites*; Lamiales- ID: Satureja-thymbra BT: Origanum; Lamiaceae; dicotyledons; angiosperms; Spermatophyta; plants; Satureja

EMBASE – DE Medical: chemical-analysis*; extraction*; article; autooxidation; high-performance-liquid-chromatography; mutagenicity; thin-layer-chromatography. Drug: antioxidant*; herb*; hexane; spice; tocopherol

FSTA – DE: antioxidants-; flavourings-; lipids-; spices-; tocopherols-; vitamins-; oregano-

MEDLINE – DE: *Antioxidants/an [Analysis]; Chromatography, High Pressure Liquid; Chromatography, Thin Layer; *Herbs/ch [Chemistry]; Hexanes; Support, Non-U.S. Gov't; Vitamin E/an [Analysis]

Document Title 2: *Independent effects of palatability and within-meal pauses on intake and appetite ratings in human volunteers*

AGRICOLA – DE: eating-patterns. palatability-. food-intake. appetite-. men-

AGRIS – ENDE: *mankind-; *pasta-; *tomatoes-; *spices-; *duration-; *palatability-; *food-intake; *appetite-; *feeding-habits ENC: behaviour-; cereal-products; feeding-habits; flavourings-; nutrition-physiology; organoleptic-properties; physiological-functions; plant-products; processed-plant-products; processed-products; quality-; time-; vegetables- ID: *oregano-*

CAB Abstracts – DE: pasta-; tomatoes-; spices-; duration-; effects-; palatability-; food-intake; appetite-; feeding-behaviour OD: man-; Lycopersicon- ID: *oregano* BT: Homo; Hominidae; Primates; mammals; vertebrates; Chordata; animals; Solanaceae; Solanales; dicotyledons; angiosperms; Spermatophyta; plants
EMBASE – DE Medical: appetite*; food-intake*; palatability*; adult; article; controlled-study; eating; human; human-experiment; male; normal-human; satiety; time; tomato
FSTA – DE: consumer-response; sensory-properties; eating-habits; foods-; palatability-; satiety-
MEDLINE – DE: Adolescence; Adult; *Appetite; *Eating/px [Psychology]; *Food Preferences/px [Psychology]; Human; Male; Satiation; *Taste; Time Factors

Document Title 3: Bactericidal activity of carvacrol towards the food-borne pathogen B. cereus

AGRICOLA – *No descriptors*
AGRIS – ENDE: *spores-; *survival-; *temperature-; *inhibition-; *monoterpenoids; *pathogens-; *mortality-; *growth-rate; *antimicrobial-properties; *chemical-composition; *population-density; **Thymus*-genus; **Bacillus-cereus*; **Origanum-vulgare*; *essential-oils; *essential-oil-crops ENC: antibiotic-properties; bacillaceae-; *Bacillus*-; bacteria-; biological-development; cells-; crops-; growth-; isoprenoids-; labiatae-; *Origanum*-; processed-plant-products; processed-products; terpenoids- ID: carvacrol-; antibacterial-properties; thymus-vulgaris
CAB Abstracts – DE: spores-; survival-; susceptibility-; temperature-; inhibition-; monoterpenoids-; pathogens-; mortality-; growth-rate; antibacterial-properties; plant-composition; population-density; essential-oils; essential-oil-plants OD: *Thymus-vulgaris*; *Bacillus-cereus*; *Origanum-vulgare*; Lamiales- ID: carvacrol BT: *Thymus*; Lamiaceae; dicotyledons; angiosperms; Spermatophyta; plants; *Bacillus*; Bacillaceae; Firmicutes; bacteria; prokaryotes; *Origanum*
EMBASE – DE Medical: *Bacillus-cereus**; bacterial-growth*; disease-transmission; growth-inhibition; concentration-response; population-density; growth-rate; temperature; membrane-fluidity; bactericidal-activity; nonhuman; article. Drug: carvacrol*/drug dose; plant-extract
FSTA – DE: bacillus-; flavour-compounds; food-safety; inhibition-; phenols-; terpenoids-; carvacrol-
MEDLINE – DE: *Anti-Infective Agents/pd [Pharmacology]; **Bacillus cereus*/de [Drug Effects]; *Bacillus cereus*/gd [Growth & Development]; Food Microbiology; Hydrogen-Ion Concentration; Spores, Bacterial; Support, Non-U.S. Gov't; Temperature; *Terpenes/pd [Pharmacology]

Document Title 4: Extraction of light filth from unground marjoram: collaborative study

AGRICOLA – DE: spices-. contaminants-. rodents-. hair-. insects-. particles-. analytical-methods
AGRIS – ENDE: *spices-; *pollutants-; *rodents-; *hair-; *insects- ENC: anatomy-; animal-anatomy; animal-fibres; animal-products; arthropods-; body-parts; fibres-; injurious-factors; integument-; invertebrates-; mammals-; plant-products; products-; tissues-; vertebrates-ID: particles-; analytical-methods

EMBASE – DE Medical: filter*; filtration*; food-hygiene+; priority-journal; short-survey; methodology; human
FSTA – DE: FILTH-TESTS; *marjoram*, light filth extraction method for unground; EXTRACTION-; SPICES-
MEDLINE – DE: Animal; *Condiments/an [Analysis]; *Food Contamination/an [Analysis]; Insects; Solvents

Document Title 5: Autumn essential oils of greek Oregano

AGRICOLA – DE: *Origanum-vulgare*. essential-oils. plant-composition. thymol-. quantitative-analysis. autumn-. summer-. seasonal-variation. geographical-variation. greece-. ID: *Origanum-vulgare*-subsp.-*hirtum*. gamma-terpinene-. *p*-cymeme-. carvacrol-

AGRIS – ENDE: *labiatae-; *essential-oil-crops; *chemical-composition; *leaves-; *stems-; *seasonal-variation; *monoterpenoids-; *aromatic-compounds; *autumn-; **Origanum-vulgare*; *essential-oils ENC: crops-; labiatae-; *Origanum*-; periodicity-; plant-anatomy; plant-vegetative-organs; processed-plant-products; processed-products; seasons-; terpenoids-; time-ID: lamiaceae-; thymol-; p-cymene

BIOBASE – DE 92–1–6–1 plant-science: biochemistry: Secondary-products: Synthesis; 92–9–1–4 plant-science: biotechnology: Biotechnology-and-Bioengineering: Production-of-useful-compounds. *Origanum-vulgare*-subsp-*hirtum*; Labiatae; *Greek-oregano*; essential-oils; seasonal-variation; gg-terpinene; *p*-cymeme; carvacrol; thymol.

CAB Abstracts – DE: essential-oil-plants; plant-composition; leaves-; stems-; seasonal-variation; thymol-; *p*-cymene; autumn-; monoterpenoids-; geographical-variation; essential-oils; chemical-composition OD: lamiaceae-; *Origanum-vulgare*; Lamiales- BT: dicotyledons; angiosperms; Spermatophyta; plants; *Origanum*; Lamiaceae; Mediterranean-Region; Developed-Countries; European-Union-Countries; OECD-Countries; Southern-Europe; Europe

FSTA – DE: agriculture-; cultivation-; essential-oils; flavourings-; harvesting-; oils-; spices-; vegetable-products; *Oregano-*

In the above example, we may observe significant differences among database indexing. With field descriptors there are so many differences that there is no point in interpreting them. The examples illustrate the varieties well enough.

With regard to *Origanum* or oregano it is possible to approximately describe some basic characteristics. The descriptor field is searchable mostly with the term *Origanum*. There are instances of oregano. Marjoram is used rarely as a descriptor term. In Medline the genus *Origanum* does not exist as a descriptor. The articles are best retrieved with free-text searching. In Embase the genus name may appear. It's presence, however, is less frequent than in agricultural databases. In FSTA the descriptor searches can be carried-out only with English terms. There, however, exists descriptor distinction between oregano and marjoram. In AGRIS, there exists an indexing rule that implies indexing of living plant-related articles with a Latin descriptor (i.e. *Origanum*), and indexing of a product, that is a plant after harvest, with an English descriptor (i.e. oregano, used also for marjoram). Example of the document no. 4 is quite interesting in that that there are almost no organism descriptors even though the organism appears in the very title.

We can see that an extended analysis would be needed to assess all principal differences. In general, however, we may conclude that there exist significant differences in structure of the descriptor field, quantity of descriptors, indexing depth, and also perspective or discipline.

DISCUSSION

Monographs on specific life sciences related topics frequently forego certain aspects of methodology. Some methodological aspects are usually incorporated into the corpus of a paper. Such is the case of the statistical analysis. Information science and documentation is presented much more rarely. It is usually not customary for scientists to explain how bibliographic data were obtained. It is thus not possible to assess if most relevant sources were consulted at all. Such particular methodology papers are usually available only in information-science publications which will probably only accidentally become known to an end-user in life sciences. With our contribution we wanted to present information-documentation aspect of medicinal plants in its own right.

Our research material was bibliographic databases and references/records on genera *Lippia* and *Origanum*. Bibliographic databases offer a much higher level of data collection consistency than enormous WWW-related corpus on medicinal plants. Even though some databases under our study were available for free via the WWW we used CD-ROM and traditional online versions. The level of uniformity of databases available via these storage media is higher and searches are easier to perform. It is also possible to obtain more bibliographic fields in the records. Consistent downloading is also easier.

We had no intention to estimate the information value of one database versus another, and thereby suggest relevance. This is for the end-users to decide. Some databases index a high number of articles published in authoritative periodicals. Some other databases, on the other hand, may boast stronger coverage of non-conventional or "gray" literature. End-users need information for specific purposes. They may seek top-rank articles but may also be interested in strictly geographically limited national publications.

Retrieval recall and precision remain an important issue in database searching. To improve retrieval recall in databases there remain several possibilities with expanding the search syntax with the use of Both English and Latin terms, end employing both oregano and marjoram terms. Unified thesauri could serve as good tools for improving recall. There, however, is still (and may never be) no single comprehensive agricultural thesaurus, which would serve the needs of the three agricultural databases. All databases under study employ particular indexing and classification schemes. Such indexing schemes have to be studied carefully prior to conducting searches. Origanum may appear as either English or Latin term with sometimes non-defined taxonomic distinction between oregano and marjoram. Sometimes there is a distinction between the green plant and the dried plant as a spice. In one instance there is no descriptor term with regard to *Origanum* or *Lippia* so records can only be retrieved by free searching in titles and abstracts. In one database there exists only English indexing term. Except for one database there exists, except for identifiers, no indexing differences between wild marjoram (oregano) and sweet marjoram (marjoram). In Lippia the situation is more clear so most records are retrievable with the term Lippia, with a few records, however, retrievable only with the term Mexican oregano.

Here it must be further stressed that our search results were established solely on the presence of characteristic words in the records. Thus a word may have appeared in an abstract only in one database whereas another database did not supply the record with an abstract so such a record had in such a case not been retrieved. The number of retrieved records in each respective database therefore merely indicates the presence of the term in a record and does not reflect the exact number of records being dedicated to either Origanum or Lippia in an absolute sense. This additionally shows the importance of exhaustive indexing and abstracting of documents.

In our contribution we showed that, in order to carryout comprehensive Origanum- or Lippia-related searches, more than one database should be consulted. In Origanum some 54 per cent of all investigated references could be retrieved by one (most inclusive) database what is not enough. And if we included some other databases, such as Napralert, Chemical Abstracts and Biological Abstracts, this percentage would probably decrease. A further analysis would enable a more exact assessment of publishing trends in the field. Our, to some extent limited, analysis, however, permits some reasonably accurate conclusions to this end. We may observe a fairly constant, if modest, yearly growth of new documents. It is difficult, however, to suggest how many databases should be used for an exact assessment of trends and for obtaining of enough relevant documents. There frequently appears the dilemma among the end-users whether some novel information retrieved from yet another database outweighs the fee costs of commercial services especially if multiple duplications are taken into consideration.

Duplications with high multiple occurrence frequency were rare. Four per cent of *Origanum* records only were retrievable in both an agricultural, a biomedical database, and in FSTA. The scatter of journal articles is also quite high. As many as 258 among 761 articles appeared as an only *Origanum*-related article in that particular journal title in the entire period. *Origanum* and *Lippia* scattered in rather different journals, however, with both genera the Journal of Essential Oil Research appears to be the highest frequency journal. These numbers can serve as an approximate illustration of scatter. It would be wrong to use such a method as a sole formula for definition of core journal titles. This is again pertinent to user needs and research preference. And finally, some journals change titles in the course of their existence so some numbers may also for this reason not be accurate.

From this analysis it is evident that single databases, or information sources, cannot serve as good examples to illustrate some principal directions in the field of medicinal plants. More sources must be used for better identification of relevant topics. Even so the conclusions can only be stated very carefully as the entire scope of certain field may never become known. The end-users must therefore, if independently employing information sources, be aware that there may always exist yet more features they are not aware of. Prior to setting up a research and conducting database searches the end-users should gather comprehensive instructions on each particular information service and get familiar with its construction and maintenance characteristics.

REFERENCES

Andre, P.Q.J. (1992) Toward a unified agricultural thesaurus. *Q. bull. int. assoc. agric. inf. specialists* 37, 224–226.

Bartol, T. (1999) International scatter of Salvia-related articles in serial publications and availability of core salvia-related serials in the libraries in Slovenia. *Res. Rep. Biotechnical Faculty – Agric. Issue* 73(1), 231–244.

Bonitz, M., Bruckner, E. and Scharnhorst, A. (1999) *The Matthew Index – Concentration Patterns and Matthew Core Journals*. *Scientometrics* 44(3), 361–378.

Brown, C.M. (1998a) *Complementary Use of the Scisearch Database for Improved Biomedical Information*. *Bulletin of the Medical Library Association* 86(1), 63–67.

Brown, C.M. (1998b) The benefits of searching EMBASE versus MEDLINE for pharmaceutical information. *Online and CD-ROM Review* 22(1), 3–8.

Chen, Q. (1989) A check on overlaping between AGRICOLA, AGRIS and CAB for tropical agriculture records. *Q. Bull. Int. Assoc. Agric. Inf. Specialists* 34(2), 67–72.

Clarke, S.G.D. (1998) Food on SilverPlatter. *Inf. World Rev.* 137, 28.

Cleverdon, C.W. (1962) Report on the testing and analysis of an investigation into the comparative efficiency of indexing systems. College of Aeronautics, Aslib Cranfield Research Project, Cranfield, 305.

Cortez, E.M. (1999) Use of metadata vocabularies in data-retrieval. *J. Am. Soc. Inf. Sci.* 50(13), 1218–1223.

Datta, V.K. (1988) Coverage of literature on mycotoxins by computer databases and comparison of this coverage with the TDRI in-house index facility. *Q. Bull. Int. Assoc. Agric. Inf. Specialists* 33(2), 61–78.

Deselaers, N. (1986) The necessity for closer cooperation among secondary agricultural information services: an analysis of AGRICOLA, AGRIS and CAB. *Q. Bull. Int. Assoc. Agric. Inf. Specialists* 31(1), 19–26.

Garcia-Lopez, J.A. (1999) Bibliometric analysis of spanish scientific publications on tobacco use during the period 1970–1996. *Eur. J. Epidemiol.* 15(1), 23–28.

Glanzel, W., Schubert, A. and Czerwon, H.J. (1999) A bibliometric analysis of international scientific cooperation of the European Union (1985–1995). *Scientometrics* 45(2), 185–202.

Gomez, I., Fernandez, M.T. and Sebastian, J. (1999) Analysis of the structure of international scientific cooperation networks. *Scientometrics* 44(3), 441–457.

Gupta, B.M., Sharma, S.C. and Mehrotra, N.N. (1990) Subject-based publication activity indicators for medicinal and aromatic plants research. *Scientometrics* 18(5–6), 341–361.

Jaccard, P. (1912) The distribution of the flora of the alpine zone. *New Phytologist* 11, 37–50.

Juliano-Longo, R.M. and Dantas-Machado, U.D. (1980) Analysis of databases in the agricultural sciences. *Revista AIBDA* 1(2), 101–134.

Kuhlthau, C.C. (1997) Inside the search process: Information seeking from the user's perspective. *J. Am. Soc. Inf. Sci.* 42(6), 232.

Lofflerova, J. and Pessrova, H. (1982) The evaluation of the international information system AGRIS using statistical methods. *Ceskoslovenska Informatika* 24(9), 238–246.

Martyn, J. and Slater, M. (1964) Tests on abstracts journals. *J. Doc.* 20(4), 212–235.

Matsubuchi, A. (1987) Comparison of MEDLINE, CAB, BIOSIS and FSTA files in retrieving veterinary literatures: a case study in Nippon Veterinary and Zoo technical College Library [in Japanese]. *Online Kensaku* 8(4), 140–152.

Matthews, E.J., Edwards, A.G.K., Barker, J., Bloor, M., Covey, J., Hood, K., Pill, R., Russell, I., Stott, N. and Wilkinson, C. (1999) Efficient literature searching in diffuse topics: lessons from a systematic review of research on communicating risk to patients in primary care. *Health Libr. Rev.* 16(2), 112–120.

Nixon, J. M. (1989) Online searching for human nutrition: an evaluation of data bases. *Med. Ref. Serv. Q.* 8(3), 27–35.

Ogawa, K. and Kangohri, K. (1989) Literature searching for herbal medicines. *Pharm. Lib. Bull.* 34(4), 234–243.

Parsons, T.J., Power, C., Logan, S. and Summerbell, C.D. (1999) Childhood predictors of adult obesity – A systematic review. *Int. J. Obes.* 23(8), S-107.

Patil, J.M. and Kumar, P.S.G. (1998) Agricultural information systems and services: retrospects and prospects. *Libr. Sci. Slant Doc. Inf. Stud.* 35(1), 37–46.

Seetharam, G. and Rao, I.K.R. (1999) Growth of food science and technology literature: a comparison of CFTRI, India and the world. *Scientometrics* 44(1), 59–79.

Small, H. and Griffith, B.C. (1974) The structure of scientific literatures I: Identifying and graphing specialties. *Sci. Stud.* 4(4), 17–40.

Sodha, R.J. and Amelsvoort, T.V. (1994) Multi-database searches in biomedicine: citation duplication and novelty assessment using carbamazepine as an example. *J. Inf. Sci.* 20(2), 139–141.

Soremark, G. (1990) MEDLINE versus EMBASE: comparing search quality. *Database* 13(6), 66–67.

Tenopir, C. (1982) Distribution of citations in databases in a multidisciplinary field. *Online Rev.* 6, 399–419.

Thomas, S.E. (1990) Bibliographic control and agriculture. *Libr. Trends* 38(3), 542–561.

van Putte, N. (1991) A comparison of four biomedical databases for the retrieval of drug literature. *Health Inf. Libr.* 2(3), 119–127.

Weintraub, I. (1992) The terminology of alternative agricultural searching AGRICOLA, CAB and AGRIS. *Q. Bull. Int. Assoc. Agric. Inf. Specialists* 37(24), 209–213.

Wilson, T.D. (1998) EQUIP: a European survey of quality criteria for the evaluation of databases. *J. Inf. Sci.* 24(5), 345–357.

Zhang, H. (1994) A bibliometric study on medicine chinese traditional in medline database. *Scientometrics* 31(3), 241–250.

Index

A. amanus Bornmüller 77
A. brevidens Bornmüller 77
A. ciliatus Briquet 77
A. cordifolius Montbr 77et et Aucher ex Bentham
A. cyrenaicus Rechinger 77
A. dictamnus Bentham 77
A. hausskenechtii Briquet 77
A. hausskenechtii Briquet var. *acutidens* Handel-Mazzetti 77
A. leptocladus Briquet 77
A. libanoticus Briquet 77
A. lirius Hayek 77
A. majorana Schinz et Thellung 77
A. majorica Sampaio 77
A. pampainii Brullo et Furnari 77
A. pulchellus Briquet 77
A. pulcher Briquet 77
A. rotundifolius Briquet 77
A. scaber Briquet 77
A. sipyleus Rafanesque 77
A. syriacus Stokes 77
A. tomentosus Moench 77
A. tournefortii Bentham 77
A. vetteri Briquet 77
A. vulgaris Hill 77
acetophenones 141
acteoside 133, 141
aglycone 45, 91, 103
Agronomical Characters 169
akhdardiol 93
akhdarenol 93
akhdartriol 93
allo-aromadendrene 83
Amaracus akhdarensis Brullo et Furnari 77
Amaracus Bentham 3, 67, 69, 72, 77, 83, 88, 96, 103, 110, 177, 180
  analgesic 112, 133, 141, 195
  Analgesic, anti-inflammatory and antispasmodic activity 195
Amaracus sp. 67, 69, 73, 77, 83, 88, 96, 103, 110–13, 177, 182

Anatolicon Bentham 3, 67, 69, 72, 83, 88, 96, 110, 177
  anthocyanins 104
  antiaflatoxinogenic 183
  antiaggregant 195
  antiasthmatic 178
  Antibacterial activity 185
anticancer 196
antifungal 6, 113, 122, 139, 179, 221
  Antifungal activity 180
antihysterial 178
anti-inflammatory 195
antimicrobial 7, 11, 22, 134, 147, 167, 173, 185, 220, 262
antimutagenic 96, 196
antioxidant 6, 12, 96, 161, 167, 179, 190, 219
  Antioxidant activity 190
antiparalytic 178
antiparasitic 197, 204
antispasmodic 112, 195
apigenin 98, 141, 179, 219
apigenin-7-glucoside 179
arbutin 95, 192, 197
aromadendrene 83, 140
arylacetaldehides 141
astringent 178

bacteriostatic activities 180
betulin 93
betulinic acid 93
BHA (butylated hydroxy anisol) 192, 219
Bibliometric analysis 245
bicyclic monoterpenoids 83
bicyclogermacrene 83, 119, 139
Biological activity *see Pharmacology*
Biotechnology 173, 235, 263
biotin 218
β-bisabolene 87
borneol 83, 122, 140, 164
bornyl 83, 141
Botany 9, 219
β-bourbonene 87

Breeding 5, 7, 8, 144, 155, 160, 163, 237
    Breeding Methods 5, 170
    Breeding Targets 5, 167
Brevifilamentum Ietswaart 3, 69, 73, 84, 90, 110, 177
    caffeic acid 95, 141

γ-cadinene 87, 140
2-caffeoyloxy-3-[2-(4-hydroxybenzyl)-4, 5-dihydroxy] 9
2-caffeoyloxy-3-[2-(4-hydroxybenzyl)-4, 5-dihydroxy] phenylpropionic acid 95
Calamintha menthifolia 28, 31
Campanulaticalyx Ietswaart 3, 70, 73, 88, 103, 177
camphene 83, 134, 140
camphor 83, 133, 139
    Capitate glandular hairs 27
carotene 191, 218
carvacrol 83, 91, 112, 139, 155, 163, 177, 183, 192, 219
carvone 130, 140, 184
caryophylla-1(12),7-dien-9-ol 139
caryophyllane-2,6-β-oxide 132
β-caryophyllene 83, 87, 139
caryophyllene oxide 134
caryoptoside 134
caryoptosidic acid 134
cerebrocides 95
C-glycosides 96, 103
    Chemotaxonomy 161
Chilocalyx Ietswaart 3, 67, 70, 73, 85, 88, 90, 110, 177
chlorogenic acid 95
choleretic 112, 179
choline 95
chrysoeriol 99, 141
1,8-cineole 5, 89, 127, 131
cineole 89, 127, 131
cirsiliol 142
cirsimaritin 141
cis-p-menth-3-ene-1,2-diol 115
cis-p-menth-4-ene-1,2-diol 115
cis-p-menthan-1,8-diol 115
cis-sabinene hydrate 83, 168, 178
cis-sabinene hydrate acetates 83
cis-sabinol 85
citronellal 134
Coleus amboinicus 180
Coleus aromaticus 180
    Comparative methods 156
α-copaene 87, 134
Coridothymus capitatus 28, 31, 177
coryophyllene oxide 83
β-cubebene 87
Cultivation 3, 151, 153, 163, 263
c-vitamin 218

Cyclone powder 7, 114
cymyl 84, 90, 122

21-α-dihydroxy-oleanolic acids 93
21α-dihydroxy-ursolic acids 93
D-(+)-galactose 91
D-(+)-glucose 91
Davanone 137
Dictamnus creticus Hill 77
7,15-dien-19-ol 139
Dietary properties 217
digalactosyl-diglycerides 95
digestive 178
dihydrocarvone 134, 139
dihydroflavones 96
dihydroflavonols 96, 141
dihydrokaempferol 192, 219
dihydroquercetin 98, 192, 219
3,4-dihydroxybenzoyl 192
3,7-dimethyl-1-octen-3,7-diol 115
dimethylsecologanoside 134
2,6-dimethylstyrene 138
1,5-dione 134, 138
diosmetin 102, 141
disaccharidic 91
Distribution 68, 88, 96, 117
diterpenes 93
diuretic 134, 178
    Economic data 6

eicosane 241
elemol 140
Elongataspica Ietswaart 3, 70, 73, 74, 88
endo-borneol 140
    Epidermal cells 15
    Epidermis 15
(−)-epihernandulcin 133
epiloganic acid 134
epoxycaryophyllene 140
6,7-epoxymyrcene 127
eriodictyol-7-glucoside 179
eucarvone 139
β-eudesmol 140
γ-eudesmol 140
eupatorin 141
eupeptic 179

fatty alcohols 95
    Fertilisation, pest and insect control 155
ferulic acid 95
flavanone eriodictyol 192
flavanones 96, 134, 141
flavone 7-O-glycosides 103
flavone apigenine 192
flavone-C-glycosides 103
flavones 96
flavonoid glycosides 96, 143

flavonoids 96, 127, 141, 179, 191, 219, 252
flavonols 96, 141
flavonone-7-O-glycosides 103
folate 218, 268
free fatty acids 95

gallic acid 95
  General 5, 6, 111
geranial 127, 132, 138
geraniol 83, 85, 134, 138, 168
geranyl acetate 83, 86
germacrene 83, 87, 119, 130, 140, 223
germacrene A 134
germacrene-D 83, 87, 119, 122, 130, 223
germacrene-D-4-ol 83
  Glandular 23
glucosides 95, 103, 127, 134, 148
glycerol 95
glycolipids 95
β-glycosidase 91
β-guaiene 131

Hairs 5, 11
(+)-hernandulcin 44, 127, 132
hernandulcin 127, 132, 144
hispidulin 102, 141
humulene epoxide II 139
Hybrids 3, 68, 73, 76, 90, 96, 110,
  122, 170
  Hybrids 3, 68, 75, 90, 110, 121–2,
    161, 170, 177
hydrodistillates 241
hydroquinone 95
α-hydroxy-6-asteriscene 136
hydroxycinnamic acid 96
  Immunostimulant, antimutagenic and
    anticancer activity 196
  In Turkey 5, 6, 109
  In vitro production of secondary
    metabolites 241
  Infrafamiliar classification 67
  Infrageneric classification 68
  Infraspecific variation of the essential oil
    yield and composition 91
(+)-4β-hydroxyhernandulcin 133
6-hydroxyluteolin 103, 141
inositol 95
insecticidal 180, 199
  Insecticidal, nematocidal and molluscicidal
    activity 199
insect-pollinating 197
  Insect-pollinating and antiparasitic
    activity 197
integrifoliane 127, 134
  Inter- and infraspecific variation 144
iridoids glucosides 4
isoacteoside 133

isoakhdartriol 93
isoprimara-15-en-3β,8β,19-triol 93
isoprimara-15-en-8β,11α,19-triol 93
isoprimara-15-en-8β,19-diol 93
isoprimara-7,15-dien-19-ol 93

jaceosidin 142

Kekik water 114

(+)-lactose 91
L. adoensis Hochst., syn. L. Multiflora 128
L. affinis aristata Schau. 128
L. affinis sidoides Cham. 128
L. alnifolia Schau. (syn. L. brasiliensis
  A.S. Muller) 128
L. americana L. (syn. L. floribunda HBK;
  L. hemisphaerica; L. pauciserrata Turcz.;
  L. pyramidata Crantz) 128
L. aristata Schau. (syn. L. arguta Mart.; Lantana
  aristata (Schau.) Briq.) 128
L. canescens Kunth 128
L. carviodora Meikle 128
L. carviodora Meikle var. minor Meikle 128
L. chamaedrifolia Stued. (syn. L. chamaedryoides
  Stued.; Aloysia chamaedrifolia Cham.) 128
L. chevalieri Moldenke 128
L. citriodora Kunth (syn. L. triphylla (L'Hér.)
  Kuntze; Aloysia triphylla (L'Hér.) Britt.) 128
L. dauensis (Chiov.) Chiov. (syn. Lantana
  dauensis Chiov.) 129
L. dulcis Trev. (syn. L. asperifolia Benth.; L. asperifolia
  Reichenb.; L. dulcis var. mexicana Wehmer;
  Phyla scaberrima (A.L. Juss. Moldenke) 4, 129
L. fissicalyx Tronc. 129
L. gracilis HBK 129
L. grandifolia et Schau 129
L. grandis Martius & Schau. 129
L. grata Schau. 129
L. graveolens HBK (syn. L. amentacea M.E. Jones;
  L. berlandieri Millsp.; L. berlandieri Schau.;
  L. graveolens Schau.; L. tomentosa Sessé et Moc.) 4
L. grisebachiana Mold. (syn. L. lantanaefolia Griseb.) 4
L. hastulata (Griseb.) Hier. (syn. Acantholippia
  hastulata Griseb.) 4
L. integrifolia (Griseb.) Hier. 4
  L. javanica (Burm. f.) Spreng. (syn. L. asperifo-
    lia L.C. Rich.; L. asperifolia var. anomala
    Moldenke; L. scabra Hochst.) 4
L. juneillana (Mold.) Tronc. 4
L. ligustrina (Lag.) Britton (syn. Aloysia ligustrina
  (Lag.) Small; Junellia ligustrina (Lag.)
  Moldenke; Verbena ligustrina Lag.) 4
L. lycioides (Cham.) Steud. (syn. L. lagustrina
  Britton; Aloysia gratissima (Gill. et Hook.)
  Tronc.; Aloysia lycioides Cham.) 4
L. micromera Schauer in DC. 4

L. microphylla Cham. 4
L. nodiflora L. Greene (syn. *L. cuneifolia* Zipp.;
  L. nodiflora (L.) Michx.; *L. repens* Spreng.;
  Phyla chinensi Lour; *Phyla nodiflora* (L.)
  Greene) 4
L. oatesii Rolfe 4
L. origanoides HBK (syn. *L. berterii* Spreng.) 6, 130
L. palmeri 6
L. polystachia Gris. (syn. *Aloysia polystachia* (Gris.)
  Moldenke) 6
L. rugosa A. Chev. 6
L. savoryi Meikle 6
L. scaberrima Sond. 6
L. schimperi Wolp 6
L. sellowii Briq. (syn. *L. affinis* Briq.; *L. spiraeoides*
  Mart.; *Aloysia gratissima* (Gill. et Hook.)
  Tronc. var. *sellowii* (Briq.) Moldenke; *Aloysia
  sellowii* (Briq.) Moldenke 6
L. seriphioides A. Gray (syn. *L. foliosa* Phil.;
  Acantholippia seriphioides (A. Gray)
  Moldenke) 6
L. sidoides Cham. (syn. *L. multicapitata* Mart.) 6
L. somalensis Vatke 6
L. stoechadifolia HBK (syn. *Phyla stoechadifolia* (L.)
  Small) 6
L. thymoides Martius & Schau. 6
L. trifida C. Gay (syn. *L. gracilis* R.A. Phil.;
  L. hispida Gay; *L. parvifolia* Gardner;
  Acantholippia trifida Clos.; *Acantholippia
  trifida* (C. Gay) Moldenke) 6
L. turbinata Gris. (syn. *L. aprica* R.A. Phil.;
  L. disepala R.A. Phil.; *L. poleo* Lillo) 6
L. ukambensis Vatke 6
L. wilmsii H.H.W. Pearson 6
Lanata sp. 153
lapachenol 134
Leaf 11
Leonotis leonurus 27, 31, 32
linalool 83, 85, 114, 120, 127, 131, 134,
  138, 140, 164, 182, 195
linalyl acetate 83, 86, 122, 178
linoleic acid 114, 192
linolenic acid 114
lipids 95, 114, 190, 261
lipophilic antioxidants 190
Lippia sp. 3, 4, 111, 127, 241, 245, 251
lippiaphenol 132
lippifoli-1(6)-en-4β-ol-5-one 136
lippifoli-1(6)-en-5-one 137, 140
lippifoliane 127, 135
lippione 132, 139, 141
loganic acid 134
loganin 134
Longitubus *Ietswaart* 3, 70, 73, 110, 177
L-phenylalanine 141
L-tyrosine 141
Luteolin 99, 101, 103, 141
luteoline-7-glucoside 179

M. crassa Moench 77
M. crassifolia Bentham 77
M. cretica Kosteletzky 77
M. cretica Miller 77
M. dictamnus Kosteletzky 77
M. dubia Briquet 77
M. fragrans Rafinesque 77
M. hortensis Moench 77
M. leptoclados Rechinger 77
M. majorana Karsten 77
M. majorica Briquet 77
M. majorica Briquet var. *lusitanica*
  Coutinho 77
M. maru Briquet 78
M. maru Briquet var. *nervosa* Briquet 78
M. maru Hayek 78
M. mexicana Martius 78
M. micrantha Briquet 78
M. microphylla Bentham 78
M. nervosa Bentham 78
M. onites Bentham 78
M. orega Briquet 78
M. orega Walpers 78
M. ovalifolia Stokes 78
M. scutellifolia Stokes 78
M. sipylea Kosteletzky 78
M. smyrnaea Kosteletzky 78
M. suffruticosa Raffinesque 78
M. syriaca Kosteletzky 78
M. tenuifolia Gray 78
M. tomentosa Stokes 78
M. vulgaris Miller 78
Majorana aegyptiaca Kostelwtzky 77
  Majorana Bentham 3, 67, 70, 73, 74,
    76, 85, 88, 90, 110, 177
Majorana hortensis 154, 178, 217, 251
Majorana syriaca 28, 31
Majoranamaracus zerniji Rechinger 78
Melissa officinalis 28, 31
Mentha longifolia 22, 32
Mentha piperita 26, 28, 31, 32
Mentha spicata 22, 32
Mentha viridis lavanduliodora 28
Mentha × *villoso-nervata* 22, 32
menthone 139
4-methoxycarvacrol 134
3-methoxy-4-hydroxybenzoyl 115
4-methoxythymol 134
3-methyl-2-cyclohexen-1-one 133
methyl 3β-21α-dihydroxy-oleanolic acid 93
methyl 3β-21α-dihydroxy-ursolic 93
6-methyl-5-hepten-2-one 132
methyl oleanol acid 93
methyl ursolic acid 93
methylarbutin 95
methylcarvacrol 134
methyleugenol 139
methylthymol 134, 139

Micromeria fruticosa 28, 31
Micropropagation 237
molluscicidal 200
Monarda fistulosa 31, 32
monogalactosyl-diglycerides 95
monosaccharidic 91
monoterpenes 121
γ-muurolene 87
β-myrcene 86
myrcene 83, 86, 118, 134, 138
myrtenol 127

n-alkanes 241
naphthalene 241
naphthoquinoids 127, 145
naringenin 100, 134, 143
nematicidal 200
Nepeta racemosa 28, 31
neral 127, 132, 138, 140
niacin 218, 268
Nonglandular 23
O. acutidens Ietswaart 23

O. aegyptiacum Savi 78
  O. akhdarense Ietswaart et Boulos 78
O. akhdarense Ietswaart et Boulos 78
O. albiflorum Koch var. *congestum* Koch 78
O. album Salisbury 78
  O. amanum Post 78
O. amanum Post 78
O. americanum Rafanisque 78
O. anglicum Hill 78
O. angustifolium Koch 78
O. balearicum Pourret ex Lange 78
O. barcense Simonkai 78
O. barcense Simonkay var. *microstachyum* Grecescu 78
O. bargyli 71, 76, 80, 89, 101, 110, 119
  O. bargyli Mouterde 78
O. bargyli Mouterde 78
O. bevani Holmes 78
O. bilgeri 70, 78, 85, 88, 110, 117
  O. bilgeri Davis 78
O. bilgeri Davis 78
  O. boissieri Ietswaart
O. boissieri 69, 78, 83, 88, 97, 101, 110, 120
O. boissieri Ietswaart 78
  O. brevidens Dinsmore
O. brevidens Dinsmore 78
O. brevidens Dinsmore var. *pubescens* Thiebaut 78
O. bucharicum Bornmüller 78
  O. calcaratum Jussieu
6'-O-caffeoyl 134
O. calcaratum Jussieu 78
O. capitatum Wildenow ex Bentham 78

O. ciliatum Boissier et Kotschy 78
O. cinereum de Noë 78
  O. compactum Bentham 78
O. compactum Bentham 78
O. confertum Savi 78
  O. cordifolium Vogel 78
O. cordifolium Vogel 78
O. crassa Chevallier 78
O. creticum auct. non L. 78
O. creticum L. 78
O. creticum Schousbou ex Ball. 78
  O. cyrenaicum Beguinot et Vaccari
O. cyrenaicum Beguinot et Vaccari 78
  O. dayi Post
O. dayi Post 79
O. decipiens Wallroth ex Bentham 79
O. dictamnifolium Saint-Lager 79
  O. dictamnus L. 79
O. dictamnus L.79
O. dubium Boissier 79
O. ehrenbergii Boisier var. *parviflorum* Bornmuller 79
  O. ehrenbergii Boissier 79
O. ehrenbergii Boissier 79
O. elegans Sennen 79
  O. elongatum Emberger ex Maire
O. elongatum Emberger ex Maire 79
  O. floribundum Munby 79
O. floribundum Munby 79
O. floridum Salisbury 79
O. glandulosum Defontaines var. *elongatum* Bonnet 79
O. glandulosum Desfontaines 79
O. glandulosum Salzmann ex Bentham 79
O. glaucum Rechinger et Edelberg 79
O. glaucum Rechinger et Edelberg var. *laxius* Rechinger et Edelberg 79
O. gracile Koch 79
  O. grosii Pau et Font Quer ex Ietswaart 79
O. grosii Pau et Font Quer ex Ietswaart 79
O. gussonei Tineo ex Lojacono Pojero 79
O. haradjianii Rechinger 79
  O. hausskenechtii Boissier 79
O. hausskenechtii Boissier 79
O. haussknecktii 120
O. heracleoticum auct. non L. 79
O. heracleoticum auct. non L. f. *trichocalycinum* Rechinger 79
O. heracleoticum auct. non L. var. *albiflorum* Halácsy 79
O. heracleoticum auct. non L. var. *creticum* Halácsy 79
O. heracleoticum auct. non L. var. *creticum* Halácsy f. *glabra* Halácsy 79
O. heracleoticum auct. non L. var. *creticum* Halácsy f. *hirsuta* Halácsy 79
O. heracleoticum auct. non L. var. *rubriflorum* Halácsy 79

O. heracleoticum auct. non L. var. *trichocalycinum* Halácsy 79
O. heracleoticum hort. ex Koch 79
O. heracleoticum L. 79
O. hirtum Link 79
O. hirtum Link f. *albiflorum* Haussknecht 79
O. hirtum Link f. *prismaticum* Haussknecht 79
O. hirtum Link f. *rubriflorum* Haussknecht 79
O. hirtum Link f. *trichocalycinum* Haussknecht 79
O. hirtum Link var. *corymbulosum* Candargy 79
O. hirtum Link var. *genuinum* Vogel 79
O. hirtum Link var. *humile* Bentham 79
O. hirtum Link var. *laxiflorum* Candargy 79
O. hirtum Link var. *macrostachyum* Candargy f. *macrostachyoides* Candargy 79
O. hirtum Link var. *oostachyum* Candargy 79
O. hirtum Link var. *prismaticum* Vogel 79
O. hirtum Link var. *subtypicum* Candargy 79
O. hirtum Link var. *typicum* Candargy 79
O. humile Miller 79
O. husnucan-baserii H.Duman, Z.Aytaç et A.Duran 79
O. husnucan-baserii H.Duman, Z.Aytaç et A.Duran 79
O. hypercifolium Schwarz et Davis
O. hypercifolium Schwarz et Davis 79
O. hyrcanum Bornmüller 79
O. illiricum Scheele 79
O. isthmicum Danin 79
O. isthmicum Danin 79
O. jordanicum Danin and Künne 80
O. jordanicum Danin and Künne 80
O. laevigatum 71, 76, 80, 89, 101, 110, 119
O. laevigatum Boissier 80
O. laevigatum Boissier 80
O. latifolium Miller 80
O. leptocladum Boissier 80
O. leptocladum 69, 80, 84, 88, 97, 110
O. leptocladum Boissier 80
O. levigatum Boisier var. *laxum* Post 80
O. libanoticum Boissier 80
O. libanoticum Boissier 80
O. lusitanicum Rouy 80
O. macrostachyum Hoffmannseg et Link var. *genuinum* Coutinho 80
O. majorana L. 80
O. majorana L. 80
O. majoranoides hort. ex Gams 80
O. majoranoides Willdenow 80
O. majoricum Cambessedes var. *lusitanicum* Rouy 80
O. majus Garsault 80
O. maru L. 80
O. maru L. f. *viridula* Bornmüller 80
O. maru L. var. *aegyptiacum* Dinsmore 80
O. maru L. var. *capitatum* Post 80

O. maru L. var. *sinaicum* Boissier 80
O. maru sensu Sibthop et Smith 80
O. megastachyum Link 80
O. micranthum Vogel 80
O. micranthum Vogel 80
O. microphyllum Vogel 80
O. microphyllum Vogel 80
O. minus Garsault 80
O. minutiflorum Schwarz et Davis 80
O. minutiflorum Schwarz et Davis 80
O. nervosum Vogel 80
O. normale Don 80
O. normale Don var. *incanum* Schmidt et Schlaginweit 80
O. nutans Wildenow ex Bentham 80
O. oblongatum Link 80
O. odorum Salisbury 80
O. onites L. 80
O. onites L. 80
O. orega Vogel 80
O. orientale Miller 80
O. pallidum Desfontaines 80
O. pampaninii Ietswaart 80
O. pampaninii Ietswaart 80
O. paniculata Spenner 80
O. paniculatum Koch. 80
O. parviflorum Dumond d' Urville 80
O. paui Martinez 80
O. petraeum Danin 6'-O-p-coumaroyl 134
O. petraeum Danin 80
O. pruinosum Koch 80
O. pseudodictamnus Sieber 80
O. pseudo-onites Lindberg 80
O. puberulum Klokov 80
O. pulchellum Boissier 80
O. pulchrum Boissier et Heldreich 80
O. punonense Danin 80
O. punonense Danin 80
O. purpurascens Gilibert 80
O. ramonense Danin 80
O. ramonense Danin 80
O. rotundifolium Boissier 80
O. rotundifolium Boissier 80
O. saccatum Davis 80
O. saccatum 69, 80, 83, 88, 97, 101–2, 110, 120
O. saccatum Davis 80
O. sardoum Nyman 81
O. saxatile Salisbury 81
O. scabrum Boissier et Heldreich 81
O. scabrum Boissier et Heldreich 81
O. semiglaucum Boissier et Reuter ex Briquet 81
O. siculum Nyman 81
O. silvestre Ortega ex Sampaio 81
O. sipyleum L. 81
O. sipyleum L. 81
O. smyrnaeum L. 81

O. smyrnaeum sensu Sibthorp et Smith 81
O. solymicum Davis 81
O. solymicum 69, 81, 88, 110, 120
O. solymicum Davis 81
O. stoloniferum Besser ex Reichnbach 81
O. strobilaceum Mobayen et Gahraman 81
O. suffruticosum hort. ex Steudel 81
O. symes Carlström 81
O. syriacum L. 81
O. syriacum L. 81
O. syriacum L. var. *aegyptiacum* Täckholm 81
O. syriacum L. var. *bevanii* Ietswaart 81
O. syriacum L. var. *bevanii* Ietswaart 81
O. syriacum L. var. *sinaicum* Ietswaart 81
O. syriacum L. var. *sinaicum* Ietswaart 81
O. syriacum L. var. *syriacum* 81
O. syriacum L. var. *syriacum* 81
O. thymiflorum Reichenbach 81
O. tournefortii Aiton 81
O. tragoriganum Zuccagni ex Steudel 81
O. tytthanthum Gomscharov 81
O. tytthanthum Gomscharov var. *seravschianum* Borissova 81
O. venosum Wildenow ex Bentham 81
O. vestitum Clarke 81
O. vetteri Briquet et Barbey
O. vetteri Briquet et Barbey 81
O. virens Hoffmannseg et Link 81
O. virens Hoffmannseg et Link var. *genuinum* Coutinho 81
O. virens Hoffmannseg et Link var. *macrostachyum* Coutinho 81
O. virens Hoffmannseg et Link var. *siculum* Bentham 81
O. virens Hoffmannseg et Link var. *spicatum* Rouy 81
O. virescens Poiret 81
O. viride Halácsy 81
O. viride Halácsy var. *hyrcanum* Bornmüller 81
O. viridulum Martin-Donos 81
O. vogelii Greuter & Burdet 81
O. vulgare L. 81
O. vulgare L. f. *albiflora* Senchovei ex Formánek 81
O. vulgare L. f. *elongatum* Formánek 81
O. vulgare L. f. *glabrescens* Beck 81
O. vulgare L. f. *grecescui* Soó 81
O. vulgare L. f. *procumbens* Jakus ex Soó et Borhini 81
O. vulgare L. ssp. *barcense* Jávorka 81
O. vulgare L. ssp. *glandulosum* Ietswaart 81
O. vulgare L. ssp. *glandulosum* Ietswaart 81
O. vulgare L. ssp. *gracile* Ietswaart 81
O. vulgare L. ssp. *gracile* Ietswaart 81
O. vulgare L. ssp. *hercleoticum* Holmboe 81
O. vulgare L. ssp. *hirtum* Ietswaart 81
O. vulgare L. ssp. *hirtum* Ietswaart 81
O. vulgare L. ssp. *prismaticum* Gaudin 81

O. vulgare L. ssp. *prismaticum* Gaudin var. *australe* Gaudin 81
O. vulgare L. ssp. *prismaticum* Gaudin var. *parviflorum* Gaudin 81
O. vulgare L. ssp. *virens* Ietswaart 81
O. vulgare L. ssp. *virens* Ietswaart 82
O. vulgare L. ssp. *viride* Hayek 82
O. vulgare L. ssp. *viride* Hayek 82
O. vulgare L. ssp. *viridulum* Nyman 82
O. vulgare L. ssp. *vulgare* 82
O. vulgare L. var. *album* Fraas 82
O. vulgare L. var. *americanum* Rafinesque 82
O. vulgare L. var. *barcense* Hayek 82
O. vulgare L. var. *bracteosum* Petermann ex Soó 82
O. vulgare L. var. *creticum* Briquet 82
O. vulgare L. var. *exile* Lamotte 82
O. vulgare L. var. *formosanum* Hayata 82
O. vulgare L. var. *glaucum* Hedge 82
O. vulgare L. var. *hirtum* Visiani 82
O. vulgare L. var. *humile* Bentham 82
O. vulgare L. var. *latebracteum* Beck 82
O. vulgare L. var. *laxiflorum* Post 82
O. vulgare L. var. *longespicatum* Post 82
O. vulgare L. var. *macrostachyum* Brotero 82
O. vulgare L. var. *magnilimbis* Boissier 82
O. vulgare L. var. *megastachya* Koch 82
O. vulgare L. var. *normale* Briquet 82
O. vulgare L. var. *pallescens* Martin-Donos 82
O. vulgare L. var. *prismaticum* Bentham 82
O. vulgare L. var. *puberulum* Beck 82
O. vulgare L. var. *purpurascens* Briquet 82
O. vulgare L. var. *purpureum* Stokes 82
O. vulgare L. var. *rotundifolium* Rafinesque 82
O. vulgare L. var. *rufuscens* Stokes 82
O. vulgare L. var. *semiglaucum* Boissier ex Briquet 82
O. vulgare L. var. *smyrnaeum* Bentham 82
O. vulgare L. var. *spicatum* Koch 82
O. vulgare L. var. *spiculigerum* Briquet 82
O. vulgare L. var. *subglabrum* Schmidt et Schlagintweit 82
O. vulgare L. var. *tauricum* Borissova 82
O. vulgare L. var. *violacea* Sennen 82
O. vulgare L. var. *virens* Bentham 82
O. vulgare L. var. *virens* Koch 82
O. vulgare L. var. *virescens* Cariot et St. Lager 82
O. vulgare L. var. *viride* Boissier 82
O. vulgare L. var. *viridulum* Briquet 82
O. vulgare ssp. *Hirtum* 30–2, 76, 88, 101, 120, 164, 168, 171, 179, 186, 194, 199, 222
O. wallicianum Bentham 82
O. wastoni Schmidt et Schlagintweit 82
O. × *adanense* Baser et Duman 82
O. × *adonidis* Mouterde 82
O. × *aplii* Boros 82
O. × *barbarae* Bornmüller 82
O. × *dolichosiphon* Davis 82

O. × *font-queri* Pau 82
O. × *hybridinum* Miller 82
O. × *hybridum* Heldeich 82
O. × *indercedens* Rechinger 82
O. × *intermedium* Davis 83
O. × *lirium* Heldreich ex Halacsy 83
O. × *majoricum* Cambessedes 83
O. × *minoanum* Davis 83
O. × *pabotii* Mouterde 83
O. × *symeonis* Mouterde 83
O.acutidens 69, 84, 88, 97, 110, 117
   O.vulgare L. subsp. *gracile* (Koch) Ietswaart
      [Syn.: *O. tyttanthum* Gontsch.] 83
   O.vulgare L. subsp. *hirtum* (Link) Ietswaart
      [Syn.: *O. heracleoticum* L.] 83
   O.vulgare L. subsp. *viride* (Boiss.) Hayek
      [Syn.: *O. heracleoticum* L.] 83
   O.vulgare L. subsp. *vulgare*
      [Syn.: *O. creticum* L.] 83
Ocimum basilicum 26, 28, 31
O-glycosides 96, 98, 101, 103
oleic acid 114
oleoresin 182
Onites tomentosa Rafinesque 78
oral antiseptic 179
Origanomajorana applii Domin 83
Origanum acutidens Ietswaart 78
Origanum bastetantum 7, 241
Origanum dictamnus 11, 13, 14, 17, 18, 21, 31, 32, 34, 53
Origanum hypericifolium 88, 110, 118, 178
Origanum onites 109, 112, 154, 172, 178, 185, 188, 191, 220, 261, 268
Origanum sipyleum 69, 76, 81, 83, 88, 101, 110, 118, 178
Origanum vulgare 4, 11, 14, 24, 28–32, 36, 49, 50, 51, 52, 56, 57
   Origanum vulgare subsp. *hirtum* 21, 32, 51, 56–7
   Origanum vulgare subsp. *viridulum* 21, 32
   Origanum vulgare subsp. *vulgare* 21, 32
Origanum × *adanense* 69, 76, 78, 84, 88, 90, 110, 119
Origanum × *dolichosiphon* 76, 82, 110, 119
Origanum × *intercedens* 12, 13, 14, 15, 21, 23, 26, 31, 32, 89, 97, 110, 120, 161, 171

palmitic acid 114
panthotenate 218
p-coumaric acid 95
p-cymen-7-ol 83
p-cymen-8-ol 83
p-cymene 83, 90, 114, 116, 127, 134, 138, 160
p-cymenene 83
p-cymyl 84, 86

pectolinaringenin 142
   Peltate (sessile) glandular hairs 27, 30
pesticidal 179
Pharmacology 177
phenol carvacrol 122
phenolics 95, 106
phenylpropanoids 127
phosphatidic acid 95
phosphatidyl-ethanolamine 95
phospholipids 95, 114
p-hydroxybenzoic 95
α-pinene 134, 140
β-pinene 134, 140
pinocembrin 134, 143
piperitone 132, 134
Plectranthus madagascariensis 28, 31
Plectranthus ornatus 31, 32
Plectrantus amboinicus
Plectrantus aromaticus 180
Pogostemon cablin 31, 32
polygalactosyl diglycedides 95
Processing 217, 227
Prolaticorolla Ietswaart 3, 71, 74, 75, 89, 110, 177
   protocatechuic acid 192
pulegone 134, 139
   Quality characters 164

Quality, commercial 5, 6, 111, 103, 163, 168, 172
Quality, sensory 224, 227
quercetin 96, 98, 100
   Root 56

rosmarinic acid 95
Rosmarinus officinalis 28

sabina ketone 83, 85
sabinene 83, 89, 117, 122, 134, 139, 160
sabinyl 84, 86, 88, 122
Safety, microbiological 224
Salvia aurea 28, 31
Salvia blepharophylla 31
Salvia fruticosa 31
Salvia officinalis 28, 31
Salvia reflexa 32
Satureja camphorata Bornmüller 83
Satureja hortensis 112
Satureja thymbra 28, 31, 32
Schizocalyx smyrnaeus Scheele 83
Schizocalyx syriacus Schelle 83
Seasoning 228
4,5-seco-african-4,5-dione 135
Sectio *Origanum* L. 83
Sideritis syriaca 27, 31
   Somatic embryogenesis 239
   Stem 54

Structural characteristics 11
  O. symes Carlström 11

(E)-tagetenone 138
(Z)-tagetenone 138
Taxonomy 67, 177
  Taxonomy 83
Teucrium 23, 26, 27
  Teucrium sect. *Chamaedrys* 28, 31
  Teucrium siculum 31
  Teucrium silicum 28

The following hybrids are mentioned in page 76
O. × *adanense* Baser et Duman, *O. bargyli* × *O. laevigatum*
O. × *adonidis* Mouterde, *O. libanotcum* × *O. syriacum* var. *bevanii*
O. × *aplii* Boros, *O. majorana* × *O. vulgare* ssp. *vulgare*
O. × *barbarae* Bornmüller, *O. ehrenbergii* × *O. syriacum* var. *bevanii*
O. × *dolichosiphon* Davis, *O. amanum* × *O. laevigatum*
O. × *font-queri* Pau, *O.grosii* × *O. compactum*
O. × *hybridinum* Miller, *O. dictamnus* × *O. sipyleum*
O. × *indercedens* Rechinger, *O. onites* × *O. vulgare* ssp. *hirtum*
O. × *intermedium* Davis, *O. onites* × *O. sipyleum*
O. × *lirium* Heldreich ex Halacsy, *O. scabrum* × *O. vulgare* ssp. *hirtum*
O. × *majoricum* Cambessedes, *O. majorana* × *O. vulgare* ssp. *virens*
O. × *minoanum* Davis, *O. microphyllum* × *O. vulgare* ssp. *hirtum*
O. × *pabotii* Mouterde, *O. bargyli* × *O. syriacum* var. *bevanii*
O. × *symeonis* Mouterde, *O. laevigatum* × *O. syriacum* var. *bevanii*
O. amanum × *dictamnus*, *O. amanum* × *O. dictamnus*
O. calcaratum × *dictamnus*, *O. calcaratum* × *O. dictamnus*
O. micranthum × *vulgare* ssp. *hirtum*, *O. micranthum* × *O. vulgare* ssp. *hirtum*
O. sipyleum × *vulgare* ssp. *hirtum*, *O. sipyleum* × *O. vulgare* ssp. *Hirtum*

$2\beta,4\beta,9\alpha$-2,6,6,9-tetramethyltricyclo[6.3.0.0^{2,4}] undec-1(8)-en-7,11-dione 136
4-terpenylacetate 134
α-terpinene 139
γ-terpinene 84, 91, 116, 139, 160
α-terpineol 83, 92, 122, 134, 140
α-terpinolene 134, 139
α-thujene 134
α-thujone 139
Thymbra spicata 177
Thymus majorana Kuntze 83
Thymus mastichina 154
Thymus serpyllum 112
Thymus vulgaris 28, 31, 32, 38
  Toxicity 201
1,6-*trans*-lippifolian-1α-ol-5-one 136
Trade of *Origanum*
Trichomes 22

Volatalile oils 83
Vulgare ssp. *Gracile* 88, 95, 121
Vulgare ssp. *Viride* 89, 101, 121
Vulgare ssp. *Vulgare* 76, 88, 101, 121